MARKING TIME
The Epic Quest to Invent the Perfect Calendar

DUNCAN STEEL

John Wiley & Sons, Inc.
New York • Chichester • Weinheim • Brisbane • Singapore • Toronto

This publication is designed to provide accurate and authoritative
information in regard to the subject matter covered. It is sold with the
understanding that the publisher is not engaged in rendering professional
services. If professional advice or other expert assistance is required, the
services of a competent professional person should be sought.

ISBN   0-471-29827-1

Printed in the United States of America

10   9   8   7   6   5   4   3   2   1

For Helen,
with love and appreciation

CONTENTS

PREFACE

A ll journalists know that any story needs a hook to catch the reader, or, in media parlance, a *peg*. In one's domestic world physical pegs come in all sorts of forms, like the single peg on the back of the bathroom door on which to hang your towel, the multiple pegs on a hat and coat stand, or the myriad clothes-pegs (clothes*pins* on the west of the Atlantic) with which one hangs out the wet laundry to dry.

To a writer a peg is something on which to hang a story, some focus which will ensure a readership. For sports columnists the pegs vary seasonally: no one much wants to read about ice hockey when the mercury is soaring, nor about tennis when the courts are cloaked in snow. Recent events such as scandals in the purportedly private lives of politicians, or upcoming events like elections, lead to political writers capturing more front-page acreage even than is their usual overallowance. After some years of public yawns over the space program, the decision by the National Aeronautics and Space Administration hierarchy to send John Glenn back into space in 1998 on a space shuttle flight was a masterpiece of public relations, because it provided a veritable cornucopia of pegs, equivalent to all the cloakrooms in Carnegie Hall, bringing the work of NASA firmly back into the public eye and appreciation.

The transition between the second and third millennium has provided an unparalleled peg for journalists, as anyone who can read will surely be aware. Our newspapers are daily filled with stories about such matters as the so-called Millennium Computer Bug, the meaning (if any) of the millennium, and so on. Clearly this provokes an upswell of public interest in the calendar and its origins.

About the middle of the 1990s I toyed with the idea of writing a popular-level book which would cater to this potential market, but then I held back, realizing that others would already have made a start on a similar quest. Such books soon started to appear, but in perusing them I found to my horror that they are largely a sorry bunch, misinforming the public on this significant matter. Let me give no names, no pack-drill, so as not to feed the ravenous libel lawyers, but one best-selling recent calendar book I scanned has nine purported "facts" on its back cover, of which five are glaringly wrong, while others are dubious. This is loony-tunes stuff, a cartoon image of the science and the history. Writing for the interested public is a great endeavor, but it needs to be based upon facts, not fallacies. Articles written for good newspapers, like the major broadsheets, are at least subject to some later correction if wrong, but there are many popular-level books around now which give an entirely false impression to the reader, in terms of both the scientific basis and the evolutionary history of the calendar, with no chance of quick redemption.

That is not to say that it is those authors who should bear all of the guilt. In this book I will be highlighting various errors made over the years in calendrical matters, and not all of them occurred long ago. A recent egregious example is contained in the *Explanatory Supplement to the Astronomical Almanac*, which is prepared by the astronomers employed by the governments of the United States and the United Kingdom at the U.S. Naval Observatory and the Royal Greenwich Observatory to provide absolute knowledge of the astronomical cycles upon which any calendar may be based. In the *Explanatory Supplement* an incorrect definition is given for the length of the year against which the Gregorian calendar (or most other solar calendars) should be judged. The difference is only in the fourth decimal place of a day length, or about sixteen seconds of time, but the error is hugely significant once one uncovers the real reason why Pope Gregory XIII instructed the Catholic Church in 1582 to start using the leap-year cycle which we employ today, that being very much a second-best solution in spite of various modern-day commentators claiming that it provides an excellent remedy to the problem. Those writers misunderstand that problem, aided and abetted by the mistake in the *Explanatory Supplement*. This is not merely inconsequential pedanticism; the Persian calendar may well be altered shortly based upon the erroneous belief of Iranian clerics that the definition given by the U.S. and U.K. government astronomers is correct.

All these things will become clear as you take your trip through the meandering path laid down in this book. When I started writing I had thought that it would be possible to tell the full story in perhaps forty thousand words, but soon I realized that a rather longer treatment would be necessary. There are so many fascinating facets which need to be covered if one is to appreciate fully the remarkable story of the calendar. For example, think of the first English attempts to found a colony in North America, the forays of Sir Walter Raleigh in and around Chesapeake Bay. These are usually explained on face value as part of a quest to obtain gold and novel products such as tobacco and coffee through the establishment of a British presence across the Atlantic, in the face of Spanish domination to that time, but in reality a strong case may be made for the first colonization attempts being prompted by calendrical matters and Protestant-Catholic rivalry on that front.

In preparing this book I have used a wide variety of disparate sources. A few detailed books are listed in the bibliography, but there were some hundreds of other volumes to which I referred. The Internet has been a valuable resource, and I downloaded over a thousand distinct web pages in doing my background research. Of particular note in this regard has been the Internet calendar discussion group hosted by Rick McCarty through East Carolina University. Many of the members of that group have very kindly assisted me, and some who have made special contributions are contained in the listing below.

During my labors the following have kindly helped in some way or answered queries that I have posed to them: Brian Marsden, John T. Ramsey, Michael Alexander, Richard John, Tony Beresford, Colin Humphreys, Jonathan Tate, Syuichi Nakano, Rob McNaught and Soo Tan, Bill Napier, Geoff Pope, David Asher, R. E. Asher, Ian Elliott, Viktor Shor, Mike Baillie, Hans Rickman,

Andrea Carusi, Peter Brown, John Newbury, Dagmar Heller, John Campbell, George Williams, Vic Gostin, Malcolm Walter, Daniel McCarthy, Karri Muinonen, Alain Maury, Rodolphe Audette, Amos Shapir, Toke Nørby, Mario Hilgemeier, Claus Tøndering, Bill Thayer, Tom Peters, Chris Carrier, Peter Meyer, Marcos Montes, Bill Hollon, Dik Winter, Robert van Gent, Steve Sohmer, Ivan Van Laningham, Alex Livingston, Rudy Limeback, Bill Becker, Kevin Tobin, Karl Palmen, Dean Tiegs, Branislav Skrobonja, Clinton Anderson, Doug Ewell, Ken Pizzini, John Ward, Jay Gary, Tom Buckley, Robert Poole, Mark Bailey, and Lance Latham. My thanks to them and all others who have assisted by whatever means.

It is appropriate that I specifically express my gratitude to Graeme Waddington, who has been an indefatigable correspondent, answering even my most foolish questions.

A special mention must also be made here of Simon Cassidy, an English mathematician who lives in California. The idea of the calendrical motivation behind the initial British attempts to colonize part of North America was suggested to me by Cassidy. His erudite work has illuminated this and various other aspects of the broad subject about which I have written.

In preparing and writing this tome I have of course incurred a great debt to my family, and so I must express my gratitude to Helen, Harry, and Elliot for enabling me to get it done. The other two individuals whose support has been invaluable are Paul Davies and Chris McKay.

Now I have written enough in this preface to foreshadow the matters which are covered herein. I have tried diligently to get my facts correct, but I would not be surprised if in twists and turns along the convoluted route I have allowed myself to be misled. Nevertheless, nothing ventured, nothing gained. Any errors I have propagated will, I hope, be corrected by later writers, and so the understanding of our intriguing but infuriating calendar will move ahead.

DUNCAN STEEL
Adelaide, South Australia
January 1999

Chapter 1

❧

GEORGE WASHINGTON'S BIRTHDAY

Fifteen miles south of Washington, D.C., abreast the Potomac River where it flows through the verdant countryside of Virginia, stands Mount Vernon, the home and tomb of George Washington, first president of the United States. Many a calendar has, under the date February 22, a note saying something like *Birth of George Washington, 1732.* But the Washington family Bible, preserved at Mount Vernon, has the following entered within its covers:

> George Washington, son to Augustine and Mary his wife, was born ye 11th day of February 1731/32 about 10 in the morning, and was baptized on the 30th of April following.

Throughout his life, Washington celebrated his birthday on February 11, and he was not mistaken. Nor is that calendar hanging on your wall incorrect: the date known as Washington's Birthday is indeed February 22. In this book we are going to see how this paradox can be reconciled. But in doing so we will open up a whole can of worms (or Pandora's box, if you prefer the allusion).

George Washington's birthday is as good a starting point as any for our tortuous trip through the history (and future) of the calendar, because it introduces several points which we need to explain if we are to understand the workings of our systems for keeping time and dating events. Let's face it, if there is one thing other than health that controls our lives, it's time.

In his best-seller entitled *A Brief History of Time,* British physicist Stephen Hawking wrote that he had been advised to avoid using any equations in his book because every equation was reckoned to cut potential sales in half. I'm going to ignore that advice here, as you'll see—I have confidence in you—and start off with an equation. And here it is: $T = \$$.

If you need that spelled out in words, it says "time is money," and most will realize the truth of that. One could claim that to be the best-known equation in modern society, ahead of $F = ma$ and even $E = mc^2$. (My publisher will now be having fits of apoplexy—three equations in the first few paragraphs!) So, time is money, and in a capitalist society we see in every minute of every day how the logic of that affects all that we do, as Adam Smith's famous "invisible hand" guides us in our actions: make a buck wherever and whenever we can.

Except that we don't. Imagine the following scene. One morning you are lying asleep in bed when your alarm rings, demanding attention. Switching off the obnoxious clarion call, what do you do? Leap from bed, shower, shave and dress for the office, grab a cup of coffee, and rush for the subway? Or slumber a while, then rise at your leisure, dress in sloppy joes, and amble down to your favorite greasy spoon for two eggs scrambled, bottomless coffee, and a perusal of the sports scores in the voluminous weekend newspaper?

Obviously, I'm comparing a working day with the weekend. But when (and how?) did that familiar seven-day cycle begin? Strange though it may seem, the course you take after tackling that damned alarm clock actually depends upon the date of Augustus Caesar's triumphal march into Alexandria after defeating Cleopatra and Mark Antony: because that's the day (August 30 in 30 B.C., a Sunday) from which our weekly cycle begins, unbroken over the two millennia since.

As we will see, Augustus made several other contributions to the calendar, so it is hardly surprising that a month is named for him. His adoptive father and predecessor as supreme ruler of Rome, Julius Caesar, is also remembered in a month name, as *he* should be, since it was he who gave our calendar much of the form which it has today. But there are other famous (and some not-so-famous) men who have left their mark on the calendar, even if they are month-less. They may be long-since dead, but how do Pope Gregory XIII, Lord Chesterfield, and Constantine the Great all continue to affect your life? And who was Denis the Little? As we'll see, it was Denis's work which led to our year numbers, and hence the instant at which a new century and new millennium is celebrated. Trouble is, he got it wrong.

Resolving Washington's Birthdate

As I wrote, there are several points arising from that inscription in the Washington family Bible which need explaining, if we are to proceed knowledgeably along the road bringing us to an understanding of our complicated calendrical system.

First, look at the date given in the inscription: February 11. That seems simple enough, except that a baby born at the same instant in a Catholic country (in Rome, or Madrid, or Mexico, say) would have been said to have entered the world on February 22. Perish the thought, could America have been *behind* Mexico, at least in terms of the date? Yes, indeed. At the time Virginia was still a British colony, and as a knock-on effect from the schism between Britain and the Roman Catholic Church in the first half of the six-teenth century, when the pope refused to grant King Henry VIII an annul-ment of his marriage to Catherine of Aragon, such places were still using the calendar introduced by Julius Caesar, rather than the reformed calendar which Pope Gregory XIII substituted in 1582. In the eighteenth century, Britain and its colonies were quite literally behind the times, still using the Julian rather than the Gregorian system, even though most of the Protestant nations in Europe had already fallen in line with the Catholic reckoning of dates. But take heart: it took Russia and several other countries until the early twentieth century to abandon the Julian calendar, which is why the so-called October Revolution in 1917 actually occurred on November 7. Britain and her dominions were late in bowing to the inevitable, but not the last. It took Turkey until 1925 to adopt the calendar which is now the global standard (notwithstanding the fact that there are dozens of alternative calendars still in parallel use), which is rather surprising because the builder of Istanbul played a major part in the story of our dating system.

Second, what is the meaning of "1731/32" in that inscription? In calendrical terms that's called "double-dating" and it's meant to avoid confusion. So let me confuse you. The actual date was February 11, 1731. But nowadays we would call it February 22, 1732. From the previous paragraph you may have come to terms with the idea that two calendars may have been in use in different countries with a few (actually eleven, at the time) days deviation between them, but now it looks like the year numbers were different, too. Well, yes: some of the time. The thing is that, although back in 46 B.C. Julius Caesar had legislated for the year to start on January 1, and of course we celebrate New Year on that date nowadays, for many centuries Britain and the colonies had a system in which the year officially began on March 25. Going back to Washington's time, the sequence of dates on the calendar used by his parents was this:

March 25, 1731

. . . then dates as usual until . . .

December 31, 1731

January 1, 1731

. . . then dates as usual until . . .

March 24, 1731

. . . and then celebrations since the next day was . . .

March 25, 1732: *Happy New Year!*

When Washington was born it was still 1731 on that system, although down Mexico way they thought it was 1732. And now one can read some significance into the use of "on the 30th of April following" for the date of his baptism. At that time young George was about eleven weeks old, but the event occurred in the year following his birth (1732, or 1732/33 using double-dating), even though in retrospect we have placed his birth and baptism in the same year.

Third, how many days between his birth and his baptism? Even though there are more days than we have fingers and toes, I'm sure we can all count them: just so long as we know whether we should include a February 29 or not. So, was there a leap year day in there? If Washington was born in 1731, surely that must have been a common year (the converse of a leap year), so that February had only twenty-eight days? Well, no. The rule employed was that the second year number in the double-date was used. Since 1732 is divisible by four, there were twenty-nine days in the month that George Washington was born.

The British Calendar Reform

In that era people were quite used to such convolutions. If you look at Samuel Pepys' *Diary* (the full unabridged version, most editions having been severely censored because the original was licentious enough to bring a blush from even Lady Chatterley's Lover), you'll find that he wrote in "New Year" on January 1, but did not change the year numbers until March 25. With other countries already working on the Gregorian calendar, but the British being unwilling even to mention the name of the pope (the "Anti-Christ of Rome" was how they referred to him), those engaged in trade and commerce, or diplomatic exchanges, had to find some way of unambiguously dating all documents. For many years the British would need to add "N.S." (for New Style) after a date given on their amended calendar (New Year on January 1, dates ten days different in the seventeenth century, eleven in the eighteenth), and "O.S." (Old Style) for their previous system, which they refused to abandon until 170 years after the papal revision. Except that the Scottish (perhaps just to spite the English) went halfway in that they adopted January 1 for New Year from 1600.

Well, we seem to have understood the date now. As we'll see later, when George Washington was aged twenty an edict from London changed the calendar in Old Virginny, and throughout what was to become the fledgling United States of America; indeed, throughout Britain, and Ireland, and all the colonies. Eleven days were struck from the calendar, with September 2, 1752, being followed by September 14. This meant that there were only 355 days in 1752 (11 less than 366, because it was a leap year). But 1751 was even shorter than that, because it lacked January 1 through March 24: the calendar reform by the British Parliament meant that 1751 was curtailed simply because the legislation called for the civil New Year to be moved to January 1. Again it's best to look at the sequence of dates (and we'll ignore the double-dating for simplicity):

March 25, 1750–March 24, 1750	(365 days)
March 25, 1751–December 31, 1751	(282 days)
January 1, 1752–December 31, 1752	(but with September 3–13 missing: 355 days)
January 1, 1753–December 31, 1753	(365 days)

So, where did January 1 through March 24 of 1751 go? If it's confusing to us now, what was it like for the people of the time? One way to answer that question would be to look back at the public response to this upset of the calendar, except that this was some decades before the *Times* of London (or, of course, the *Washington Post*) was founded. But the calendar was of great importance to people then, and almanacs proliferated. One of the major almanacs in the American colonies was written and published by Benjamin Franklin, and as we'll later see he had a few things to say about this imposition.

March 25 was abandoned as the date of the New Year in the civil calendar in the eighteenth century, then, although as we will be discovering its importance echoes on in other spheres of life, such as the ecclesiastical calendar (depending upon your church). The British have never totally abandoned it, as follows. While the financial years in the United States have occasionally been shifted around, the income tax year in Britain still ends with April 5: in the eighteenth century, when the calendar was reformed, April 5 was the Gregorian equivalent of March 25 on the Julian calendar. But don't let me fool you with a subterfuge like that in my previous sentence: I wrote that the British tax year *ends* with April 5, whereas that date would be the *start* of the year in a straightforward change between the Gregorian and the Julian calendars, so that there might be thought to be an anomaly of one day. Even today this causes confusion for people, and explanations for the apparent anomaly range from it being a fix by tax accountants to confusion in the 1750s over whether 1752 would be a leap year or not. In fact, although March 25 was the *first* day of the legal year it was actually counted as the *last* day of the financial quarter, so that an income tax year to start on April 6 (Gregorian) was the correct eleven day jump in dates. To make it even more confusing, the British adopted a *financial* year (for the government bookkeeping rather than the income tax returns of individuals) starting on April 1 from 1854.

The Old Style March 25, then, has not given up its grip on the lives of the British, and we'll be learning that it affects all of us in one way or another. As an astronomer I have to say that March 25 actually has a lot going for it, illogical as it may seem: its origin as a significant point in the year arises from the fact that it was near the date of the vernal equinox (and hence the start of spring) at around the time that Christ was born. It would seem logical to start the year with some astronomically defined juncture, even if we gave it some other label than "March 25." But the whole story of this book is of human responses to multifarious concerns, many of them being religious rivalries, and as the Vulcan Mr. Spock in *Star Trek* would tell you, humans rarely follow logic.

So When Is Washington's Birthday?

So what of the Washington's Birthday public holiday in the United States? We have seen that in effect his birthday was changed during his lifetime, from February 11 to the twenty-second, although he persisted in celebrating it on the eleventh. Nowadays one thing is clear: February 22 is listed in reference books as the date when his birth occurred.

Ah, but that is not quite the same as the public holiday which is scheduled in the official U.S. calendar for each year. Public holidays are funny things:

some occur on the same date regardless of the day of the week (like Independence Day on July 4), while others occur on a certain day of the week regardless of the day of the month (like Thanksgiving, on the fourth Thursday in November). But people often like to have public holidays on Mondays, giving three-day weekends rather than odd days off work midweek. Thus Canada has *its* Thanksgiving on the second Monday in October; then again, by having Thanksgiving on a Thursday many in the United States manage to end up with a four-day weekend break. In 1968 the official U.S. government statutes were changed to define Washington's Birthday (the public holiday, that is) as the third Monday in February. That means that it falls between February 15 and 21, inclusive, which means that it is *never* commemorated on either of the days one might legitimately recognize: February 11 (Julian) or February 22 (Gregorian).

What Time of Day Was Washington Born?

The final thing to discuss concerning the inscription detailing Washington's birth is the time of day: "about 10 in the morning." What does that mean? In those days a quoted time had a rather different implication from our understanding of the time nowadays: there were no railway trains to catch, no time signals on the hour over the radio, no continuous transmission of precise reference points from satellites far above. There was no nationwide coordinated time system, since there was no telegraph to carry the signals. There was no standard time for the whole of the east coast; from Virginia to New York was several days by ship or horse, and rather further to Boston. The time was set by the available technology, and the local need.

That just meant that the time followed the sun: midday was when the sun crossed the meridian, reaching its highest point in the sky, and George Washington entered the world about two hours before that, about forty miles south of Mount Vernon, in Westmoreland County. If his father possessed a reasonably accurate watch or clock, and the same sort of astronomical equipment as Charles Mason and Jeremiah Dixon used to survey their eponymous line dividing Maryland and Pennsylvania a few decades later, then he would have been able to tell the time more accurately. But to what benefit? One can define midday as being the time of meridian crossing, but with our modern definition of the day (which makes all days the same length, whereas in reality they're not) the sun is on the meridian at midday on only four days during each year.

There are ways to answer the question "What time is it?" too numerous to list. The only sensible way to answer is, "It all depends . . ." It all depends on whether you're worried about missing a television program, about how long it is until your next birthday, or what the time is to the nearest nanosecond for some complicated physics experiment. George Washington was born at about ten in the morning, and that means a couple of hours prior to the time when the sun passed closest to overhead. No more accuracy is needed. No more accuracy would have been possible without major effort.

Astronomical Cycles

Astronomy, however, has been introduced into our story. By making precise observations of the Sun, or perhaps other celestial bodies, the time of day *could* have been deduced more accurately. The length of the day depends upon the rate at which Earth spins on its axis. The year depends on the time taken to circuit the Sun, and it was a refinement of the measurements of that duration which eventually led to the calendar reform which moved Washington's birthdate. Once upon a time the month was determined by the time taken by the Moon to orbit Earth, as the word *month* suggests, but that linkage was abandoned, at least in our dominant Western calendar. Other societies and religions, such as Islam, persist in using dating systems based on the Moon rather than the Sun. As we will see, these various astronomical factors define and control our calendar, although it's not as simple as one might think. And the week? Well, that's a different matter—a peculiarly human one, involving religion and astrology—as we'll also discover.

Later I will outline how the cycles of the sky are defined, and how they have contributed to the ways in which we keep time in our calendar. Throughout, the comparisons I will be making are with the dominant calendar in human affairs, which is usually termed the *Gregorian calendar,* the origin of which we glimpsed above. As we will see at length, this might be termed the *Western calendar* since it is the calendar of what is usually termed the *Western world*. Terminology can be confusing. For example, the country where I live (Australia) is part of this Western world even though its name indicates it to be a southern land, and one would also count a country with a very similar title in English (Austria) as being part of the West despite its name in the tongue of its inhabitants, Österreich, actually meaning "eastern nation"!

The story of this book is the story of the calendar which is the world standard for business and communications, and which I presume is familiar to all readers, even though they may live in countries which use other dating systems, or may be members of religions which use parallel but independent calendars. What is *not* familiar to the vast majority of people is why that calendar—indeed, *any* calendar—is the way it is. By the time you finish reading this tome, my intention is that you will have a much greater appreciation of the intricacies of astronomy, history, and human foibles that have shaped our system of labeling the days. Many questions will be answered, but be warned: you may be left with more puzzles than you started with, because there are many things we just do not know about the calendar's evolution. What will be clear to you is the fact that the calendar has had a much greater effect upon human destiny than you may have fondly imagined. It's not just about numbering the days: it's about the rises and falls of empires and religions, about internecine squabbling and one-upmanship, about controlling the masses and subverting your rivals. The calendar story is one of human strife and aspirations.

To give you some idea of the breadth and spectrum of the matters to be discussed, in the next chapter I will give a summary of the topics we are going to cover. Following that I will give a succinct account of the various astronomical factors which control the lengths of the units of time—the year, the month, the day, and the second—which shape all calendars. After that, the calendar story will begin in earnest.

Chapter 2

❧

THE COUNTRY PARSON'S FORMULA

The so-called Country Parson's Formula will be well known to anyone who has ever given a public talk. Or at least it should be.

This formula is taught to a novice rural vicar by the elders, when worrying over the preparation of his or her first sermon. It says simply: *First of all you tell them what you are going to tell them; then you tell them; and then you tell them what you have told them.*

In less convoluted language, one starts by giving a succinct introduction to the matters that are going to be discussed in detail, so as to make the audience familiar with the subjects to be covered. One then presents the main discussion, taking up the majority of the available time in this way. Finally, one summarizes the main points, that summary often being the only substantive information which the audience will remember later.

I am going to try to employ the Country Parson's Formula here, to some extent. In this chapter I am going to outline quickly what I am going to tell you at length in the other chapters and the appendixes of this book. Rather than waste space with a formal summary, when you are done with the epilogue, having read the four appendixes along the way, I suggest that you come back to chapter 2, if a reminder of the whole story is required.

No Apologies

Another basic rule of public speaking is that usually ascribed to Michael Faraday, the nineteenth-century British pioneer of chemistry and electromagnetism; if you open up a television or radio set you will find that the capacitors therein have their characteristic behavior quantified in *farads,* a unit named for that great scientist.

Michael Faraday was a renowned master of the art of presenting popular lectures, his public talks at the Royal Institution in London being immensely well-regarded. One rule he applied to himself was: *never start a presentation with an apology.*

That seems a simple enough instruction to the budding public speaker, but it is astonishing how often it is disobeyed. Just note for yourself how often the talks you attend, whether at some science congress, an art gallery, or the local tennis club dinner, begin with the speaker apologizing in some way. They set themselves up for failure by telling their audience that they are sorry to be so ill-prepared, that their photos are too dark to be clear, and that they know comparatively little about their subject, anyway. The listener immediately decides that he or she is in for an hour of boredom and ineptitude.

So, I must not start with an apology. Rather, it is incumbent upon me to make clear that the astronomical considerations upon which our calendar is based are not simple, and do take some understanding. Nevertheless I see no reason why a person of average intelligence should not be able to comprehend how each of the astronomical cycles is defined and behaves. Unless you make the effort to properly grasp those matters, the story of the calendar will forever be a mystery.

Baiting the Hook

In describing the confusion surrounding George Washington's birthday in chapter 1, I have tried to bait a hook to catch you, persuading you of the intrigue of the calendar. To mix my metaphors, a spoonful of sugar helps the medicine go down, and we will meet Washington again in a later chapter. Another great figure of eighteenth-century America, Benjamin Franklin, also makes repeated appearances in the story.

The first chapter, then, was to get you interested, like a public speaker walking onto the stage wearing a bright yellow bow tie, or with an iridescent flower tucked behind the ear, while the present episode is designed to reel you in, to persuade you that the story to be told is indeed one which will entertain, illuminate, and maybe even fascinate you.

In chapter 3 it is necessary to swallow just a little drop of medicine. Therein I will summarize in very brief form the astronomical considerations that lead to the units of time which we commonly term the *day*, the *year*, the *second*, and the *month*. But beware that these are not as you might presently believe: the day is not the time it takes Earth to spin on its axis; the year is not the time taken to orbit the Sun; according to the stars, the second is getting longer; and the month is growing, too.

In that chapter, then, I will summarize these things, but several larger spoonfuls of medicine are needed if you are to appreciate fully the intricacies involved, and these figurative spoonfuls appear as appendixes A through D. All I can do is to hope that you will consume these at some stage in your reading of the book, preferably as early as possible. I have tried to sprinkle in a generous quantity of sugar lumps so as to make those appended essays as palatable as possible.

The Dawn of Calendrical History

That done, the story of the calendar can begin in earnest. The main text of the book follows the chronological development of the calendar with which we are all familiar, the Gregorian system, although I will be making a case for the dating scheme which is the international standard properly being distinct from the religion-based Gregorian, and so I prefer to think of that standard as being the *Western* calendar, it having been developed in the Western world from knowledge originally built up by various civilizations in the Middle East.

The Middle East is the place to start then, although also mentioned are independent calendrical developments, like that at Stonehenge in ancient Britain, where the people appear to have charted the movements of the Sun and the Moon from at least the third millennium B.C.

In the Middle East the first calendars were constructed by the various civilizations which thrived in and around the Fertile Crescent of Mesopotamia, the Babylonians being the best-remembered of these, although it is important to note that in nearby locations in different eras the dominant cultures of the region may not have been connected with Babylon itself.

While the Mesopotamians were following the cyclic motions of the Sun, the Moon, the planets, and various groups of stars, in Egypt the ancients had other concerns. A year may be defined by any annual phenomenon, such as the call of the first cuckoo of spring, although it makes sense to employ celestial observations once one realizes that these repeat with great regularity. The Egyptians watched eagerly for the reappearance of the star Sirius, associated with their deity Sothis, after a period during which it is invisible behind the Sun, eventually appearing again in the morning sky shortly before sunrise.

But their interest was not only driven by the desire to see the star again. Rather, they knew that when Sirius rose again in the east, soon the River Nile would rise again in flood, the remarkable annual inundation of the Nile being largely responsible for the wealth of the nation as a consequence of its delivery of water and nutrients to the wide valley and delta through which the river flows into the Mediterranean Sea.

By keeping a watch for Sirius, and counting the days between its annual reappearances, the Egyptians determined one length for the year, the Sothic year. But by similarly counting the days between the peaks of the Nile floods, another evaluation of the natural year was derived.

The Cycles of the Moon

To the ancient Egyptians, the Nile, Sirius, and the Sun were all more important than the Moon in many respects, but that was because of the peculiarity of their geography.

Elsewhere, the Moon was tracked with enthusiasm by most civilizations, and it was soon apparent that the lunar cycle (the time from one full moon to the next, say) was not an even divisor of the year. That is, a year defined by the apparent motion of the Sun does not contain an exact number of lunar months.

On the other hand, it is possible to find a longer period over which a certain number of years contains close to an integer number of months. The best-known, and much-used in various forms of calendar design, is the Metonic cycle of nineteen years, which contains very close to 235 lunar months.

That cycle gets its name from an astronomer living in Athens in the fifth century B.C. In chapter 5 we discuss this development, and other longer and more precise cycles discovered by later Greek investigators.

The Moon Abandoned

Although many features of disparate calendars still depend upon the Moon (for example, the calculation of the date of Easter in the Christian churches, and of Chinese New Year), the Western calendar bears no relation to the Moon in that the months, despite the origin of that word, are independent of the lunar phases.

This calendrical abandonment of the Moon, in the thread of history leading to our contemporary calendar, occurred in the Roman republic era, in the fifth century B.C. Thus, while the Greeks were paying much attention to the Moon, the Romans were disregarding it.

The fundamental reform of the calendar rendering the month lengths which we employ today was due to Julius Caesar. This man, effectively the first emperor of Rome, introduced the calendar we know by his name in 46 B.C., although it was another half-century before the leap year rule he stipulated was properly obeyed.

Julius Caesar was a pivotal influence in the history of the calendar, and chapter 6 details his contributions.

The Seven-Day Week

Originally the Romans used an 8-day week, and the familiar 7-day rhythm came about through the reinforcement of two 7-day cycles of quite distinct cultural origins, but of geographical coincidence.

Sometime after the start of the Exile to Babylonia in the sixth century B.C., the Jews began to use a Sabbath cycle in which the seventh day was held to be holy. Also in the Middle East, adherents to various celestial religions developed an astrological week based upon the planets, which in this sense included the Moon, the Sun, Mercury, Venus, Mars, Jupiter, and Saturn. It was the identity of these cycles which produced a profound effect upon the Romans and upon the members of the fledgling religion of Christianity.

Although the seven-day week had gradually grown in importance in Rome from the last few decades B.C., whence it had been imported after the start of the Roman era in Egypt, it was not until the fourth century A.D. that the concept of the Christian Sabbath (as opposed to the Jewish Sabbath) was legislated for, that being occasioned by the conversion of Emperor Constantine the Great. Constantine and the week are the major subjects of chapter 7, an appropriate enumeration in the circumstances.

Numbering the Years

A continuing theme throughout this book is the importance of Easter. The shape and phasing of our calendar owes a great deal to the computation of when Easter should be celebrated in the Christian churches. We first get the gist of this in chapter 8, where the work of the sixth-century monk Dionysius Exiguus is discussed along with the beginnings of the continuing disputes over when Easter should occur. For example, we see that the rules for Easter written into the British act of Parliament reforming the calendar from 1752 (the provisions of that act being inherited by the United States) are actually nonsense.

Although it was not his aim, Dionysius's tabulations of the past and future dates of Easter led to the eventual adoption of the chronology which we use to number our years. That is, he is primarily responsible for the appropriation of the juncture in history which, after the fact, we recognize to be the meeting point between the years 1 B.C. and A.D. 1 (note that there is no year zero), although he got it wrong in that Jesus Christ seems to have been born some years earlier than this dating convention indicates.

The Early British Role in Calendar Development

The Western calendar as we have received it owes its origin to the Middle Eastern civilizations, not least because Julius Caesar's principle adviser was a learned Egyptian. Over the following six centuries or so the major developments in this connection were through the burgeoning Christian religion within the Roman Empire.

While Christianity was also responsible for the subsequent ascendancy of that calendar, the development it underwent in the following several centuries occurred in Britain. The adoption of the Roman Church's calculation of Easter throughout much of the British Isles following the Synod of Whitby in A.D. 664 kept Dionysius's system alive during a time when there was mayhem on the Continent. Early the next century the historical work of the Venerable Bede led to the *anno Domini* dating system eventually becoming dominant.

Bede, although he was not born until later, is implicated in the continuation of the subterfuge which was conducted by the Roman Church party at Whitby, an apparent bluff in which they misrepresented the perfection of the Dionysiac system for predicting the lunar phase and thus when Easter should be celebrated. If it were not for that subterfuge, and Bede's cover-up, then our calendar would be very different.

Bede's promotion of the *anno Domini* chronology came about from a quite different motivation. In the eighth century there was a widespread expectation that the end of time and the Day of Judgment would come in about the year we now refer to as A.D. 800, because of the use of an *anno mundi* (year of the world) system counting from a zero point around 5200 B.C. The end of the six thousandth year would mark the end of the sixth millennium on that scale, and thus the dawn of a sabbatical millennium to be accompanied by the Second Coming of Jesus Christ and the Apocalypse. Bede's uptake of the *anno Domini* scale meant that the Apocalypse, if it were to come with A.D. 1000, was postponed for another two centuries. A useful tool with which the Church could control the masses.

In the last few decades before A.D. 1000 there was a new calendrical development. Although January 1 had been used as New Year from the era of Julius Caesar and before within the Roman republic, Bede had introduced annalistic years counting from a date in September. At the end of the tenth century a cult of Saint Mary grew in England, with the result that her feast day (Lady Day or Annunciation Day, March 25) became accepted as the appropriate date for New Year. This habit continued until the calendar was changed from 1752 onward.

An interesting consequence of the adoption of Lady Day from about A.D. 980 is that the year numbers got shifted. Dionysius's calculations were based upon the Incarnation/Annunciation having occurred on March 25 in 1 B.C. (the traditional date of the equinox). This leads to a Nativity on December 25 in that year, the year count ("the year of the Lord," *anno Domini*) quite correctly following the Jewish custom of reckoning from the circumcision of the child on the eighth day following, hence January 1 in A.D. 1. The later British year count, though, effectively started from March 25 in A.D. 1, a full year after Dionysius's date. All the above British episodes in the story are covered in chapters 9, 10, and 11.

Astronomical Markers of Time

The British were using one calendrical system, as mentioned above, while others were employing alternative calendars, even though they were merely a few tens or hundreds of miles away on the Continent. Even now different societies use different dating conventions. I write on January 6, which is sometimes called *Old Christmas* (more correctly, *Epiphany*). Tomorrow is January 7, and that could be called *Old Style Christmas,* because it is Christmas on the Julian calendar, the Old Style used in Britain and its American colonies before 1752, and still employed today in various churches. There exists the possibility for confusion over dates, then.

Over the period from which we have extensive written records, and especially since the invention of printing over five hundred years ago, it is possible to tally disparate dating systems and thus work out what happened when. Further back in time, matters become more difficult.

One way in which chronologies from distinct sources have been calibrated is through the use of astronomical time markers. For example, if the ancient Chinese recorded bright comets in the third, sixth, and tenth years of a certain dynasty, also mentioning the seasons, then a similar pattern discovered on some ancient Babylonian clay tablet allows the latter to be dated. Periodic comets like Halley's allow an independent assessment, in that we are able to calculate with some precision when and where it would have been visible, back to the fifteenth century B.C. It would have appeared about every seventy-six years; so far we have confirmed records dating back to 240 B.C. With modern knowledge of the apparent motion of the Sun and the Moon we can compute the occurrences of eclipses in the distant past, and thus date records of such obscurations. Various major battles, or the deaths of kings and emperors, have been dated in this way.

This modern ability we have developed to perform retrospective dating is discussed in chapter 12.

The Gregorian Reform

Earlier I mentioned that it was appropriate that chapter 7 concerns the ascendancy of the seven-day week. Similarly, it is fitting that chapter 13 should cover the reform of the Julian calendar promulgated by the Catholic Church in 1582, because the pope responsible was Gregory XIII.

Most popular accounts give a totally false impression of how and why the Gregorian reform took place. The illusory fable often told simply involves the date of the vernal (or spring) equinox having slipped by ten days, the pope being persuaded to delete ten days from the almanac for 1582 so as to correlate again the dates with the position of the Sun, continued agreement being afforded by the uptake of a leap year scheme in which three days are dropped from every four centuries, compared to the Julian rule, which has a leap year *every* fourth year.

This is nonsense, on several grounds. The Church had long been aware of this discrepancy, but had been delayed in acting by considerations of the Easter computation. The Metonic cycle as employed by Dionysius had led to the Moon being absent from its calculated position by an amount which in part compensated for the solar position, a complication unrecognized by most commentators. In fact the ten-day correction was suboptimal, eleven days (or more) being better, but most significant is the fact that the four-hundred-year leap cycle introduced was very much a second-best solution, and the pope and his advisers knew this very well.

The Perfect Protestant Calendar of the British

It seems that the Catholic Church adopted a second-best calendrical solution largely to head off any possible counterreform by members of the Protestant churches, perhaps in the belief that the latter did not realize that the use of a thirty-three-year cycle could lead to the vernal equinox being kept on a single day. This could lead to the regularization of Easter along the lines envisioned by the early fathers of the Church, before the various schisms which have wreaked havoc within Christianity.

The British, however, were on the ball. Keeping their activities a secret, they had noted that in order to maintain the equinox on a single date it was necessary to use a prime meridian at the longitude we nowadays label seventy-seven degrees west. Where was that? It ran through the regions of the Caribbean and South America claimed by Spanish Catholics, but also passed across the coastline of unclaimed North America in the regions now known as North Carolina and Virginia.

In chapter 14 it is argued that the fundamental motivation behind the early British attempts to colonize the Atlantic coast at Roanoke Island and Jamestown has not been recognized until now. If this interpretation is correct then the real reason for Sir Walter Raleigh's adventures in Old Virginny was simply to found a city on the meridian which, once claimed, would allow the pronouncement of a Perfect Christian (and Protestant) Calendar, a weapon with which to attack the Church of Rome and to convert others to the Protestant faith.

The Age of the World

In chapter 15 we turn attention to another vociferous British Protestant from about a half-century later, Archbishop James Ussher. Nowadays Ussher is remembered as a figure of derision, the man who claimed that God created the world in 4004 B.C. In this regard, the joke is on those who poke fun at him. He may well have been wrong, but judged against the circumstances of the era there was nothing aberrant about his methodology. He was a fine scholar, not part of the lunatic fringe.

Many different dating systems employed by unconnected civilizations have assumed initial epochs around that time, a handful of millennia B.C. This makes one wonder how this apparent synchroneity came about.

We now know that the Earth was formed closer to 4 *billion* years B.C. rather than 4 *thousand,* but that knowledge has only come about from various contrasting techniques brought to bear over the past century or so, such as measuring the accumulating level of salt in the oceans and also the radioactive dating of rocks, especially meteorites. In chapter 15 the spectrum of high-technology methods through which we have ascertained the age of the world is outlined, and the implications discussed.

The British (and American) Calendar Act

After the Roman Church amended the calendar in 1582, most Catholic nations quickly changed their dating systems, and many Protestant countries followed, in particular in a wave around 1700.

Britain, however, held out. As we see in chapter 16, it was not until 1751 that the calendar used there was changed by an act of Parliament taking effect from 1752, which was declared to begin from the following January 1 (as opposed to March 25). This act also applied to all British dominions, such as the American colonies. Since the United States was formed, there has been no formal statute declaring the official calendar, so that the calendar employed owes its authority to the inheritance from the British act, plus of course common assent.

The United States being the dominant power in the world, effectively controlling much of the business, trade, and scientific developments, it follows that the Western calendar universally regarded as the standard for international transactions stems originally from that British act. Which happens to contain several errors of fact.

Poor Richard

The calendar reform was imposed upon the American colonies by edict from far-off London. Given knowledge of events soon to come, such as the Boston Tea

Party, one might imagine that this instruction to amend the calendar would not have been well received. On top of that, many recent authors have assured their readers that even in England there were riots provoked by resentment over eleven days being deleted from the almanac for 1752. One might therefore guess that there must have been uproar in Philadelphia and Baltimore.

As a matter of fact, the reception of the revised calendar seems to have been totally muted (as it was in Britain itself: the stories of riots are fallacious, an early day urban myth). A good source to consider is Benjamin Franklin who, under the pseudonym Richard Saunders, was writing his famous annual publication *Poor Richard's Almanack* at that time. In chapter 17 we see that Franklin simply gives an erudite description of the history of the calendar, and why the adjustment was necessary. No mention of any great civil unrest over the change in the calendar is made in any subsequent editions.

Later in this book, though, in chapter 20, I make the case for Franklin having been responsible for the first suggestion of a radical calendar reform involving one day a year (two in a leap year) being no part of any week or month, with a jump of four in the year numbering of the *anno Domini* system so as to accord with Jesus having been born in about 5 B.C.

Standard Longitude, Standard Time

It is surprising how recent is our system of time zones counting from the standard meridian passing through Greenwich in London. In chapter 18 we discuss how this convention eventually came about, after an international conference held in Washington in 1884, prompted largely by the problems of railway timetables based on local sun-based time, which alters from city to city, and the need for some standard longitude system for navigational charts.

From this stems a whole host of other matters which are now familiar to us, but did not crop up until the advent of telegraph communications, transcontinental railway transport, round-the-world travel, and radio time pips. For example, the international date line is made necessary so that we know which day it is as we crisscross the Pacific (although some island archipelago nations have unilaterally decided to deviate the IDL to the east of them, so as to be able to claim that they start the day earlier than others). Similarly, daylight saving time as enacted in many developed nations is another manifestation of time according to the Sun being deserted in favor of time according to some artificial clock. Such things have only been a concern within the past century or so.

Other Calendars Than Ours

Although the main theme of this book is the calendar which is the standard for international communications and timekeeping, it does not follow that there are no other systems used perhaps within one country alone, or within a specific religious or racial group. The obvious examples thrown up are Chinese New Year, in January or February, and the Jewish festival of Hanukkah, occurring in November or December.

In chapter 19 a wide variety of other calendars are sampled. Examples include many which are still in use, like the Islamic, Hindu, and Iranian systems, and some which were tried but then abandoned, such as the French and Russian revolutionary calendars. In this way we see why Ramadan shifts through our solar year, how the Coptic Church still clings to the era definition which Dionysius wanted to drop, and how the Soviet Union tried to make people work on five-day weeks.

Who Will Rid Us of This Illogical Calendar?

Most people do not think much about changing the calendar: it is like a comfortable old pair of slippers, in need of replacing but not worth the effort of breaking in new ones. Consider how complicated the annual round is in actuality, with New Year's Day, July 4, and your birthday shifting by one or two weekdays each year, but Easter having a mobility covering thirty-five possible dates.

There are those among us who would like to remedy this, producing a calendar which is more logical (although that does not necessarily mean *better*). One form of revised calendar would contain thirteen months each of four weeks, so that the first of the month is always a Sunday (or a Monday, or whatever). This would require one day (two in a leap year) to be counted as extraneous to the seven-day week cycle.

Another proposal calls for four quarters each of three months containing thirty-one, thirty, thirty days; again there are blank days that are part of no week. Both concepts meet difficulties with the various religions for which one day in seven is holy (Friday for a Muslim, Saturday for a Jew, Sunday for a Christian), so that there is great resistance to such reforms. The unbroken weekly cycle seems sacrosanct, inviolable.

In chapter 20 we discuss these and various other calendar reform suggestions, including one from 1745 which I suggest may have been anonymously invented by Benjamin Franklin.

Dating the Nativity and Crucifixion

It has been common knowledge for at least four centuries that the reference point for the B.C./A.D. dating convention is misplaced, Jesus having been born a handful of years earlier. Elsewhere in the book we see that astronomical phenomena such as eclipses, comets, and conjunctions (comings-together in the sky) of the bright planets provide useful markers from which historians can derive time frames for specific events in the past.

The Nativity is no exception, in that several distinct phenomena have been proposed for the identity of the star of Bethlehem. My favored explanation is that a thrice-repeated conjunction of Jupiter and Saturn in 7 B.C., followed by a three-way conjunction of those two planets plus Mars in 6 B.C., sensitized the Magi and others to expect some third celestial event. Chinese records indicate a bright comet in 5 B.C. which was visible for seventy days, rising in the east and perhaps moving in an appropriate direction to lead the Magi to Judaea. This is the correct time frame for the birth of Jesus (before Herod the Great died in 4 B.C.), and there are various other aspects which allow a refined assessment of the date of the Nativity (to mid-April in 5 B.C.).

This cometary hypothesis for the star of Bethlehem is detailed in chapter 21. Having discussed there the birth of Jesus, attention is then turned to his death, and a dating for the Crucifixion (April 3 in A.D. 33) is derived on the basis of the phases of the moon (which control the calling of the Passover) and the fact that a lunar eclipse may be computed as having occurred around moonrise that evening, a few hours after Jesus died on the Cross and in accord with biblical accounts of the moon turning to blood.

Long-Term Changes in the Length of the Day and the Month

Over the past several millennia the rate of spin of Earth has slowed such that a day is now almost a tenth of a second longer than it was when the Great Pyramids were built. Our rotation rate is dropping because of the drag force imposed by the tides raised by the gravitational attraction of the Moon. In a related way, the Moon is gradually receding from Earth, by about an inch and a half per year, making the lunar month longer.

This means that although our year (which is itself of uniform absolute duration apart from some minute oscillations) currently lasts for about 365¼ days comprising nearly 12⅓ lunar months, far in the past there were more than 400 days spread over in excess of 13 months in every orbit of our planet about the Sun.

This leads to the idea that the calendar would have been quite different millions of years ago if there had been anybody around to invent it, and similarly any future calendar developments may need to reflect cycles of the heavens which are different to those which we experience now. Certainly there is no point in trying to define calendars for use more than a thousand years into the future.

In chapter 22, then, I ask "How many days in a dinosaur year?" and then answer that and various related posers.

Do We Need More Days in the Year?

Finally, in chapter 23, I consider whether the calendar we use is based upon sound principles. Those principles stem from religious considerations based upon certain beliefs about the natural seasonal cycle, and have never really been doubted. In this summary I must not give away the punch line; so let it suffice to say that what you have read about the length of the seasonal year in practically every textbook or other work ever published seems to be wrong, in a critical regard.

First Step Complete

Now I have finished the first step of the Country Parson's Formula: I have told you what I am going to tell you. Next I have to tell you. In chapter 3 the basics of the cycles of the sky are outlined, although I urge strongly that you read and comprehend the fuller accounts in appendixes A through D before proceeding beyond.

If you do choose to bash on full steam ahead, chapters 4 through 23 are waiting, eager to tell you the story of our calendar and associated timekeeping systems. It's a riveting tale, in my opinion, and I've had great fun winnowing out the details and discovering where others have (deliberately or otherwise) gone wrong in the past. I can only hope that my own errors are few.

Not only is our calendar imperfect and illogical, but it was set up that way for reasons connected with religious and social conflict. Will it ever be put right? I think not. Like the reversed retina, vermiform appendix, and other suboptimal peculiarities of the human body, the calendar came about through an evolutionary process and so retains vestiges of past concerns which have long since ameliorated or completely disappeared. The computer age makes it unlikely, in my estimation, that the calendar will be changed soon, although I do show the way ahead for would-be reformers. Please do read, puzzle, and enjoy.

Chapter 3

❧

THE CYCLES OF THE SKY

The lengths of the three divisions of time which we call a *day*, a *month*, and a *year* were each originally based upon an obvious astronomical cycle. Having read that opening sentence, you will likely now be thinking that the reason I used the word "originally" is simply that we have a calendar in which the months, despite the derivation of the term, are no longer tied to the brightness cycle of the Moon. If that is your belief, let me say that you are incorrect, because *none* of those three are nowadays defined strictly on the basis of an astronomical cycle. The day, for example, is *not* the length of time between the sun making consecutive passes overhead.

If you were to be asked what defines a *day* you might well reply that it is the amount of time which it takes Earth to spin on its axis. Wrong, wrong, wrong: it never has been set in this way, as we'll soon see. More complicated is the length of the year: "the time it takes Earth to orbit the Sun," you might guess. That is too simplistic, too inexact, for a calendar, as we will eventually learn. But we'll begin our examination of the cycles of the sky by looking at the day: surely nothing could be simpler than that?

What Do We Mean by a "Day"?

First we must deal with what may appear to be a triviality: what do we mean by *day*? The answer is not trivial because of the ambiguity of the English language. We use the word *day* to refer to both a period of twenty-four hours *and* the part of that period during which there is sufficient sunlight to function outdoors without artificial light; that is, the hours of *daylight*. When there's insufficient sunlight we call it *night*. According to our ambiguous usage of the word *day*, the arithmetic implies that a day equals a day plus a night, so a night must be zero. Correct? Obviously not, but it does show that we must be precise about that which we have in mind when we are discussing how we mark time.

Surprisingly few languages have different words distinguishing between those two meanings of *day*. One of them is Swedish, in which for most purposes the direct translation of *day* is *dag*, but when a twenty-four-hour period is implied the word is *dygn*.

In fact I have come across an English word in nineteenth-century documents whose meaning is the same thing: nychthemeron (sometimes spelled with a *u* instead of a *y*, or other variants). If you're ever asked for a term which unambiguously defines a twenty-four-hour period, that's about the best you can do, but obviously it is not in general usage and it appears in few dictionaries: I challenge you to use it in a game of *Scrabble*. Literally it means "night and day," the meaning of *nycht* being obvious, as is that of *hemeron* when one thinks of words like *ephemeral*. In some erudite scientific tomes *nychthemeral* is used as an adjective, and quite correctly: many scientists use *diurnal* when they wish to imply something happening on a twenty-four-hour cycle, but that is ambiguous in that it is also the converse of *nocturnal*.

There are other ways of saying the same thing, however. When you are watching some corny old cowboys-and-Indians movie, and an army major asks the friendly Apache chieftain how long it is since the renegade Indians rode off, the scriptwriter will often have had him reply something like, "Three sun-ups have passed." Now that may sound quaint and old-fashioned, but it makes the point that the Native Americans were (and remain) a culturally distinct race from the Europeans and others coming from across the ocean. And the ways of the wave of immigrants to North America, from the times of Christopher Columbus and the Pilgrim fathers, may not necessarily have been superior: as we can see, "sun-ups" is actually better than "days" because it is not ambiguous. It is written in the Bible that Jesus rose from the dead on "the third day" after the Crucifixion (e.g., Luke 24:46), but that was *not* seventy-two hours later; in fact the elapsed time in the biblical account was only half that, around thirty-six hours. Clearly the use of the term *day* can lead to some confusion.

When Does the Day Start?

This also brings up another related matter: when does a day start and finish? To define a length of time obviously one has to give it a beginning and an ending. Having *sunrise* as the start (and the end) of the day is one solution.

In fact, many cultures have their marker points at *sunset;* for example, the Jewish Sabbath begins at nightfall on Friday evening, meaning that it is incorrect to assume that it coincides with the period defined as *Saturday* in most usages, although there is an overlap of about eighteen hours. If Jesus had died on the Cross after about 6 P.M. on the day which we call Friday, midnight to midnight, then it would already have been the Jewish Sabbath, and so the Bible might say that Jesus rose from the dead on the *second* day, the Sunday which is commemorated as Easter.

Muslims also use sunset as the marker point between days. Each new month on the Islamic calendar begins with the sighting of the new moon, and this can only occur soon after sunset. Because a new month must start with a new day, it follows that the Islamic day begins and ends at dusk. That makes a lot of sense. What may seem not so sensible to a reader familiar only with the world's dominant calendar is the fact that the Islamic "year" is defined as being twelve of these *lunar* months long, which is only 354 or 355 days, and as a result the Islamic calendar slips through all the seasons in about thirty-four of our (365-

or 366-day) years. And because the declaration of each month (and hence each year) depends, as I wrote, on the *sighting* of the new moon, which may be inhibited by weather conditions and other vagaries, the Islamic calendar strictly speaking cannot be declared or printed ahead of time, with the months having irregular beginnings and lengths. Of course indicative calculations are carried out, but the official declaration of the sighting of the new moon each month is the cause of much concern and angst in the Muslim world.

It can also affect others. For example, the military strike by the United States and the United Kingdom upon Iraq in December 1998 was timed with a plan of completion before the start of Ramadan, but in the event Ramadan was called a day earlier than anticipated because the new moon was sighted from Mecca a day sooner than had been expected.

Before non-Muslim readers begin to sneer with perceived superiority, let me point out that in the Christian world the developments of the calendar over the past two millennia have hinged to a large extent on the determination of the date of Easter, and that depends upon the date of the first full moon after the March equinox, except that the Church gave up using the real (astronomically defined) equinox and moon a long time ago, instead using artificial constructs for the sake of convenience. Thus the Islamic calendar may be said to be scientifically based, whereas the calendars of the Christian churches have, to varying extents, lost their connections with the cycles of the heavens.

One could raise other objections to our calendrical scheme on the basis of science and logic. The convention of using midnight as the end of one day and the start of the next is illogical in that there is no phenomenon which may be directly observed (like sunset or sunrise) so as to define the instant: one cannot actually watch the sun passing directly underfoot. If *midday* were used, as defined by the time at which the sun transits the meridian (the line passing directly north-south above your head), that might make more sense scientifically speaking; but for human usage, the day ending at midnight is more convenient.

As an international standard the concept of the day starting at midnight was adopted at the International Meridian Conference in Washington, D.C., in 1884. While many people regard that conference as having merely defined the prime meridian for the reckoning of longitude as that passing through the Greenwich Observatory in London (hence *Greenwich mean time*, or GMT, being the time standard), in fact many other matters of import regarding dates and time and its measurement were decided there by the two dozen nations who sent representatives, and in the century or so since, the conventions ratified in Washington have been adopted by most other countries.

That is not to say that the Washington conferees got everything right, and I will argue later in the book that they might have done better to have adopted the *Washington* meridian as being the prime reference for longitude and timekeeping, but the important thing is that everyone have some standard system which they employ in international dealings. The Japanese have a calendar whose year number is reset with the accession of each new emperor, but for ease of business dealings with others they use the calendar introduced by Pope Gregory XIII in 1582 and time conventions laid out in Washington in 1884, although with later slight amendments made necessary by subsequent advances of science and technology.

Astronomical and Nautical Time

I wrote above that the day ending at midnight is convenient. That is true for *most* humans, but until 1925 astronomers, as you might expect given that they observe at night, used a day beginning at twelve noon, and some of their time-keeping systems are still defined in this way. If that were not confusing enough, at one time a day in nautical parlance *ended* at noon. "Is that not the same thing?" you might ask. The answer is *no*. The nautical day began and ended at 12:00 hours on a twenty-four-hour clock. The dates and days of the week were out of step. This is easiest seen with an example. Consider how different time systems would record two instants of time, 6 A.M. and 6 P.M. on a Friday, in common usage:

	Civil time	*Nautical time*	*Astronomical time*
Instant 1:	06:00 Friday	06:00 Friday	18:00 Thursday
Instant 2:	18:00 Friday	18:00 Saturday	06:00 Friday

That may be confusing enough, but to top it off it was the custom when a ship entered harbor to switch over to civil time. Such a hybrid system has led to some difficulties—for example, in interpreting the diaries of the astronomers who went with Captain Cook on his Pacific voyages. Just which time system were they using on any particular day?

The Royal Navy abandoned the nautical day in 1805—in fact, ten days before the battle of Trafalgar—and most other nations followed suit by the middle of the nineteenth century, but it was still necessary for those at the 1884 Washington convention to pass a motion calling on all others to convert to a universal system with days beginning and ending at midnight. Conformity took another few decades, with astronomers being the last holdouts.

Again this leads to difficulties in interpreting old records, because it was astronomers who defined GMT. Before 1925, zero hours GMT implied noon, but from 1925 onward zero hours GMT meant midnight, the start of the day. Greenwich mean time has since been superseded by coordinated Universal time (UTC). One should not imagine, though, that the Royal Navy conforms totally to this timekeeping system, because it still divides up the day into *watches*, the first of which begins at 8:00 P.M. (20:00 hours). Similarly other organizations, military or otherwise, may reckon by their own particular time systems.

Pause for Breath

Many and varied have been the considerations touched upon so far in this chapter. These are intended merely to give a flavor of the subjects, with the main courses yet to come, and by the time that the book is ended you will doubtless have a better grasp of the sense, nonsense, and technicalities of our calendar.

Let us now just pause for breath. The discussion so far has concerned the day—when it starts, when it ends, and what we mean by the term. Since weeks, months, and years all begin with a new day, the definition of the initiation and termination of a day may equally well apply to those units of time.

We have not, though, said how long a day lasts. Nor a month, nor a year, both of which will be recognized to be based upon astronomical phenomena. A week lasts for seven days, and so its duration will depend upon the lengths of the days it contains. Do they vary?

Having gotten our breath back, we must next consider the durations of these units of time. A proper discussion of the questions at hand warrant an essay each, and these appear as appendixes A through D. In appendix A the question asked is "How Long Is a Day?" while appendix B queries "How Long Is a Year?" Do not imagine that the answers are simple. In addition, the slowing down of the terrestrial spin rate means that the day is getting longer and thus the number of days in a year is changing, so that some absolute definition of time duration is required, independent of astronomical vagaries. In the modern world the length of a second is defined not by the rotation of Earth, but through the intercomparison of atomic clocks. Appendix C asks, then, "How Long Is a Second?" Finally, although our calendar has month lengths set up independent of the motion of the Moon, the lunar phase cycle is important for many lunisolar calendars, including the Christian calendrical schemes used to define when Easter and the other movable feasts will occur each year. In appendix D, then, we ask, "How Long Is a Month?"

Those four appendixes provide the details needed to comprehend just why our basic units of time are defined the way they are. If you are so inclined, I would encourage you to read them now, or perhaps at some later juncture when you feel that you have missed some vital point in the main text through not understanding the subtleties of the celestial rhythms.

It is anticipated, however, that most readers will want to proceed immediately to discover the history of humankind's response to the cycles of the sky in chapters 4 through 23. With that in mind I supply below concatenated descriptions of what defines those four units of time, under section headings identical to the titles of the four appendixes. These will at least provide you with the basic tools, if not the finesse.

How Long Is a Day?

The *day* in its common meaning is not the time taken for Earth to spin once on its axis, because during such a period our planet has moved along its orbit such that to get the Sun back to more or less the same place as it was twenty-four hours earlier, an additional rotation of Earth by almost a degree is required.

To look at it another way, if a year were exactly 365.25 days long, then during that time the planet would need to spin precisely 366.25 times, once extra for the orbit around the Sun. Thus, in a year the Sun crosses one's noon meridian on average 365.25 times, whereas the stars cross the meridian 366.25 times. The time between solar transits renders a *solar day*, and it is this which is most often meant by the diminutive *day*. Stellar transits define the *sidereal day* necessarily employed by astronomers in their observations, which is almost four minutes short of a solar day.

The Sun would transit the noon meridian (alternatively, one could say that the Sun would rise, or set) precisely every twenty-four hours if the terrestrial orbit were circular, and our equator were not tilted against the plane of the orbit. In reality neither is the case, with the effect that the time between transits wanders away from twenty-four hours during the year. In times past people used the local solar time—the time effectively using the Sun in the sky as a clock—but for some centuries the availability of mechanical (and now electronic) clocks has meant that instead an artificial definition of time has been used. Instead of using the real Sun, an imaginary *mean sun* is effectively employed (hence the term *Greenwich mean time*). The mean sun is defined upon the supposition that Earth's orbit is circular and its spin axis perpendicular to the plane of that orbit. The difference in the time indicated by the real Sun and the mean sun is called the *equation of time*, and it may be as much as sixteen minutes at some times of year.

Because our spin axis is in reality tilted against the pole of our orbit by just over twenty-three degrees, there is an angle equal to that measure between the orbital plane (called the *ecliptic*) and the extrapolation of the equator out into space (the *equatorial plane*). The points on our orbits where Earth comes to the junctions between those two planes are called the *equinoxes*. One, called the *vernal* or *spring equinox*, occurs around March 20; the other, called the *autumnal equinox*, is reached around September 21 each year. At the equinoxes Earth passes through the equatorial plane, and the Sun rises due east and sets due west. Although their collective name suggests that daylight and nighttime hours are equal at the equinoxes, this is not quite the case.

Close to midway between the equinoxes are the *solstices*, when the sun reaches its furthest point north (the *summer solstice*, near June 21) and south (the *winter solstice*, near December 22). Although it is common to say that these are the longest and shortest days of the year, in reality the equation of time interferes with such a simple half-truth, although to a very small extent.

How long is a day? The answer is that originally it was the time between sunrises or sunsets (or midnights or middays as defined by the Sun itself), and that meant that the length of the day varied during the year due to our noncircular orbit and tilted spin axis. Since clocks have been available humankind has defined the length of the day artificially so that all the days in the year contain twenty-four hours. More precisely, a day is now defined to last for 86,400 atomic seconds, the meaning of which we will meet below. Because that final unit is divorced from the rotation rate of Earth, which is actually slowing, every so often a leap second needs to be inserted between one day and the next so as to keep the time according to the heavens in step with that according to atomic clocks.

How Long Is a Year?

It takes about a year for Earth to orbit the Sun, but there are many different ways to define what is meant by a *year* and so it is vital to differentiate between them. In appendix B it is shown that one could come up with a dozen or more distinct definitions; here we will limit ourselves to those which are pertinent for the calendrical matters which are the major focus of this book.

Because the terrestrial orbit is noncircular, the Earth-Sun distance varies, reaching a minimum around January 4 each year at a point called the *perihelion*. The time between perihelion passages is called the *anomalistic year*, and that lasts for about 365.2596 days. That January date is currently correct, but the perihelion point is gradually swiveling around such that it moves later by about one day every 57 or 58 years. Obviously our calendar is not predicated upon the anomalistic year: if it were, then the date of perihelion would be held steady. This swiveling of the orbit, which takes about 110,000 years to complete relative to the distant stars, is called the *precession of perihelion*. Which direction is the movement? Looking down upon the solar system from the north, the perihelion point would appear to be moving anticlockwise, our egg-shaped orbit systematically shifting around in that direction. As noted above, this makes the date of perihelion gradually get *later*, in that Earth's annual movement around its orbital path proceeds in the same anticlockwise direction.

A quite different type of precession is that undergone by the equinoxes. If we ignore the orbit of our planet for a minute, we can think about its spin in isolation. The spin axis points to a particular point in the sky, which currently happens to be near the star we call *Polaris* for that reason. Just over twenty-three degrees away on the celestial sphere is the pole of the ecliptic, and the spin axis direction is gradually rotating around the ecliptic pole, taking about 25,800 years to complete the circle. Thus Polaris is only temporarily the polestar, and over the millennia our spin axis revolves so as to point instead towards other stars. This rotation is called the *precession of the equinoxes* because it moves the times of the equinoxes. Again looking down on the solar system from above, the direction of the motion may be stipulated, but now it is clockwise, in the opposite direction to the precession of perihelion.

The timescale for perihelion to complete a full turn is 110,000 years compared to the stars; that for the equinoxes to do likewise is 25,800 years. But we are not interested in the stars, because they do not affect terrestrial affairs, such things as the climate. What we require is the *relative* rate of precession of the equinoxes and perihelion, and because these are moving in opposite directions the overall figure is near 21,000 years. One may expect climatic cycles to follow such a periodicity.

Earth's orbit being noncircular, our speed varies during the year and is highest at perihelion passage in early January, only a couple of weeks after the winter solstice. Because of this changing speed, the time taken to move from that solstice to the same solstice the following year will not be the same as the time between, say, two autumnal equinoxes. The differentials will be small, only amounting to a couple of minutes at most, but they are of critical importance for calendrical purposes, as below.

Averaging for *all* start-and-end points around our orbit, and also figuratively averaging over some decades so as to smooth out slight variations from year to year due to tugs from the other planets, one deduces a parameter known as the *mean tropical year* with a current value of about 365.2422 days. If instead the specific start-and-end point of the vernal equinox only is used, the period deduced is near 365.2424 days, and may be called the *vernal equinox year*. The distinction is pivotal.

The Gregorian reform of the calendar was a religious reform with the avowed aim of keeping the vernal equinox on about the same date. (That it fails to do so is another matter entirely.) This was because the date of Easter depends upon that equinox. The target year length for the Gregorian reform should therefore be regarded as the vernal equinox year of 365.2424 days. The leap-year cycle of the Gregorian calendar results in there being only 97 leap days added to each four-hundred-year period, rather than the full 100 days of the Julian calendar. The average duration of the year in the Gregorian calendar is thus in excess of 365 days by 97 divided by 400, or 365.2425 days in all. That is three times closer to the 365.2424-day vernal equinox year than to the 365.2422-day mean tropical year. Unfortunately many, many commentators have missed this fundamental point and therefore misjudge the Gregorian calendar from this perspective.

How long is a year? It all depends on the phenomena which are of concern to you. If you were a pagan sun worshiper for whom the summer solstice was the most important time of year, you would want a calendar with a mean year length of 365.2416 days. Two millennia ago the sect of Mithra and associated sun cults regarded the winter solstice as the fundamental turning point of the year (and it is from that festival that the date of Christmas derives); under that circumstance you would nowadays want to design a calendar with an average length of 365.2427 days. If you wanted a calendar simply to keep the seasons in step so far as is possible, then one might imagine that a year length equal to the mean tropical year would be the target, but this is a matter we will leave aside until chapter 23.

For the Gregorian calendar the appropriate yardstick is the time between vernal equinoxes, 365.2424 days, and judged against that standard the Catholic Church's 97/400 leap-year cycle does quite well. On the other hand, an 8/33 leap-year cycle, producing a mean year duration of 365.242424. . . days does even better, and the implications of such a realization we will meet at length.

How Long Is a Second?

To some extent the answer to this question was given above when we considered the duration of a day. I wrote there that a day is defined to last for 86,400 *atomic seconds*, but one might wonder what that term means. More correctly, the fundamental unit of time (and indeed of *length*, because the meter is defined in terms of the speed of light, a certain number of meters per second) is called the *SI second*, the SI standing for *Système International d'Unités*, or International System of Units.

The atomic second is based upon the spin rate and orbit of Earth in an epoch some decades before the atomic age began. For reasons which will be skipped here, the atomic second is predicated on the duration of the tropical year back in 1900, although its specific definition is rather more long-winded than that, and is spelled out in appendix C.

Since 1900 the rotation rate of Earth has been slightly curtailed by tidal drag such that a day now lasts for about 1.7 milliseconds longer than a day back then, a century ago. This means that there are *not* precisely 86,400 seconds in a day, if by "seconds" one means "atomic/SI seconds." The average day actually lasts for around 86,400.0017 seconds, although our spin rate varies slightly with both erratic and seasonal wanderings.

In the more distant past, time was defined in quite different ways. For example, before the first mechanical clocks were built, although there were twenty-four hours counted in the day (each subdivided into sexagesimal minutes and then seconds), the lengths of the daylight and nighttime hours varied seasonally. Dependent upon one's latitude, in the summer a daylight hour might last for seventy of our minutes with a nighttime hour containing only fifty, the converse being true during the winter: both day and night were allotted twelve hours, and their actual durations would vary with the length of time which the sun was above the horizon. These, then, were *unequal hours,* and it was not until the invention of clocks that *equal hours* started to be used; hence our term *o'clock,* as opposed to the time according to the Sun.

The transition from the sun time to the clock time represents one revolutionary era in timekeeping practice, but the clocks were still calibrated according to the stars and thus the rotation of Earth. It has only been with the electronic and atomic revolution that humankind has been able to construct clocks more precise than the spin of the planet, such that we now implicitly employ the atomic second for most timekeeping, this being independent of all astronomical considerations. Physics laboratories, not astronomical observatories, now define time.

How Long Is a Month?

Our Western calendar months have lengths defined independent of the time taken by the Moon to orbit Earth. In this section we are not interested in those calendar months, only the months as defined by the motion of the Moon and Earth. Note that some other calendars, such as the Hebrew and Islamic schemes, *do* have month lengths which track the lunar phases.

The time taken for the Moon to complete one orbit about Earth is twenty-seven days plus a fraction, that fraction depending upon whether one defines the orbital period according to the lunar equinox, or perigee, or some other juncture. From our present perspective this is not of great concern, though, because the essential feature we are interested in is the cycle of the Moon's brightness—for example, the time from one *full moon* to the next.

After 27 to 28 days the Moon may have completed one orbit about our planet, but in the meantime Earth has moved substantially along its own orbit about the Sun. For a full moon to occur, our natural satellite must reach the position where it is again 180 degrees away from the direction of the Sun, which we call *opposition*. The extra distance which must be moved means that there is rather more than 27-and-a-bit days between full moons. On average the time between full moons is close to 29.5306 days. This is also equivalent to the mean time between new moons, although with a complicating twist summarized below.

This length of time is termed the *synodic period* of the moon, or a *lunation*. One should be clear, though, that it is an *average* value which I have quoted there. Due to the eccentricity (noncircularity) of the lunar orbit, the actual interval between successive full moons varies between about 29.2 and 29.8 days, a very substantial difference.

The variation in the interval between *new moons* can be even greater. The important point I need to make transparent here is that one can calculate the time between full moons, even if you cannot see the strict full moon because your geographical location happens to be on the opposite side of the planet when fullness is achieved. This is also the case for the *dark of the moon,* when it is at the same ecliptic longitude as the Sun (that is, the angles as measured in the plane of the terrestrial orbit are the same; if the Moon happened to be crossing that plane at that time then there would be a solar eclipse, but most often this is not the case because the lunar orbit is inclined by just over five degrees to the ecliptic, putting the Moon above or below the ecliptic when the longitudes match).

Dark of moon is otherwise known as *inferior conjunction*. But *new moon* is not the same thing. New moon is when the thin lunar crescent can first be sighted just above the western horizon soon after sunset. For this to be possible the Moon must have moved in its orbit a sufficient distance away from the solar direction, the required separation being critically dependent upon one's location on the surface of the planet, the season, the atmospheric conditions, and so on. The new moon will be visible one to three evenings after conjunction has occurred, the typical delay being thirty hours. But if it is just missed on one evening, the new moon will not be seen for almost another twenty-four hours, and so clearly the time between successive new moons can vary by a much greater amount than the range of intervals between full moons. Nevertheless, if averaged over many years the mean time between new moons is the same as the mean time between full moons, the average synodic period.

This consideration is pivotal in calendar matters, because many cultures live under calendrical systems based upon the sighting of the new moon resulting in the declaration of a new month. For the Islamic calendar this is the case. Two millennia ago it was also the case for the Hebrew calendar, but nowadays the lunar phases are instead subject to advance computation using assumed lengths for the synodic period, and the other divisions of time.

The Alligator and the Red Tab

The rhythms of the skies are deceptively easy to describe in the simple way above. The details are more complicated, as you will see from the appendixes, but even those descriptions only cover the most rudimentary details. For example, the lunar motion is extremely complicated to define completely, and whole books have been written on the subject, whole careers have been spent working toward ever-better theories of its orbit.

Leaving such complications aside, we are going to learn in this book that the tiny difference in duration between the vernal equinox year (near 365.2424 days, but varying) and the mean Gregorian year (365.2425 days precisely) has in effect been the subject of great disputes between various factions of the Christian church, in particular between the Roman Catholic and Protestant sides, although the Eastern Orthodox churches have also beaten their metaphorical chests in this regard.

That seems to be a mind-boggling concept when you first hear it, and I pondered long and hard how to persuade you to take such an idea seriously. How could such a trivial difference be a fighting ground between warring factions? I discovered my rhetorical device while Christmas shopping, trying to find some suitable summer blouse for my wife.

In perambulating through various ladies' fashionwear stores, a pastime which I prefer to minimize, I noticed that there was a huge price differential between essentially equivalent products depending upon the prestige of the shop involved, and this was portrayed for all to see by means of some distinct labeling on the exterior of the clothing. Again I will not name any specific companies for fear of libel action, but I think that my message should be clear enough.

To transfer the tale to menswear, I note that the cost of polo shirts of seemingly comparable quality could be tripled by the simple addition of a little cloth alligator sewn over the left breast of the garment. If one were concerned only with the utilitarian functions of such a shirt, it would seem that the alligator adds nothing: it does not keep the wearer warmer, shield him from harmful sunlight, or any of the other uses which could be ascribed to the apparel.

What the little alligator does do, though, is to serve as a status symbol. When considered in an abstract sense, this story seems almost too fantastic to be true. People will scour shops for bargains, boasting how they got 30 percent off the price of their socks in a sale, but then they will proudly wear polo shirts embossed with a tiny decoration that declares to all and sundry that they paid some ludicrous amount more than a shirt lacking the alligator would have cost!

This is not to say that brand marking is not a useful phenomenon. I have always found that denim jeans with a little red nylon tab sewn into the seam of the right buttock pocket tend to be of uniform higher quality than any other on the market, and so I tend to buy that brand only. But I tend not to worry too much whether people notice the absence of an alligator on my polo shirts, or three stripes on my tennis shoes. For others, this seems to be a matter of overriding concern, sufficient to cause them to dig deep into their wallets.

If one cut out and weighed that red tab alone and expressed the result as a proportion of the weight of the trousers as a whole, it might well be that the slender piece of nylon represents about the same fraction as that fourth decimal place in the year lengths quoted above. The simile is apt. There may be tiny differences between the average year lengths resulting from the leap-year rules espoused by different religious sects, but they have been major fighting grounds as a consequence of the fact that they act like brand labels. Small parts of the whole, but the feature which everyone notices.

Chapter 4

❧

STONEHENGE AND SOTHIS
(THIRD MILLENNIUM B.C.)

The First Cuckoo is the title of a compendium of letters written to the *Times* of London. That newspaper, which is now well into its third century of publication, was renowned in earlier years for its bombastic editorial comment, often being sharply critical of the British government, earning itself a nickname of the *Thunderer*. While many of the letters it published also thundered about some contemporary matter, often the most interesting contributions were the brief, whimsical notes which terminated its letters pages. These showed various seasonal trends, one of which gave the book in question its appellation. Each spring a host of retired colonels, country parsons, and village schoolmarms would hasten letters to the editor, trying to claim precedence as the first hearer of the call of a cuckoo for that year.

Let us regard these letters as a plausible rudimentary data source for assessing the length of the year. Ignoring the dates on the newspaper, which are of course based upon a prior determination of the year length, one could just count the days between appearances of the annual champions' names, the claimants for hearing the first cuckoo. If you did this for ten years, the irregular skips from one spring to the next due to the vagaries of the weather and so on would be averaged out somewhat, and you would know the year length to better than a day. Extending the length of your study to a century into the future would try your patience and longevity, so you could count back through past issues, adding one-sixth to the number you reckoned (the *Sunday Times* is a different newspaper), in order to deduce a year length precise to a few hundredths of a day providing the cuckoo-noting custom has persisted for long enough.

Elsewhere, disparate civilizations have used quite different animals to define the starts of their years. In the western Pacific, the Trobriand Islanders begin with *Worm Day*. At the dark of moon following the full moon between October 15 and November 15, the palolo worms begin to spawn over the coral reefs in that part of the ocean, turning the sea into a writhing mass. The precise date will vary from one island to another, and so may be different in Samoa or Tonga. Even though the spawning depends on the Moon and may vary from year to year by some weeks in consequence, if the days between eruptions are counted for a few decades an evaluation of the year length will result.

Similar cycles are seen in many marine creatures, largely due to their dependence upon spring (highest) tides, when the attractions of the Moon and the Sun act in consort. This is useful information if you wish to see turtles laying their eggs, and also if you want to avoid the seaward migration of the millions of red land crabs which inhabit Christmas Island in the Indian Ocean (not the same-named island 1,200 miles south of Hawaii, made famous when the United States and the United Kingdom tested hydrogen bombs there in 1957). Plants provide year markers, too: count the days between the cherry blossoms blooming in Japan, or the tulips erupting in Amsterdam, or the Rocky Mountain aspens turning russet and rufous.

The point here is that *any* event which recurs annually may be used to determine the length of the year with some accuracy, provided that a long enough sequence of records is maintained. The phenomena studied may be terrestrial, like those mentioned above, or celestial. We know very well the duration of a year as defined by astronomical considerations, but that knowledge is based upon many centuries of accumulation by observation and testing. Let us step back some thousands of years and wonder about how ancient humans determined the year length. Let us begin with Stonehenge.

The Stonehenge Year

Put yourself in the place of Neolithic Britons. The length of the day is obvious: it is the time between sunrises (or sunsets). Without a digital watch on the wrist the intrinsic variations in the day length were not noticeable, although the contrast in the durations of the daytime/nighttime hours between summer and winter would have been. That, with the concomitant change in the weather, would have provided some motivation to chart the annual course of the Sun, and it is easy to see that the places where it rises on the eastern horizon and sets in the west alter from day to day. To measure the year in this way some sophistication is needed: one needs to erect horizon markers and keep a consistent watch each sunrise or sunset.

On the other hand, the *month* is easy: not only is it relatively brief, but also one can ascertain the Moon's phase to within a day from simply looking at it, with a little experience. A day after full moon the disc looks rather different from when it's at its brightest with complete illumination. Similarly, the days between spottings of the new moon can be counted. A sequence of counts of twenty-nine and thirty days (broadly alternating) would result, and after a few years the average duration of the synodic month would be known to better than a tenth of a day. Any interested party could have done that. There is even evidence—hotly disputed—that animal bones dated to about thirty thousand years ago show scratch marks inscribed by humans which may be interpreted as being records of waxing and waning moons continuing for an extended period. Maybe those were the first calendars.

The Sun does not, of course, show changing phases like the Moon. Its visual appearance does not really alter during the year, but the places of rising and setting do. The range of azimuths (angles measured clockwise from due north) at which the sun rises depends upon the latitude of the site in question. For the latitude of Stonehenge the sun rises at an azimuth near 50 degrees at the summer solstice, and near 130 degrees at the winter solstice. It may be better to think of these directions as being 40 degrees north and 40 degrees south of due east. Between these extremes the sunrise position varies, passing due east at the equinoxes.

If the sunrise azimuth moved consistently it would change by 0.44 degrees from one day to the next, which is almost the whole angular diameter of the Sun, so that measuring off its motion would be easy. The problem is that the sunrise position does *not* move steadily, but rather sinusoidally. Around the equinoxes that position changes substantially from day to day, whereas approaching the solstices it merely creeps along the horizon before turning around to move back toward the east again. This would make it very difficult to determine the length of the year if one merely counted days between returns to either the most-northern or the most-southern sunrise location: the diurnal movement is much less than the solar diameter, and it is difficult to judge just when the extreme point is reached. It would be more accurate to count days between risings at due east. A drawback is the fact that you only get one chance a day to make your observation, and with the rapid movement near the equinox, sunrise might be some distance south of east on one morning, but well north of it at the next daybreak. Over some years, though, this influence averages out and the year length may be calculated to a small fraction of a day.

The above is one of the reasons I believe that many of the modern astronomical hypotheses for the *original* purpose of the developments at Stonehenge by Neolithic people are incorrect. The axis of symmetry of the various phases of construction does seem to be toward the summer solstice sunrise azimuth of fifty degrees mentioned above, but that is not the best way to determine the year length, sunrise tracking close to the equinoxes being more efficacious. I believe that the orientation towards fifty degrees began in the first, simple phase of construction (3200–2800 B.C., many centuries before the familiar stones were dragged to the site), the purpose being predictions of phenomena associated with a giant comet and its associated trail of dust and meteoroids which (not by chance) had an orbit close to the ecliptic and therefore near to the course of the Sun. After this comet had decayed away the awe of the Stonehenge people was transferred to the Sun and a religious cult surrounding it and the Moon began, the construction of a mighty temple oriented toward the summer solstice sunrise following in various stages between 2500 and 1200 B.C. Later utility does not necessarily follow directly from initial purpose, a lesson learned from evolutionary biology: for example, many fish have organs allowing them to sense the presence of prey through electrical field disturbances, but in the case of the electric eel that sense developed further into an ability to stun other fish.

Notwithstanding my objections above, it is a fact that if you stood at the center of Stonehenge for several years, weathering the damp English climate and the attempts of security guards to remove you, and you counted the days between risings of the sun over the heelstone, then eventually you would arrive at an evaluation of the length of the year. I have suggested that there are simpler ways to do better, and I will put forward others below, but one could certainly accomplish the basic aim in this way.

The Sothic Year

At about the time that the Neolithic Britons were beginning their labors at Stonehenge, in the Middle East several great civilizations were starting to flourish. The first written records of the division of the year and the day seem to have been made by the Sumerians around 3500 B.C., and one may wonder what marker points they and other societies used in order to delineate the year. There are several possibilities one might choose apart from the sunrise azimuth. For example, in the tropics the Sun passes overhead at some stage during the year, so that if a monolith were erected vertically then the Sun would first shine on its northern face, putting the southern face in shadow, at a specific time each year (when the Sun reached a declination north of your latitude); by keeping track of these times, the year would be determined. Further north (in Sumeria and Babylon) the day of summer solstice each year could be identified by seeing when the length of the shadow on the northern side of a gnomon was minimized.

Rather than using the Sun itself as an indicator, one could think of other factors it imposes. One cannot see stars during the day because of the sunlight flooding the sky, but at nightfall they reappear. Are there any stars which cannot be seen at any stage during the night? Obviously from the north one cannot see the Southern Cross, the constellation which appears on the flags of Australia, New Zealand, and various other Southern Hemisphere countries, and we Down Under cannot see Polaris, the North Pole star, but what about near-ecliptic stars (those along the zodiac) which may be observed from either hemisphere? At any time of year several zodiacal constellations cannot be seen at any time of night because they are in the part of the celestial sphere beyond the Sun: this is called *superior conjunction*. As the Sun moves through the zodiacal constellations during the year, eventually a particular bright star which has been hidden for some time will rise just before dawn over the eastern horizon and shortly before the Sun's rays drown it out. Its renewed visibility is called *heliacal rising*. In general, over the eons the heliacal rising of Sirius has been used by many Northern Hemisphere civilizations as a year marker, while the nebulous cluster of stars known as the Pleiades has been used in the south.

In many native tongues of South America the words for "year" and "Pleiades" are the same, impressing upon one the fact that this was their sign of the annual cycle. The Pleiades are at the tip of the western horn of the constellation Taurus and may be viewed through to temperate northern latitudes. In Hawaii, before the arrival of Captain Cook the dry and the rainy seasons were named and regulated according to the visibility of the Pleiades. Being nebulous, they were called *zappu* (the hair) by the Mesopotamians and the Greeks named them *kometes* (the long-haired), this also being the origin of the word *comet*. In English the group is often called the Seven Sisters, although only six are visible (in mythology one of these seven daughters of Atlas is said to hide in shame or grief), which is why the symbol for the automobile manufacturer Subaru has only six stars (*Subaru* is the Japanese word for the Pleiades).

Rather than the Pleiades, the Egyptians were more interested in the helical rising of the star Sirius. A modern-day astronomer might refer to it by that name (which is the Latinized version of the Greek name *Sothis*) or as *Alpha Canis Majoris,* the alpha implying it to be the brightest star in the constellation Canis Major, and it is often called the Dog Star. In fact it is the brightest star in the whole sky; for the pedants among my readers let me note that I am neglecting our local star, the Sun. In the Northern Hemisphere the hottest time of year follows the reappearance of Sirius soon after the summer solstice, around mid-July, hence the term *dog days.*

The Mesopotamians visualized the constellations differently, calling this the *Bow* with Sirius being *gag.si.sa,* meaning "tip of the arrow." The Egyptians called it *Sophet,* hence *Sothis* above, while the Greeks themselves actually called it *Seirios,* meaning "the scorcher." The dynastic Egyptians had another reason for regarding this star so highly: it appeared at the time of the annual inundation of the Nile.

The Nile Flood Year

It would be difficult to overstate the importance of the flooding of the Nile for these people. At the time of year that most other civilizations would be preparing for the summer heat and dryness, they would be readying themselves for a huge flood, with attendant benefits for agriculture. The proliferation and wealth of Egyptian civilization, and the emergence of the first government bureaucracy set up anywhere so as to allow the exploitation of a resource for the common good, may be ascribed to the annual flooding of the Nile, bringing with it not only water for irrigation but also rich nutrients in its plentiful sediments, making the Nile valley immensely fertile.

To the ancient Egyptians the annual inundation must have seemed like a gift sent from heaven: no wonder they worshiped Sirius/Sothis as a deity, Osiris being the fellow god of the flood. Although they could not have known it at that time, the details only comparatively recently being understood, the waters of the world's longest river actually swell and burst their banks due to rain derived from the southern Atlantic Ocean. The rain does not fall in Egypt, or even Sudan to the south, so to the early Egyptians the floods seemed to be a matter of divine providence.

The White Nile rises in the central African highlands, on a six-thousand-foot plateau north of Lake Tanganyika, although with its tributaries it drains much of equatorial Africa. The Blue Nile rises to the northeast, in the mountains of Ethiopia, meeting the White Nile near Khartoum. These are later joined by the Atbara, which rises in Eritrea. The three together form the 4,160-mile-long Nile system. When the rain-bearing winds beat in from the

Atlantic in March, the monsoon begins and huge volumes of water are dropped on these mountains. The flow of the White Nile is buffered by Lake Victoria and Lake Albert, and it disperses much of its water in the swamps of Sudan. But there is nothing to slow the Blue Nile. By June it is in flood, and by late July or early August the wave of water reaches Cairo, persisting through September. Paradoxically, with the flooding the White Nile turns *green* (due to vegetative detritus), while the Blue Nile turns *red* (from the mud which it carries downstream).

At the inundation the overall flow is twenty to fifty times the volume at low water, making the flood a remarkable sight. One could wonder why civilization did not thrive earlier in Egypt. Part of the answer might be that the floods were too severe: from 20,000 to 6,000 B.C. the peak level, from geological evidence, was up to thirty meters above the present floodplain, whereas since then it has moderated to only a five meter enhancement. For more recent times we know a great deal about the flood levels due to the records from seven separate nilometers. The most famous of these is one dating from A.D. 711 at Roda, while there is also one on Elephantine Island near the first cataract, and another called the House of Inundations in Old Cairo.

Shakespeare wrote in his play *Antony and Cleopatra* (where else?) that

> . . . they take the flow of the Nile
>   by certain scales in the pyramid.
> They know by the height, the lowness, or the mean,
>   if dearth [scarcity] or foison [abundance] follow.
> The higher Nilus swells,
>   the more it promises;
> As it ebbs, the seedsman upon the slime and ooze
>   scatters his grain,
> And shortly comes to harvest.

This does not refer to the familiar Great Pyramids at Giza, standing high and dry, but rather the pyramidal building at Roda which houses a large graduated column, water being allowed in from the river to indicate the flood level. The centuries of records derived from this nilometer have been used to search for periodicities in (and perturbations of) the climate. Actually the levels of the floods were followed also for earlier millennia, even if we no longer have the figures from the dynastic era.

In the present instance the implication is clear: if one kept a record of the number of days between the highest water levels, or just the date when the rise began, then an evaluation of the year would follow. In the case of the nilometers good records were kept, and for an excellent reason. The ancient Egyptian state bureaucracy was established to enable the Nile floods to be exploited for the greater good, but someone had to pay for such work: farmers were taxed at a rate which depended upon the peak level indicated by the nilometers.

A civil year lasting 365 days seems to have been ascertained by about 3000

B.C., based on the average time between arrivals of the flood at Heliopolis, just north of Cairo, where the Nile begins to break into its delta. It has been claimed (on somewhat dubious evidence, outlined below) that as early as 4241 B.C. the people of Egypt were using that year length. They did not recognize initially that a fraction of a day was required.

The Egyptian Adaptation of the Babylonian Calendar

For their civil calendar the Egyptians inherited some features from the Babylonians, but gradually altered them to fit their own needs. (Here I am writing "Babylonians" to cover all the ancient Mesopotamian civilizations, rather than mentioning each individually.) The Babylonians had used a lunisolar calendar with the first day of the month being when the crescent moon was observed (as persists in the Islamic calendar, the crescent appearing in the flags of many Muslim nations and in the name of the Red Crescent, the Turkish equivalent of the Red Cross), but they also wanted to keep a year length in step with the seasons.

There was an early evaluation of 360 days in a year, which was soon recognized to be incorrect, but one can see how confused they might have been: 360 is five or six days longer than the approximately 354 occupying twelve lunations, but also five or six shorter than the seasonal cycle. In the end the Babylonians bowed to the Sun, recognizing that the year was about 365 days long on average, and so arranged a calendar such that many years contained twelve lunar months totaling 353, 354, or 355 days, and every so often a thirteenth (intercalary) month was inserted making a year lasting 383, 384, or 385 days. The actual length would depend upon the varying numbers of days between new moons. The Hebrew calendar used today is very similar to this, with Deficient (353-day), Regular (354-day), and Complete (355-day) years, each increased by 30 days in a Leap (thirteen-month) year; this is why the date of Hanukkah jumps around so much on the Western calendar.

For the Egyptians the Babylonian calendar system was too complicated, and neglected their major consideration: the annual cycle producing the flood, governed by the Sun and unaffected by the Moon. They abandoned the lunar months in their 365-day calendar. This calendar had twelve months, each of 30 days precisely, and at the end of every year five supplementary, or *epagomenal,* days were included. In Latin this design was called the *annus vagus,* or "wandering year," because the solar year wandered relative to the Moon. Our words *vague* and *vagabond* have similar roots. The Egyptians abandoned the Moon because to them, but not the Babylonians, it was the rising of the Nile which was the singular dominant factor in arranging the year, and this is the underlying reason for our calendar having months divorced from the Moon. The Egyptian use of a round 365-day year led to a peculiarity, however.

At the start of the third millennium B.C. the initiation of the Nile flood at

Memphis (this is the original Memphis, not the one in Tennessee associated with Elvis Presley), just south of Cairo near the Giza pyramids, occurred at much the same time of year as the summer solstice and the heliacal rising of Sothis. Textbooks will tell you that in about 3000 B.C. the solstice occurred on July 17, but beware: such dates are usually given on the Julian proleptic calendar (the Julian calendar extrapolated backward), rather than the Gregorian calendar. On the Gregorian calendar the date must be around June 21, but again one needs to be careful because even if the Gregorian reform had been successful in keeping the vernal equinox on the same date, the summer solstice date would still have shifted a little.

Leaving such things aside, the point is that the Egyptians would have a triple celebration because of the summer solstice, the reappearance of Sothis, and the rising of the Nile occurring at the same time—at least around 3000–2500 B.C.—but such times would actually be spaced by about 365.25 days, whereas their civil calendar contained only 365 days. To us the cyclical insertion of an extra day to make a leap year seems obvious, but it was not to them. The result was that the date of the celebrations, set by the natural phenomena, slipped backward through their calendar by a quarter-day per year, and they did not correct it. To give a parallel with our present-day calendar, this would be equivalent to shifting the date of Christmas backward by a day every four years—December 25 in the year 2000, November 30 in 2100—but still at the same time of year in terms of the climate (so that a white Christmas would be equally as likely).

Because of this quarter-day in 365 anomaly (one part in 1,460), the Egyptian civil calendar followed a grand cycle with duration equal to what is known as the *Sothic period* or *cycle,* this being 1,461 Egyptian years (of 365 days) long, or 1,460 Julian years (of 365.25 days). This meant that every 1,460 years the great celebration would occur on the same date, the major reference date (like Christmas eventually returning to December 25) being the first day of the month of Thoth. A century after the fact, the Roman writer Censorinus reported that in A.D. 139 the cycle phase had been such that the heliacal rising of Sothis was on the first day of Thoth, and a coincidence around 2773 B.C. is also indicated. By assuming that the civil calendar was defined at the time of a similar coincidence, various writers have suggested that the Egyptian calendar was instigated in 4241 B.C. (three 1,460 year cycles earlier than A.D. 139), so that the summer solstice in that year would be the earliest recorded date in history, and thus it might be regarded as the beginning of *history* itself because that term is usually applied to the period from which written records are available. Anything earlier is termed *prehistory.*

This is not only an academic exercise. While the evidence for the 4241 B.C. date is poor—for example, if one used the correct interval between the risings of Sirius at Memphis, rather than the Julian year, then the date gets moved by about a decade, explaining a coincidence in 2773 rather than 2780 B.C.—it is interesting that many cultures and religions have traditions which point toward an origin of the world within a few centuries of 4000 B.C. In a later chapter we will meet the famous 4004 B.C. of Archbishop Ussher, and the year-numbering system of the Hebrew calendar starts in 3761 B.C.

Why No Leap Years in Egypt?

I have mentioned that the period between heliacal risings of Sirius (the Sothic *year*, as opposed to Sothic *period*) is not equal to the Julian year of 365.25 days, and the reader might wonder why I have been so lax in stipulating year lengths here. A handful of millennia back the sidereal year was about 365.2564 days long, but this is determined for a star on the ecliptic. For a star some distance from the ecliptic, as is Sirius, the annual round will be slightly different. The center of Egyptian astronomy was at Memphis (and it is still nearby, at the modern Helwan Observatory), and at that latitude (near thirty degrees north) in that epoch the average interval between heliacal risings of Sirius was 365.2507 days, which is not much different from the Julian year. On the Julian proleptic calendar—it is tedious, but absolutely essential, to define which calendar one is using—the date of the heliacal rising of Sirius at Memphis changed only from July 17 to July 19 between 3000 B.C. and the Roman era in Egypt (the time of Christ): 3,000 times the 0.0007-day difference renders 2 days. With observations stretching over eons the Egyptians actually determined the correct 365.2507-day Sothic year, but did not reform their calendar to bring in quadrennial leap years. Why?

Part of the answer seems to be that the summer solstice and the rise of the Nile were behaving differently with respect to the repeated arrivals of Sirius. Over three thousand years the star's heliacal rising arrived later by only 2 days (against the artificial Julian year construct in use in this discussion), whereas the summer solstice progressively came earlier by a rather larger amount. For the reasons expostulated in appendix B one must be wary about what year length one uses, but for present purposes I will assume that it is sufficiently accurate to use the average tropical year length of 365.2424 days appropriate for the period 3000 to 1 B.C. (it happens that both the vernal equinox year and the summer solstice year had values near this in that era).

This is 0.0076 days short of a Julian year, so that the summer solstice would shift backward by 23 days over three millennia, separating by 25 days from the heliacal rising of Sirius. Around 3000 B.C. the summer solstice arrived and the Nile began to rise about July 16–17, coinciding with the reappearance of Sirius from the glare of the sun. By 46 B.C., when Julius Caesar introduced his calendar, the summer solstice was on June 24, but Sirius did not show up until July 19. When was the flood? There is evidence that between 2000 and 1000 B.C. it moved earlier in the year, following the movement of the summer solstice, but then jumped later again, due to climatic vagaries, such that at present the real flood begins in mid- to late July. Over the centuries these shifting dates must have mightily confused the Egyptians, who had come to believe that the gods Sothis and Osiris were the beneficial deliverers of their essential inundation. In the final chapter of this book we will revisit this problem and consider its possible implications for our understanding of the current seasonal cycles of the whole planet.

The First Leap-Year Proposal

Whatever the reason, despite their advancing astronomical knowledge the Egyptians persisted in using a civil year 365 days long, leading to the inundation and all other seasonal variations cycling through their whole calendar in 1,460 years. Rather belatedly realizing that an extra day every fourth year could pretty much fix this drawback, in 238 B.C. Ptolemy III Euergetes decreed that with such a frequency the year should contain a sixth epagomenal day. This Ptolemy—not to be confused with the famous Greek astronomer-mathematician who lived some centuries later—was king of Egypt in the Hellenistic period, during which rulers from Greek Macedonia governed from about 332 B.C., when Alexander the Great conquered the region, until the demise of Cleopatra in 30 B.C.; collectively these rulers are known as the Ptolemies. Ptolemy III had *Euergetes* (meaning "benefactor") added to his name because of the many scholarly things he accomplished—for example, the major development of the famous library of Alexandria. In essence he looked at the ancient Egyptian calendar with a learned Greek eye—perhaps almost literally, because there is evidence that he was advised by the great astronomer Aristarchus of Samos—and made the first proposal of what we now call a *leap year*.

It is another of the quirks of calendrical history that Ptolemy's instruction is called the *Canopus Decree*. The peculiarity derives from the fact that while the decree obtains its name from the coastal town near Alexandria where an inscription proclaiming it was discovered in modern times, the appellation "Canopus" is most often seen in connection with the second-most brilliant star in the sky, Sirius being the brightest. One should not, therefore, be misled into thinking that the Canopus Decree represents the abandonment of one star for another.

In the event, Ptolemy's order was largely ignored, and it was not until Augustus Caesar imposed the Julian calendar on Egypt over two centuries later that the civil calendar there with its regular 365-day years was superseded.

The Meteor Shower Year

Let me conclude this chapter by pointing out that there is another astronomical technique for determining the length of the year with some accuracy, which has the advantage of producing a good result within a few years, but which has been neglected in all discussions I have read concerning year measurements by early civilizations.

Every year there are several distinct meteor showers which occur on basically the same dates, each being characterized by an apparent direction of arrival (the *radiant*) which sets it apart. Indeed these meteor showers are conventionally named according to the constellation in which their radiants lie. For example, the *Geminids* have a radiant in Gemini; they peak on December 14, with a slow buildup in activity over a week, and then a rapid drop-off. The *Lyrids*, emanating from the direction of the constellation Lyra, are briefer, being visible for only a few nights with a distinct peak on April 22. There are at least half a dozen other prominent showers one might mention, but those will suffice for present purposes. The actual dates may differ by one day on our calendar, depending upon the phase of the leap-year cycle and one's longitude, but putting yourself in the place of an ancient human, if you watched for the peaks of each of the showers and counted the days between, then the year length would quite quickly be determined.

Chapter 5

✑◡◠◡↩

METON (432 B.C.), CALLIPPUS (330 B.C.), AND HIPPARCHUS (130 B.C.)

In the previous chapter I wrote that the ancient Egyptians abandoned the Moon in adopting a calendar with 365 days, because they were most concerned with the annual cycle of the seasons and in particular the inundations of the Nile. But that is not the complete truth. The 365-day cycle was their *civil* calendar. Today it is common for several intermeshed annual rounds to be used in parallel, such as the familiar calendar beginning on January 1 for civil purposes, a financial calendar starting on some other date, and a liturgical year—which may or not have the same year length as the civil calendar—commencing on another date again. The list might be expanded: school years, legal years, and so on. It should not be a surprise, then, that the Egyptians had several different calendars: they looked to the appearance of Sirius to herald the awaited flood, but they also counted a year starting with spring, which was signaled by the heliacal rising of the constellation Taurus, and in particular the Pleiades.

This is still an annual cycle, though. In the same way as our calculation of the date of Easter stems in part from an original need to provide a full moon for pilgrims traveling to Jerusalem, for their festivals the Egyptians needed to consider the phases of the Moon. This led to a parallel lunar calendar being used, based on a cycle containing a certain number of lunations (durations of 29.5306 days). Because of their nonastronomical year of precisely 365 days, the sums are quite simple. A period lasting 309 lunations is on average just an hour short of 9,125 days, and that is divisible by 365, providing a cycle of 25 years. This provided another argument for the Egyptians not shifting from their apparently illogical 365-day year: every quarter century the lunar phases repeated, and so their festivals returned to the same sequence of dates, even if they slipped against the seasons.

Intercalary Month Cycles

How could one fit the lunar-defined months with the solar-defined years? There are about 12.37 lunations in a year, the exact figure depending upon the value one uses for the year and the precise characteristics of the sequence of lunations (because the time between full moons can vary by some hours). Those civilizations which used a lunisolar calendar would tend to assume at first that there would be 12 months in a particular year, but then when it became apparent that the full or new moon (depending upon their favored marker) was to occur at an impropitious time a decree would be issued, intercalating an extra (unlucky 13th!) month; such years are said to be *embolismic*. As their understanding of the celestial cycles advanced, a rule specifying some cycle of intercalations would result. For example, in the last century B.C. the Celts in Gaul used a cycle with two embolismic or intercalary months every five years, the resultant average of 12.4 months per year producing between four and five days too many in the five-year round. A solution to this, which we will meet below, would be to drop a day from some months, but for the time being we are discussing solely the number of months in a year.

That average of 0.4 intercalary months per year is not a very good approximation to the 0.37 (or 0.3683. . .) set by the celestial realities. The Celts mentioned above had a relatively unsophisticated understanding of the astronomical cycles, and in much earlier times other civilizations had developed a more advanced knowledge of the lengths of the year and month. In the centuries around 700–800 B.C., the learned Greeks had adopted a cycle known as the *octaeteris*. This provided for 3 embolismic months in each eight-year cycle, hence its name. The average of 3/8 or 0.375 extra months per year was still not very precise, resulting in a day too many being included every five years or so.

A more complicated cycle is needed, but before we turn to that we should mention that there is a derivative of this Greek eight-year cycle which is still used today: the quadrennial Olympic cycle. The original Olympic Games were held around the time of the first full moon following the summer solstice, beginning in 776 B.C. Although we now count four *years* as being the cycle, as the games were tied to the lunar phase in fact they were spaced alternately by forty-nine or fifty *months* (because there were ninety-nine months in each octaeteris: eight times twelve plus three embolismic months).

The Metonic Cycle

To attain greater accuracy one must lengthen the cycle. One possibility is a nineteen-year round, called the *Metonic cycle*. The reason that an uppercase letter is used there is that, unlike in the derivation of the word *octaeteris* above, this cycle is named for a man: Meton of Athens. Meton enunciated this general scheme in 432 B.C., with seven embolismic (thirteen-month) and twelve ordinary (twelve-month) years in a cycle of nineteen.

We can now introduce the notion of varying the number of days in a month in order to converge on an agreeable calendrical solution. Meton's 235-month calendar cycle contained 110 months of 29 days and 125 of 30 days, thus averaging just over 29.53 as required, and producing a total of 6,940 days. In fact, those 235 lunations total 6939.69 days on average, whereas 19 tropical years have a duration of 6,939.60 days, so that the Metonic cycle wanders by a fraction of a day against either of those measures. That fraction, we will see at length, is hugely significant.

There is an important distinction to be made here. Many lunisolar calendars in use today employ regular repeats of 13-month years in a 19-year cycle, and are said to use the Metonic cycle, even though they may not be the same as the calendar devised by Meton. The Metonic cycle is a cycle of precisely 235 lunations lasting for 6,939 days plus a fraction; Meton's calendar design contained exactly 6,940 days. Thus there is a difference between them, although one should realize that Meton himself did not have access to measurements of the year length of sufficient accuracy to make this a matter of concern.

This distinction shows itself especially if one is wanting to compare the accuracy of different intercalation cycles. The difference between 6,939.69 and 6,939.60 days is about two hours, which would accumulate to a full day in a little over two centuries. On the other hand, the difference between 6,940.00 and 6,939.69 results in a one-day deviation after about half a century, so that it would soon be realized that the full moon was slipping away from the regularized date sequence which Meton had planned. To look at it another way, the calendrical scheme defined by Meton results in an average year length of 365.2632 days (6,940 divided by 19), compared to a tropical year which was actually 365.2423 days long. Note the significance of the difference in those year lengths: if one took Meton's calendar and, in an evolutionary scheme, decided to abandon the Moon and instead adopt a purely solar calender, then due to the fact that *according to Meton's calendar* the year length is longer than 365.25 days a leap-year cycle with *more than* one leap year in every four would result, rather than a cycle with *less than* one in four as the orbit of Earth actually requires.

The point I am making here is that the calendar we have received from antiquity is contingent upon the path actually taken through a large number of possible branches, and so the use of the word *evolution* is very appropriate. Our calendar has evolved in a similar way as all species of animal and plant have evolved, with a contingent history which is impossible to comprehend fully. In biology one sees many examples of *convergent evolution,* resulting from the fact that there are only so many physical solutions to a particular problem. For example, ichthyosaurs, sharks, and dolphins (representatives of reptiles, fish, and mammals) all assumed the same basic form, as set by hydrodynamical constraints. In the case of the calendar, the constraints are now well known, but in antiquity the knowledge was more hazy. Convergent evolution caused all calendars developed by disparate civilizations to be broadly similar, especially after understanding had accumulated from centuries of observation, but significant differences remained due to the differing evolutionary history. Similarly, the eyes of the octopus and the human are similar, except that our retina is back to front.

Better Monthly Cycles

Let us pause here for breath, and remind ourselves of the central question we are considering: how many months should be in a year? In any calendar the days are quantized: each month and each year contain an integral number of days. Despite the fact that one might adjust the relative number of twenty-nine- and thirty-day months, attempting to get the lunar month to rhyme with the year is like trying to find a rhyme for orange, purple, and silver, as discussed in appendix D.

In 330 B.C. another astronomer, Callippus of Cyzicus (a Greek city in what is now Turkey), announced a more accurate version in which he assumed that the year length should be 365.25 days. On that basis he combined four Metonic cycles of 19 years into one cycle of 76 years (the *Callippic cycle*) from which he dropped one day. This resulted in 27,759 days spread over 76 years which individually varied in length by 30 days but averaged 365.25 days precisely. Note what is going on here in these Greek calendrical refinements: our present-day calendar produces an average year length containing a fraction of a day by inserting a single *day* into some years; the Callippic cycle achieved the same thing by inserting a whole extra *month;* small adjustments were made in the latter by balancing between 29- and 30-day months.

Two hundred years later Hipparchus, one of the truly great astronomers of ancient times, measured the year from solstice to solstice as being slightly short of 365.25 days. Hipparchus derived a length 4.8 minutes, or one part in 300 of a day, below that value. We now know that he was 6.3 minutes off in his evaluation. On the basis of his measurements Hipparchus suggested a revised intercalary cycle. If the year were one part in 300 of a day less than Callippus's assumed value, then one day in three centuries would need to be deleted from the latter's cycle. Conveniently, 4 × 76 = 304 years, and so Hipparchus proposed a cycle involving four Callippic cycles with a single day being subtracted from one of them.

Given the precision of the measurements possible at the time, the second century B.C., Hipparchus's cycle was the best that could be achieved. But many persisted in believing that the assumed year duration of Callippus was correct; that in some way (perhaps divine design) the annual round had attained a length of exactly 365.25 days. Hipparchus certainly knew better: his eclipse observations were sufficiently precise for him to discover, in about 130 B.C., the precession of the equinoxes (an important phenomenon discussed at length in appendix A). That is, he managed an accuracy good enough to differentiate between the sidereal and the tropical years, a remarkable achievement. There have been claims that a Babylonian, Kidinnu, discovered the precession of the equinoxes 250 years earlier, but that is not provable; Kidinnu certainly did, however, make remarkably accurate determinations of the lengths of the tropical year and the synodic month.

The Story So Far

We can summarize the important points from the preceding discussion as follows. Meton suggested a cycle of 19 years in which 7 years would be lengthened by the insertion of an extra month, the month lengths being fixed so as to produce 6,940 days/235 months in the 19 years. Callippus thought that this resulted in a year length which was longer than the natural cycle; he suggested that the year should be precisely 365.25 days long on average, and invented a cycle of $4 \times 19 = 76$ years from which one day was deleted, the 76 years thus lasting for $(4 \times 6,940) - 1 = 27,759$ days spread over 940 months. Hipparchus thought that the year length should be less than that by 1/300th of a day, and so proposed a cycle of $4 \times 76 = 304$ years lasting for $(4 \times 27,759) - 1 = 111,035$ days, spread over 3,760 months.

Meton of Athens gained immortality in two ways. One was through being a subject of ridicule in Aristophanes' play *The Birds* (414 B.C.). This seems unreasonable of Aristophanes, because he knew of the problems inherent in trying to construct a calendar following the cycles of the heavens: in his earlier comedy *The Clouds* he had the Moon complaining that the days did not follow in precise step with her phases. It is difficult to have too much sympathy for Meton, however, since he was a draft dodger, having feigned insanity so as to avoid being sent to fight in the Greek invasion of Sicily.

The second way that Meton is perennially commemorated is through his eponymous calendrical cycle. It seems certain, however, that he was not the first person to have recognized the nineteen-year repetition, although it bears his name alone. (Even in Athens he was assisted by Euctemon in his calculations, the latter being largely forgotten.)

The Babylonians appear to have anticipated his discovery, and the Chinese were well ahead in matters calendrical during the first millennium B.C. As early as 1300 B.C. they knew that the year was about 365.25 days long and the synodic month 29.53 days. By two millennia ago the Chinese had improved their measurements, differentiating between the tropical and the sidereal years with values known to better than one-thousandth of a day, thus deriving a reasonably accurate value for the rate of equinoctial precession, while their measure for the synodic month was accurate to within a single second. The early Chinese calendar had alternating 29- and 30-day months, with intercalary months being inserted into seven of nineteen years, a period which they called *chang*. Later, during the first millennium A.D., they refined this using the seventy-six-year cycle which they called *pu*, one day being dropped from each set of four *chang*.

The Chinese thus independently discovered both the Metonic and the Callippic cycles; but the subject of this book is the thread which leads to the calendar we use today, and the development of the Chinese calendar is a parallel but unattached thread. If you object to Meton's fame, then use the Latin term: the *circulus decemnovennalis*.

Meton's calendar was never actually employed in ancient Greece, nor was that of Callippus, although they provide benchmarks against which any suggested "improved" cycle might be judged. Similarly, Hipparchus's determination of the year length (1/300th of a day short of 365.25; in reality the year is about 1/128th of a day below that value) provides the standard against which one might judge later measurements. In the history of our calendar Callippus's revision of Meton's theory is most significant because it assumed a year exactly 365.25 days long, and that became the basis of the calendar reform instigated by Julius Caesar in 46 B.C.

Chapter 6

❧

JULIUS CAESAR (46 B.C.)

Julius Caesar was introduced among the dramatis personae early in our story, since when I have implicitly referred to him through the use of the *Julian year* (of 365.25 days precisely) and the *Julian calendar.* It would be easy to think, especially from many popular-level accounts, that the only substantive calendar legislation for which Julius Caesar was responsible was the leap-year cycle (one every four years, rendering a quarter-day-per-year average), but in fact there are several other features of the present-day calendar for which we owe him a debt of gratitude (or otherwise, depending upon your opinion). The acceptance of January 1 as New Year stands largely upon his calendar reform, as do the lengths of our months. Having written that, I note that the contrary is also true: many descriptions overstate his contribution to the calendar.

Before going further, let me note that I am aware that some will find it a mouthful that I repeatedly refer to this character as "Julius Caesar." His proper name was Gaius (or Caius) Julius Caesar. Merely writing "Julius" does not seem quite right, and I cannot use just "Caesar" because that might lead to confusion with his great-nephew and adopted son Augustus Caesar, or other members of the dynasty. The word "Caesar" is also used with other connotations; for example, to refer to *any* ruler or tyrant with variations on the spelling (such as "kaiser").

As it is, confusion over the name of Augustus (who took the appellation Gaius Julius Caesar Octavianus, having earlier been known as Octavian, although later he changed again and nowadays he is usually referred to as Augustus Caesar) led to a four-year error in the dating of the birth of Christ, and hence the numbering of the Christian Era, as we will see. To add to the bewilderment, later in the history of the Roman Empire (around A.D. 300) a four-way split in the leadership, the tetrarchy, was instigated with one Augustus and one Caesar in each of the east and west domains; these are *titles,* not *names.* Precision can lead to problems. Let me refer to these two men as "Julius Caesar," but mostly just "Augustus." They are both giants of the calendar.

Changing the Calendar

Julius Caesar introduced the basic calendar which bears his name in 46 B.C., the restructured format thus beginning in usage on the first of January of 45 B.C.

(a year in which he was sole consul). Note that one can introduce a calendar at any juncture, but not actually start using it until a later year. For example, the reform of the calendar used in Britain was legislated for in 1751, which is the year in which the calendar should be said to have changed. The first actual alteration in the dates was when the following January 1 was taken to be that date in 1752 rather than a holdover from 1751, and the deletion of eleven days bringing Britain into line with the nations already employing the Gregorian calendar occurred another eight months later. But the new calendar was definitely introduced in 1751, because that's when the legal statutes were amended.

This distinction between *changing the calendar* and *the actual alteration of dates* is not just pedanticism. Opponents of present-day calendar reform falsely claim that we cannot change the calendar simply because we are now too dependent upon the accepted system through computer software and so on. In fact international legislation *could* be passed in, say, the year 2000 or 2010 to alter the calendar, with the actual date changes to take effect in 2050 or 2100. An example is that members of various Eastern Orthodox churches sometimes claim to use a calendar—the Revised Julian—distinct from the Gregorian in part because their leap-year rule would lead to a deviation in 2800 and a rejoining in 2900, although it is not clear that such a calender has yet been adopted either ecclesiastically or legally in any of the relevant nations.

Simply for interest, let us consider a possible reformed global calendar. There are many suggestions for the shape of such a calendar. For example, it might exclude January 1 and February 29 from being *any* day of the week, such that all other dates are on fixed days (because 364 is divisible by seven, giving an even fifty-two weeks), and one could then arrange for July 4 always being on a Saturday while fixing the specific date of Thanksgiving. Just an idea. If the legislation had fifty years or more of lead time, then there would be a couple of generations to get used to the idea and the programmers could make all new software compliant with this computationally simplified calendar as the old software was superseded. It is one of the great advantages of modern technology (I write ironically) that Julius Caesar in 46 B.C. and Lord Chesterfield in A.D. 1751 could reform the calendar with effect from the subsequent years, whereas we would need to plan a half-century ahead.

Counting the Years

I wrote above that Julius Caesar is a giant of the calendar. He also made 46 B.C. a giant year: it was 445 days long. This is often called the "Year of Confusion" (*annus confusionis*), although that is hardly fair: Julius Caesar's calendar reform, and the lengthening of 46 B.C., were made necessary by the disorder which had preceded, so maybe it should be labeled the "Year of the End of Confusion." On the other hand, Julius Caesar was largely responsible for that disorder. To see why, we need to take a look at the earlier calendars of the Roman republic.

Obviously Julius Caesar and his cohort did not think of themselves as living in the years "before Christ." As we will see, the introduction of the *anno Domini* (A.D.) dating system did not occur until many centuries after this time, and the use of *before Christ* (B.C.) for earlier dates came remarkably recently, only being extensively used after the seventeenth century. So what year numbers *did* the Romans use?

As a matter of fact, they did not use year numbers at all, at least for the first seven hundred years after Rome was founded. This was the case with almost all ancient civilizations. Rather than numbering the years from some initial reference year, nations tended to use regnal years, if they had a sovereign, or some similar dating based upon whatever rulers they lived under. This system still persists, for example, with British acts of Parliament being dated such as "In the forty-fourth year of the reign of Queen Elizabeth II," with Latin often being used so as to make the system even more anachronistic. Traditions die hard. In Japan the Gregorian calendar has been used since 1873 for ease of dealing with other parts of the world, but domestically the country also uses a regnal dating system, restarting with each new emperor. For example, 1989 was both the sixty-fourth year of the Showa era and the first of the Heisei.

This illustrates one drawback of such a system: unless a ruler has the foresight to die on December 31-January 1, the year of death is effectively counted twice and this must be allowed for in assessing chronologies using lists of rulers. But it does happen. Queen Elizabeth I, the last of the Tudors, died on a date we now refer to as March 24, 1603. This was actually the last day of 1602 for the calendar in force in England at the time.

An added degree of confusion may arise when a regent obtains more territory during his or her rule. The succession after Elizabeth I provides an example. Having no children (she is often called "the Virgin Queen," and the state of Virginia gets its name from her) or apparent English heir, King James VI of Scotland marched south to claim the monarchy and thus unify the crowns of England and Scotland. He had been king of the latter since 1567, when he was just thirteen months old, his mother, Mary, Queen of Scots, having been captured by the English; she was executed twenty years later on a trumped-up charge of treason. From 1603 he was James VI of Scotland *and* James I of England and Ireland, most history books referring to him succinctly as James VI and I, the first of the Stuarts. Documents dated according to his regnal year would need to account for *which* of his thrones was significant.

(This amalgamation of the countries has continued repercussions. Even today the legal systems of England and Scotland are distinct. For example, in a Scottish criminal trial the jury may, apart from verdicts of guilty or not guilty, bring down a decision of not proven. One might wonder what the result would have been in the O. J. Simpson murder trial if California law had allowed the jury to return a verdict of not proven, this having a quite distinct implication as compared to not guilty.)

With the boom in documentary records since the Industrial Revolution two centuries ago, or even the invention of printing 540 years back, such considerations of dating from regnal lists are trivial; but if one were working on the chronology of some eastern Mediterranean region from 2,500 years ago, it would prove very difficult to date events, given that the year of the reign of some king might jump as he conquered new territory.

The Roman Year Count

Having recognized that the Romans did not use year numbers, we *do* need to define some chronological standard in order to provide a reference against which to work, in the same way that we casually use A.D./B.C. dates which were defined long after the events under consideration.

The reference point used for Roman chronology is the traditional date of the establishment of the city. This seems to have been around the middle of the eighth century B.C., and several centuries later (when the emerging pre-eminence of Rome made it a matter of newfound significance) various scholars suggested different dates for the event. The year which is normally used is 753 B.C., its calculation being due to Marcus Terentius Varro (116–127 B.C.), a scholarly senator and general. This reference is often termed the *Varronian system*. In this framework years are labeled A.U.C., an abbreviation for *anno urbis conditae* or *ab urbe condita,* meaning "in the year of the foundation of the city" or "from the foundation of the city."

One of the ways in which Varro arrived at this date was by making a case for it being in the fourth year of the sixth Olympiad, meaning that twenty-three complete years (it was in the twenty-fourth year) had elapsed since the first Olympic Games in 776 B.C. although that year in itself is uncertain; it is just the earliest date for which lists of victors are known. Eratosthenes, a Greek mathematician of Alexandria who lived in the third century B.C., numbered the Olympiads from that year, and from his time onward the Greeks used the Olympiad number, with the addition of the year count (one to four), in order to date events. (Note that *if* 776 B.C. *was* the year of the first Olympics, then we currently hold them at the wrong times: the cycle of four years would lead to the summer games needing to be held in 2001, 2005, etc., there being no year zero in our A.D./B.C. dating scheme. Then again, the winter games are now misphased, anyway.)

Before Eratosthenes and from at least 683 B.C., the Greeks used a system in which the year was referred to by the name of whoever was archon (chief magistrate) in Athens. After becoming a republic, Rome used a similar system. But for the first two-and-a-half centuries of its existence, Rome was a monarchy.

This means that king lists are used to position events during this period. For example, one king who is significant in terms of the calendar was Numa Pompilius, who is supposed to have died in 673 B.C., although the chronology, actions, and very existence of various characters in the early history of Rome are quite uncertain. It follows that our knowledge of the Roman calendar in this era is very hazy, with later Roman writers giving quite different accounts of its evolution. Some wrote that Romulus, who founded Rome with his twin brother Remus, at least in mythology, introduced a calendar having 304 days, including ten months of widely different lengths (and a long, intra-annual un-dated period over winter), while others maintained that Romulus had used a

360-day year with regular month durations. The weight of evidence seems to favor the former, with Numa adding 50 days to make a 354-day calendar of twelve months, but then adding another day to make 355 due to the Roman preference for uneven numbers. The later mathematical work of Pythagoras—in the sixth century B.C.—was instrumental in making this superstitious regard for odd numbers a continuing theme in calendar formulation.

In this version of history Numa then intercalated, late in February in every second year, a 22-day month called *Intercalaris* or *Mercedonius,* from *merces,* meaning "wages" (hence our word *mercenary;* workers were paid at around this time of year). This resulted in a year averaging 365 days, and thus it was he who added two new months (Januarius and Februarius) to the previous ten. Note that throughout I will write Januarius rather than Ianuarius, as did the Romans; similarly for Junius, and Julius rather that Iulius Caesar. They also wrote the letters *u* and *v* identically, the phonetic purpose being apparent from the context.

The initial ordering of January and February is a matter of scholarly dispute; nor is it clear whether these two months were regarded at this stage (around 700 B.C.) as being at the *beginning* or the *end* of the year, a matter we will discuss later. Right now our concern is just to show that the Roman calendar seems to have started to attain its basic shape some 650 years before Julius Caesar reformed it, and that in those days the years were numbered according to the reign of succeeding kings.

The Birth of the Roman Republic

The monarchy ended with the downfall of the despotic Tarquinius Superbus in 509 B.C. (245 A.U.C.), when the Roman republic was established. The years were then referred to with a system similar to that used in Athens, with the names of the two consuls for the year being employed. Men serving more than once as consul had subsequent consulships defined with the application of a numeral. Thus, for a Roman, the year we call 52 B.C. or 702 A.U.C. was adequately stipulated by saying that it was the third year in which Pompey was consul (after his tenures of 70 and 55 B.C.); the name of his co-consul was redundant, although useful in avoiding any misunderstandings. The fateful year we would call 44 B.C. was known to the Romans as the first year of consulship of Mark Antony (he was consul again in 34 B.C.) and the fifth year of Julius Caesar's consulship (although he did not live to see his term out).

While we have the lists of consuls right down to A.D. 541, with the establishment of the Roman Empire (effectively from when Augustus beat Mark Antony and Cleopatra at the battle of Actium in 31 B.C., although Augustus did not assume that name and the mantle of emperor until 27 B.C.) a parallel system for numbering years—according to the regnal year of the emperor, or of his tribunate—became available. This was a time of great change in terms of political power: some officials were elected annually, but the emperorship was virtually hereditary, as such, the new emperor often being the adopted son of the previous incumbent who put pressure on the Senate to approve his designated heir.

The major system for reckoning years continued to be the names of the consuls until A.D. 537, when Justinian decreed that the regnal year of the emperor should be used, and only the emperor could hold the consulship. Well before this time the original Roman Empire had been split in two, with the seat of power in the west being Rome and that in the east Constantinople. The latter city (known as Istanbul since 1930) had been developed from the ancient town of Byzantium by Constantine the Great (A.D. 280–337), the emperor who made Christian worship acceptable, changing forever the character of the empire.

From several centuries B.C. the head priest of the Roman religion was called the *pontifex maximus,* the leader of the Pontifical College, a title which has endured to this day: it actually means "chief bridge builder"! For example, Julius Caesar was appointed pontifex maximus in 63 B.C., and Augustus in 12 B.C. This title continued to be held by the leader of the religion of the western part of the Roman Empire, who of course was based in Rome. The percipient reader will recognize that I am telling another story about evolution here: from being the leader of a pagan religion well over two millennia ago, the title of pontifex maximus has been held by emperors who fed Christians to the lions, through many changes, until the present holder of the title is better known to us as the pope (or the bishop of Rome, the head of the Roman Catholic Church, or just the pontiff). But beware that the eastern empire did not desert the title: the headquarters of the Eastern (or Byzantine, or Greek Orthodox) Church had to be removed from Constantinople after its capture by the Ottomans in 1453, stepping across the eastern Mediterranean to Alexandria where the Coptic Church was already ensconced. The patriarch of Alexandria in both the Eastern and the Coptic Churches is also called the pope (three churches, three popes).

When Did the Year Begin?

Having understood, then, that in the days of the Roman republic people did not use year numbers, but rather denoted each year in the records by the names of the consuls, we need to consider how the years were actually defined. The simplest question we could ask would be, "When was New Year?" But there isn't a simple answer.

Even in terms of actual dates for the start of the year the situation is complicated and not definitively known by modern-day scholars: the basic outline is known, but precisely when changes to the calendar and political system took place is disputed, there being few extant records. Several times we have noted that nowadays we use many parallel "years" (calendar, financial, school, religious), and the Romans did likewise.

In the early phases of the Roman republic the start of the civil year was March 1. That's simple enough, one might think, and actually obvious from the names of September, October, November, and December, which proclaim them to be the seventh, eighth, ninth, and tenth months (as they would be if the year began with March). But I have already explained that the years were reckoned according to the consuls, and in the early second century B.C. they did not enter office until March 15 (a date which we will meet with again), various other starting dates having been used after 509 B.C., so the potential exists for some confusion. On the other hand, we do not have great problems with the years of U.S. presidents, even though they are not inaugurated until January 20; for example, we say that Ronald Reagan was president from 1981, even though Jimmy Carter was actually the (lame duck) president for the first nineteen days of that year.

As a matter of fact this parallel with the U.S. presidency is very apposite. The Twentieth Amendment to the U.S. Constitution moved the date of inauguration back to January 20 from March 4, in order to minimize the duration between the election in early November and when a new president would assume office. When the Constitution was written, affairs of state moved slowly and transport slower still, necessitating some months between election and entry to the presidency. With the successive invention of the steam train, the automobile, and the airplane, the length of the lame duck period became absurd, and so the Twentieth Amendment was enacted. The immediate spur was the Great Depression of the 1930s, with the public perceiving a need to have the newly elected president in the Oval Office and doing something positive as quickly as possible.

Now look back at the Roman republic. As time passed it grew from a small city into a powerful nation-state, its borders stretching north and south as it conquered and subsumed the other regions of the peninsula we now call Italy. Despite the fame of Roman road construction, which mostly came much later in any case, transit times were long: for example, it would take eight to ten days to travel by land from Rome to Calabria, the province comprising the "toe" of Italy. Military campaigns would begin with the clement weather in the spring, and in the early days, with wars being fought against other city-states within a few hundred miles of Rome, there was no problem with the consuls taking up power on March 15 and then moving to the battlefront, the vernal equinox occurring about ten days later. It was because of this that the month was named for Mars, the god of war. But as its fledgling empire expanded, more time was needed for the consuls to enter office, make their plans, and then travel to their theaters of war. A transalpine march takes weeks, as does a sea voyage to Spain or Syria. For this reason the consular year was changed such that the newly elected leaders assumed power soon after the beginning of winter. From 153 B.C. the *dies solennis* (inaguration day) became January 1.

The Wandering Roman Year

The above sounds quite straightforward, but don't be misled. Although the *dates* of induction of consuls may have been set, the time in the *seasons* was not because the Roman calendar at this time was not anchored against some seasonal/astronomical marker. The years changed their lengths, often erratically, at the whim of the Pontifical College.

The scheme used was that there were 355 days in the basic year: the months known (in Latin) as Martius, Maius, Quintilis (or Quinctilis, the fifth month, now called July), and October had 31 days each; Januarius, Aprilis, Junius, Sextilis (the sixth month, now called August), September, November, and December had 29; and Februarius just 28. The Roman predilection for odd numbers is obvious, and it was for this reason that 30-day months were abandoned, even though a 29/30-day alternation gives the possibility of making the calendar months agree with the lunar phases. In every alternate year an intercalation of either 22 or 23 days took place (the month of Intercalaris), so that the Romans had a four-year cycle of 355, 378, 355, and then 377 days.

The place where the additional days were inserted is important. February 23 and 24 were feast days, called *Terminalia* and *Regifugium* respectively. In an embolismic year that month would prematurely end after one of those dates, and the otherwise anticipated final five (or four) days of February would be appended to the twenty-two (or twenty-three) added, to make an intercalary month consistently lasting for 27 days. The system is similar to, but distinct from, the intercalary cycle of Numa which we met earlier; this calendar cycle followed soon thereafter, at least by 500 B.C.

If one adds up the total number of days in this quadrennial cycle and divides by four, it will be found that the average length of the year under this scheme was 366.25 days, which is longer than the tropical year by one day (plus a small fraction). Of course this meant that the dates soon slipped out of agreement with the seasons, and by 450 B.C. attempts were being made to correct it. The pontifices would decide quite close to the time when an intercalary month might be needed whether to insert one or not. Of course this led to much uncertainty, and the citizens did not like it. A puzzle is why they didn't simply drop a single day from their calendar, making 365.25 days the average length of a year.

Various attempts were made to remedy matters, often when defense considerations meant that it was important to have the consuls coming into command at the right (seasonal) time of year. For example, between the First Punic War (264–241 B.C.) and the Second Punic War (218–201 B.C.) a calendar cycle much like that introduced by Julius Caesar was employed, but during their travails with Hannibal over the subsequent few decades the Romans omitted to keep things in step, such that by 190 B.C. the calendar was ahead of the seasons by 119 days.

We can be quite precise about that because Livy recorded an eclipse of the Sun as occurring on July 11 (on the Roman calendar then in use) in that year. With modern astronomical knowledge and computers we have the ability to calculate when that eclipse actually occurred, and the answer is March 14 on the Julian calendar projected backward.

This is termed the *proleptic* Julian calendar, and this date (in the second century B.C.) will be close to the correct seasonal date; the eclipse would have been about ten days before the vernal equinox (the start of spring) and *not* in the summer, which is what a naive acceptance of July 11 would suggest. Note that further back in time the proleptic Julian calendar deviates from the seasonal date because it assumes a 365.25-day year. For epochs more than a few centuries B.C. the proleptic *Gregorian* calendar performs much better in terms of agreement with the seasons because it uses an average year length closer to the tropical year. This will become clearer when we consider, in chapter 15, Archbishop Ussher's beginning-of-the-world date of 4004 B.C.

The Seasonal Wander Is Reduced, but Then Expands Again

By 168 B.C. the discrepancy between the calendar date and the seasons had been reduced, presumably through the insertion of supplementary intercalary months by the pontifices, to 74 days: Livy and Pliny recorded that a lunar eclipse was predicted by the tribune Sulpicius Gallus and seen on the night before the battle of Pydna, which the Romans dated to September 3, whereas recent astronomical calculations show that the eclipse actually occurred on June 21. The disagreement was further reduced in subsequent decades, and a nominal 24-year cycle was adopted on a de facto basis: in this cycle there were 13 ordinary years of 355 days, and 11 embolismic years containing extra months (4 of 23 days and 7 of 22 days). If this cycle had been adhered to then, the average year length would have been 8,766/24 = 365.25 days, and it would have taken centuries for any slip against the seasons to have become noticeable.

The problem was that the pontifices and other officials had vested reasons for fiddling with the years in which intercalary months were to appear. Imagine if the directors of some large corporation could control the length of the financial year so as to put good profit figures into one year (boosting the share price) and later loss figures into another (by which time they may have departed with the dividends and other benefits they had voted themselves). In effect the pontifices could do this, and more. They extended the years as they pleased to prolong the terms of office of magistrates sympathetic to their interests, for example. It is important to recognize that the declaration of an intercalary month could come very late: in 50 B.C. Cicero was complaining on February 13 that the people still did not know whether an intercalary month was going to be declared to begin ten days hence.

This situation was not entirely the fault of the Pontifical College, however. Rather, the blame might be laid specifically at the feet of the pontifex maximus (Julius Caesar from 63 B.C.). By about 140 B.C. the seasonal and calendar dates had been brought quite nicely into line by the pontifices, as witnessed by the dates recorded over the following seven decades for military campaigns, and the dates when the Roman armies wintered over in far-flung parts of Europe.

From 70 B.C. onward, though, there was a period of great expansion of the Roman hegemony. Initially Pompey and later Julius Caesar led campaigns far from Rome. The former conquered much of the eastern end of the Mediterranean, in the regions we now call Israel, Jordan, Palestine, Syria, and Lebanon. Julius Caesar was governor of Spain and later Cis- and Trans-alpine Gaul, a few years later bridging the Rhine and then conducting his expeditions to Britain in 55 and 54 B.C.

Spending years away from Rome (between 50 and 44 B.C. he was only in the city for a total of fifteen months), Julius Caesar was perforce neglecting his calendrical responsibilities as pontifex maximus. Further years of civil unrest in Rome and elsewhere, with Julius Caesar embattled against his former colleague Pompey, and concurrent military actions in many spheres of conflict with the armies overstretched, meant that the seasons slipped ahead of the calendar until 47 B.C., by when there was a discrepancy of ninety days. This was as large a disagreement as there had been 130 years before. But now the domestic and foreign situation was sufficiently quiescent for Julius Caesar to at last give the calendar the attention it required.

The Romans Abandon the Moon

Many books state that it was Julius Caesar who first made January 1 the start of the year. This is clearly not the case. Other more knowledgeable authors state that the pivotal time, with the start of the calendar year being changed from March 1 to January 1 (making September through December all misnomers), was 153 B.C. In fact, this is simply not known. As I will suggest at the end of this chapter, it is quite possible that at an earlier stage the year was reckoned to begin with the vernal equinox, which was around March 25.

What we do know is that the consular term was altered so as to start on January 1 in 153 B.C., but it seems that the beginning of the calendar year had already been set as January 1 by then. As previously noted, exactly when January and February were included in the named months is not assuredly known, but from our perspective of wanting to trace the evolution of our modern calendar the important point is that at least from 153 B.C. the calendar of the Roman republic had twelve months (plus an occasional thirteenth), with a year beginning on January 1 and ending with December. In essence this was the calendar inherited by Julius Caesar. We need to consider its structure.

The lengths of the twelve months at the time of the Julian reform were given earlier. They sum to 355 days, which is 10 days (and a bit) short of any solar year but just 0.63 days longer than a lunar year defined by a dozen lunations of average length. One may think, then, that the 355-day year was based largely upon the Moon, but an immediate mystery appears: why use 355 days rather than 354? The latter is certainly closer to twelve lunations, but the most significant point is that by having a year lasting an even number of days the Romans could have had twelve months each with their preferred odd number of days (say three of 31 and nine of 29). Part of the answer is that the length of February had earlier been reduced to 28 days, as it was considered in any case an insignificant month, and the unlucky intercalary month was tacked onto it.

Leaving such things aside, the 355 days suggests a lunar connection, and in the previous chapter we saw that the Greeks were keen on calendar cycles which accommodated months defined by the lunar phases. The peculiar thing is that the Romans had no such interest, and indeed could not fathom the determination of the Greeks to maintain a lunar-based calendar.

We do not know for sure when the Romans abandoned the Moon for calendrical purposes, but it was certainly before 400 B.C. Again the clue is given by a recorded eclipse. Cicero wrote of an account of a solar eclipse which occurred in *about* the 350th year after Rome was founded; that is, within a few years of 404 B.C. While the actual year is vague in Cicero's account, the date is not: it was on June 5 (on the calendar then in use in Rome).

Actually we can calculate the date of this eclipse using modern knowledge of celestial mechanics: on the Julian proleptic calendar the eclipse path crossed Rome on June 21, 400 B.C. The fact that the dates are different is not important: they are equivalent dates on different calendars. Nor is the actual year of great concern. The significant factor is that on the Roman republican calendar the date was given as June 5. If that calendar had months defined by the actual phases of the Moon, the eclipse *must* have occurred in the last few days of a month. The critical point there is that a solar eclipse can only occur when the Moon is in conjunction with the Sun, with the sighting of the lunar crescent (the new moon) and hence the declaration of a new month following by one to three days. Because the Roman calendar recorded this eclipse as occurring on June 5, that calendar must have been divorced from the phases of the Moon by that time.

This gives the lie to another of the common fallacies concerning the Julian reform. It is often said that it was Julius Caesar who decided to abandon the Moon, calendrically speaking, and invent a new calendar—the substance of which continues to this day—in which the months bear no relation to the lunar phases. That is incorrect. The Romans had given up on the Moon several centuries before (and the weight of evidence suggests that this occurred in about 450 B.C.). It was the Greeks who persisted with a calendar tied to the Moon.

The Cult of Mithras and January 1

While debunking fallacies about the Julian reform, let's deal with another one. Many authors have claimed that it was Julius Caesar who changed the start of the Roman civil year from March 1 to January 1 (an idea which was dismissed above), and write that the reason was involved with the sun cult of Mithras. At the time Mithraism, an ancient Persian religion, was in the ascendant, in particular in the military, and by the second and third centuries A.D. it was more popular than Christianity.

The story told is that Julius Caesar was having difficulties getting his calendar reform legislation accepted, and that he agreed to move the start of the year from March 1 to January 1 only at the behest of the followers of Mithras, as a quid pro quo for accepting the rest of the reform package. This fiction explains that they wanted this to be the start of the year because in 45 B.C. (as it became) there was a new moon due on January 1, and this was viewed as a most propitious event.

Now let us dispel this erroneous belief. Again we are able to calculate the relative positions of Earth, the Sun, and the Moon quite accurately, and such calculations show that the Moon arrived at conjunction soon after midnight on January 1-2, Rome time. The new moon (when the crescent may actually be seen) we know occurs one to three days later, in the evening after the Sun has set. An observation even on the second would have been unlikely, with it more probable that the new moon was not visible until the third or even the fourth. Further, as we'll see, the Julian reform gave January 31 days and left February with 28 (in a common year, 29 in a leap year), so that if the new moon *had* occurred on January 1 (it did not) then the second new moon following would have been on March 1 (31 + 28 = 59 = 2 × 29.5 days later). The argument concerning the influence of the cult of Mithras is therefore specious.

There is still a mystery about the way in which Julius Caesar phased the year, however. As we will see in the next chapter, the cult of Mithras and other sects practicing sun worship led to the midwinter festival at the solstice becoming hugely significant over the next few centuries, but even at the time in question (mid–first century B.C.) the winter solstice celebrations on the traditional date of December 25 constituted a major marker in the year. By adding seven days less than he did to 46 B.C., Julius Caesar *could* have put the solstice on January 1, providing a perennial astronomical marker for the start of the year. There are several counterarguments, however; for example, there would have been resistance to moving that festival from its traditional date, and also some recovery time would be necessary between the bacchanalia (to abuse the term) into which the celebrations had descended and the start of the new consular year. Many centuries later various societies regarded their year as starting with December 25, but with that date considered as Christmas rather than the winter solstice (which had actually slipped earlier in the month, on the Julian calendar).

The Peculiar Roman System of Day Counting

Although the Romans had given up on the Moon some centuries earlier, their calendar had a connection with lunar phenomena beyond merely the approximate lengths of the months. The dating scheme they used *within* the months reflected an earlier system based on the phases of the Moon.

Above I wrote that an eclipse was observed on July 11, but that involves a conversion beyond a trivial translation from Latin into English. Progessively numbering days may be our idiom, but the Romans would have found it alien. They did *not* number their days successively from the beginning of the month; rather, they counted days *down* to the start of the next month, or some other marker point within the current month, just as department stores may have huge signs showing the number of shopping days to Christmas.

This basic system of counting down is reflected in other aspects of Roman life. For example, the Roman numeral for 9 is IX (one less than ten), and indeed the Romans quite often would use "two less than ten" rather than "three more than five" to represent 8 (thus 28 might be written as XXIIX, rather than XXVIII as we would tend to assume nowadays).

The place to begin our discussion of Roman intra-month dating is with the most famous date in Julius Caesar's life; in fact, the day of his death in 44 B.C. Shakespeare has the soothsayer in the eponymous play about Julius Caesar warning him to "Beware the ides of March." This was the date (March 15 in our parlance) on which Julius Caesar was stabbed to death on the Senate steps by the conspirators. In Latin the word is *idus*, in English, *ides*, and in our language that may be construed as a singular or a plural; the associated verbs and pronouns generally sound better if the word is taken to be a plural.

In March the ides are on the fifteenth, as they are in May, July, and October. In the other months the ides are the thirteenth. Comparison with the month lengths in the pre-Julian calendar given earlier will show you a pattern here: the ides were on the fifteenth in the months containing thirty-one days, and on the thirteenth otherwise. Days within each month *after* the ides were counted down to the first day of the next month (which was called *kalendae* in Latin, *kalends* in English). Thus the date we would call the twenty-sixth of May was labeled by the Romans as *ante diem VII kalendas Junias*, meaning "the seventh day before the first of June." The Latin would usually be abbreviated as *a.d. VII kal. Jun.* or just *VII kal. Jun.* The last day of a month was called *pridie;* thus, May 31 was *pridie kalendas Junias (pr. kal. Jun.).* It is very important to note here that the Romans used *inclusive* counting. We would tend to think of May 26 as being the sixth day before June 1 (we would count May 26, 27, 28, 29, 30, and 31—six days), but the Romans reckoned also June 1, making seven.

Ignoring the maltreated month of February, the systematic thing about the ides was that they occurred on what the Romans thought of as being the eighteenth day before the start of the next month. Using our more familiar non-inclusive reckoning we would say that it was the seventeenth day before the first of the month. On a lunar calendar the first of the month is the day of new moon, which is one to three days after conjunction (for the latitude of Rome), or a couple of days on average. Taking 2 off of 17, one derives 15. This means that the ides occurred in the middle of such a (29/30-day) lunar month. In short, the ides marked the approximate date of full moon, so long as a lunar calendar was maintained.

In the early (pre–450 B.C.) Roman calendar, then, the new moon was on the kalends (the first day of the month) while the full moon was on the ides (the thirteenth of the month). Later calendar changes which divorced the calendar from the moon and gave four months 31 days each meant that the ides had to be pushed on to the fifteenth, so as to allow the correct number of days being counted down, even though the actuality was that the lunar phases now wandered through the months because their average length was more than 29.53 days.

But there is another lunar phase marker used in the Roman intra-month dating system. First quarter occurs about eight days before full moon. For the Romans, their inclusive counting calls that *nine* days, hence the name of the *nonae* (*nones* in English). For a month with the ides on the fifteenth, the nones were on the seventh; in months with the ides on the thirteenth, the nones were on the fifth (the fifth day of the month, about the seventh after conjunction). Earlier I mentioned an eclipse which was recorded as occurring on June 5. In fact the record said that it was on the nones of June (the lack of any possible ambiguity in that label adding weight to the belief that the Romans had abandoned the lunar month by 400 B.C.).

The Roman Eight-Day Week

For some reason the Romans did not have a special word for the day of last quarter, but the period between the ides and the kalends *was* subdivided. The Romans used an effective week of eight days (we will discuss this further in chapter 7). With their inclusive counting they thought of this as being nine, hence the name *nundinum* for this week, and the *nundinae* (the ninth day, which we would consider to be the eighth) were the days on which markets were held and the agricultural folk would come to the city or other marketplace with their produce.

On written almanacs the eight days were lettered from A to H, and in any year the markets opened on the days with the same letter, this changing from year to year because the eight-day cycle was unbroken (like our seven-day week); but the letter cycle restarted every year, January 1 always being an A day. On the Roman annual almanacs (or *fasti*), painted on the interior walls of patrician houses and displayed in public places for the plebeians to see, B to H would be shown in black, but all the letters A were painted in red. In addition, some days had additional notes colored red, indicating holidays or dates on which specified events (like legal action or civil voting) either could take place, or might be prohibited: hence our expression, "red-letter day." Many churches continue this usage, showing festival days in red in their printed almanacs.

The period lasting from the ides to the last day of the month (pridie) contained two of these nundinal weeks, and the nones were of course one nundinum before the ides. Because the month lengths were not divisible by eight it does not follow that the markets were always on the same day of the month (else the *dates,* rather than the *letters,* would have been painted red), but it is possible that early in the history of Rome the markets were instead held on a cycle tied to the lunar phases and the nundinal week was a later adaptation made necessary by the Republican calendar falling out of step with the Moon around 450 B.C.

Having now met the term *kalends* for the first of the month, we should examine its etymology. The word is derived from an ancient Greek term through the Latin verb *calare,* meaning "to proclaim." Originally the Greek and Roman new months were proclaimed on the sighting of the new moon (as are the Islamic months today, along with other societies which maintain a lunar calendar). Obviously we get our word *calendar* from this source.

What we have shown above is that the whole Roman dating system bears witness to an earlier lunar calendar, which they gave up on quite early while preserving its basic terminology. This is another evolutionary story: the Romans, it seems, could no more abandon the old words signifying a lunar-phase dating system than we have been able to stop using terms like *wire service, icebox,* and *telegraph pole.* (And why does the international road sign for a railway crossing show a steam locomotive belching smoke from its funnel?)

Julius Caesar's Adviser Sosigenes

Having emphasized various calendrical changes for which Julius Caesar was *not* responsible, contrary to many accounts, we can now address the question of what he *did* do. The first thing he did was to get expert advice. In the previous chapter we learned that the Greeks knew quite a few things about the cycles of the heavens, but the Romans would have no truck with the Hellenic fixation on the phases of the Moon ("Your new moons and your appointed feasts my soul hateth: they are a trouble unto me; I am weary to bear them." Isaiah 1:14). In any case Greco-Roman relations tended to be less than cordial.

For expertise on matters calendrical, Julius Caesar looked to Egypt, which had a long history of indigenous and imported knowledge (from Babylon and the East, and from Macedonia through the Seleucid dynasty beginning with Alexander the Great). And of course, through its dependence on the Nile floods and its infatuation with Sothis, Egypt had a year based upon a single Earth orbit rather than multiple lunar orbits. Rome turned to Egypt for advice, then, although in later times the Romans denied the Egyptians the respect which they were due in this sphere; days which were regarded as being unlucky for some reason were known as *dies aegyptiaci* (days of the Egyptians).

The particular adviser which history remembers is Sosigenes of Alexandria. Indeed some have argued that his name, rather than that of Julius Caesar, should be connected with the calendar reform. I use the word *reform* there advisedly. Several authors have stated that Julius Caesar *revolutionized* the calendar, effectively doing away with the entire republican system, haphazard as it was, but we have seen in this chapter that this view is incorrect. The Moon had long since been abandoned, and the calendar altered so as to try to keep it in step with the seasons.

The vital point at this juncture is that Sosigenes knew of the Canopus Decree promulgated by Ptolemy III Euergetes fully 190 years earlier, and how, despite its rejection by the Egyptian people, it would produce an average year length of 365.25 days, which was close to the actual solar year length. Similarly, Sosigenes was aware that Callippus had assumed that this was the appropriate length to use for a year. The twenty-four-year cycle used in Rome between about 140 and 70 B.C. rendered a mean year length of 365.25 days (between extremes of 355 and 378 days), but adherence had been slack.

Thus Sosigenes' advice to Julius Caesar was simply to abandon the incumbent system involving the insertion of whole intercalary months (of differing lengths) every so often on an ad hoc basis, whenever the pontifex maximus paid attention to this necessity and was not off campaigning somewhere. Rather, Sosigenes suggested the obvious—at least to us—alternative of increasing the basic year by 10 days, making it 365 days long, by spreading these days over the months, and inserting an extra day (a leap-year day in our idiom) once every four years, with no pontifical intervention being required. And this was what was decreed.

But if Sosigenes knew of Callippus and Ptolemy III Euergetes, surely he would also have known of Hipparchus, who only eighty years before had shown that the solar year was actually slightly less than 365.25 days long? Undoubtedly this is the case, but it is not a matter of great consequence. Julius Caesar was looking for some simple, straightforward rule to replace the miscellany which had gone before: having 365 days in most years, but an extra single day in one in four, was about as simple as one could get. A leap-year rule based on Hipparchus's (inaccurate, we now know) evaluation of the year would have meant that once every three centuries a scheduled leap-year day would need to be skipped. If Julius Caesar could legislate to insert 90 extra days into one year, then surely some ruler several centuries hence would be able to delete a single day? Simplicity was the key to ending the confusion which reigned.

The Erroneous Triennial Leap Years

I wrote above that the one-leap-year-every-four-years rule is about as simple as one can get. But the pontifices managed to get it wrong: no wonder the verb *to pontificate* is often used in a pejorative manner!

The problem arose from the Roman inclusive-counting scheme: what to us is every fourth year would have been every fifth year to them. When the Egyptian Sosigenes stipulated an extra day every fourth year, this was interpreted by the Romans as being one year in three. The supposed Julian calendar was introduced, starting in 45 B.C., and in 44 B.C. Julius Caesar (who, one presumes, understood what Sosigenes had prescribed) was assassinated. And for the following few decades Rome mistakenly employed a cycle of two common years followed by a leap year, thus allowing the calendar to lag progressively behind the seasons.

As a matter of fact we do not know for sure *which* one-in-three years were leap years in this period, but by about 9 B.C. the problem was obvious, at least to the astronomers, with twelve rather than nine leap years having been deployed since the Julian reform.

As a result Augustus declared a moratorium to let the dates catch up with the seasons, and there were no leap years again until over a decade later. Again we are not sure whether the next leap year was A.D. 4 or A.D. 8, but from then through to A.D. 1700 every fourth year was a leap year in all countries which inherited the Julian calendar.

How Were the Roman Months Numbered?

Let me here add an aside which may be of interest, with regard to the Roman inclusive-counting system.

Many authors have noted that the the names of September, October, November, and December indicate them to be the seventh, eighth, ninth, and tenth months. Indeed, I wrote precisely this earlier. But an inclusive-counting system would have meant that they would be called the eighth, ninth, tenth, and eleventh months if the year began on March 1. One might claim that the names are all nouns so that September implied month number seven rather than the seventh month, but actually in Latin they are all adjectival in form with the qualifying *mensis* (month) usually being omitted but always understood. Thus I believe that the oft-stated opinion that the names of September through December implied them to be the seventh through tenth months in an early Roman calendar is another fallacy, at least in terms of our exclusive-counting system.

A possible solution is that the Roman year, when these months got their names, began with April. This makes sense, astronomically speaking, because the vernal equinox (and hence the start of the seasonal year) was occurring close to the end of March in the early part of the first millennium B.C. Additional support comes from the fact that the Romans believed their city to have been founded in April (of 753 B.C.). My comments in this connection are somewhat speculative, but I think worthy of consideration. I am not aware of this apparent anomaly having been pointed out elsewhere.

The Lengths of the Months Post-Reform

How long were the months after the Julian reform? I wrote above that the ten additional days (eleven in a leap year) were spread over the months, but this was done unevenly: Julius Caesar added two days each onto the ends of January, Sextilis (now August), and December, and one to April, June, September, and November; in a leap year an extra day was inserted into February (my terminology there is important: that day was *not* just tacked onto the end of *that* month, as we will see). The pre- and post-reform month lengths were therefore as follows:

Month	Before Julian reform	After Julian reform
Januarius	29	31
Februarius	28	28/29*
(Intercalaris*	22 or 23)	
Martius	31	31
Aprilis	29	30
Maius	31	31
Junius	29	30
Quintilis	31	31
Sextilis	29	31
September	29	30
October	31	31
November	29	30
December	29	31
Total	355 (plus* 22 or 23)	365/366*

*In a leap year

After Julius Caesar's assassination, as a mark of gratitude for his services to the Roman state, the Senate voted to rename Quintilis as Julius (our July); it was the month of his birth in 100 B.C. I am sure that he would have appreciated the sentiment, if not the knives in the back. Similarly, during Augustus's lifetime, the Senate later changed the name of Sextilis to Augustus, perhaps after some prodding from the emperor himself. Later Caesars such as Nero and Claudius thought that this was a good way to gain immortality, and merrily appropriated months for themselves, but except in a few sycophantic, far-flung Roman colonies there was a quick reversion to the traditional month names.

In all the modern erroneous accounts of what was involved in the Julian calendar reform, the most-repeated mistake concerns the lengths of the months which Julius Caesar defined. A variety of stories are told, with February being given 29 days (30 in a leap year), and Sextilis through December alternating between 30 and 31 (i.e., the opposite way in which we have those month lengths). This fiction continues with Augustus insisting to a rather obsequious Senate that Sextilis be renamed for him and extended to 31 days so as not to be shorter than Julius Caesar's eponymous month. Actually, if this had really happened, he would have been more concerned with the number of days being odd and therefore lucky, rather than the comparative length. The extra day to supplement August in this fabrication was pinched from February, and then to avoid having three consecutive 31-day months (July, August, September) the 30/31-day alternation of the last four months was inverted. The invention of this account seems to date back to at least thirteenth-century England. It is false.

On the other hand, various writers have discussed why it was July and August which were named for the pair, rather than any other months, but they have missed an interesting point. Quintilis was allotted to Julius Caesar largely because he was born in that month. Augustus was born in September, but he viewed Sextilis as being his lucky month, having had many victories and triumphs then. In view of the preceding paragraph, one might also note that if it were Julius Caesar who had given Quintilis and Sextilis thirty-one days each, and left September with thirty, then Augustus would have another reason for regarding Sextilis as being auspicious. The point which I have not previously seen noted, however, is that in the Julian calendar Sextilis/August became the *eighth* month; it is therefore appropriate that it be associated with someone whose name once was *Octavian*.

Where Was the Leap-Year Day Added?

Let us turn now to where the leap-year day was inserted under the Julian rules. First we must understand precisely where the other 10 days, bolstering the basic year from 355 to 365 days, were placed: they were added to the ends of their respective months. This is not a trivial statement. The Julian reform changed the face of the Roman calendar in various ways, and this is an important one, especially if one wants to date accurately events in the pre- and post-Julian periods.

The best way to understand the significance of this is with an example. It is thought that Augustus was born on September 23 in 63 B.C. To the Romans that was *VIII kal. Oct.* because September had twenty-nine days. But after the Julian reform September was thirty days in duration, meaning that the twenty-third was now *IX kal. Oct.* To avoid confusion, dates around this epoch often have *anni Juliani* (years of Julius) added.

Actually, it is not known for sure whether Augustus was born on September 23 or 22 (which would itself have been known as *IX kal. Oct.* in 63 B.C.), but that is not surprising. Birth dates are often vague, and it is not known whether Beethoven was born on December 15 or 16, although the latter is normally assumed. We do know that in later years Augustus celebrated his birthday on September 23, or *IX kal. Oct. anni Juliani,* perhaps because this was an auspicious date, being the dedication day of the Temple of Apollo.

On the other hand, feast days were not moved, so that a celebration which had earlier been held on December 21, or *X kal. Jan.,* was still held on the ninth day (inclusive) after the ides (the thirteenth) even though its notation had been changed to *XII kal. Jan.* because December now had thirty-one rather than twenty-nine days. Note that I have given away another implication of the Julian reform there: even though December now had thirty-one days, the ides still occurred on the thirteenth, not the fifteenth as in the four original thirty-one-day months (March, May, July, and October). All the months preserved the pre-Julian placing of the nones and the ides, the extra days merely being added to the ends. This represented a further distancing between the Roman calendar and the phases of the Moon.

The above does *not* apply to the day added to February in leap years, however. Although comparatively recent legislation in some countries has altered matters, from the Julian reform and virtually through to the present the extra day was *not* February 29, because the day was *not* just tacked onto the end of the month.

In the pre-Julian calendar the intercalary month was inserted after the twenty-third or twenty-fourth, as mentioned earlier. Remember this was late in winter while people were rather indolently hanging around waiting for spring, when agriculture and military campaigns could begin again, so that February (and even January) was a rather vacuous month regarding constructive activity, with many festivals to while away the days.

Julius Caesar decided to insert the additional day after the twenty-fourth, which to give its full name in Latin was known as *ante diem sextum kalendas Martias,* or "the sixth day before the kalends of March." The difficult point to understand here is that the inserted twenty-four-hour period was considered to be a "second sixth day before . . . ," and was termed *ante diem bis sextum kalendas Martias.* That is, the composite forty-eight-hour period was counted as being a single day in the calendar: in effect the day occurred twice. From that Latin we derive the word *bissextile,* which is the correct adjective to describe such an extended year; we will examine the origin of the more-often used term *leap year* in a later chapter. Note that bissextile contains the letter *s* twice; it has no connection, then, with one's sexual proclivity.

There is also another Egyptian connection in the stipulation of this date. Recall that this was the end of the pre-Julian year, with five days following. In the Egyptian calendar there were five epagomenal days ending the year, which were part of no month. The effect of the Julian reform was always to close February, previously the end of the Roman year, with five days echoing the Egyptian convention.

How Many Days in a Leap Year?

A bissextile year contains 366 twenty-four-hour periods, but legally speaking, dependent upon where you live, it may be that there are only 365 days because two of those 366 are counted as being one only. For example, in England a statute of King Henry III in 1236 entitled *De Anno Bissextili* states that "the day of the leap year and the day before should be holden for one day." If this is taken at face value, then one would have to admit that legally there are only 365 distinct days in a leap year.

This statute was inherited by other legal systems, and various repercussions from this have led to court cases in the United States and elsewhere. Imagine some of the possible implications: if your legal system says that February 24 and 25 should be counted as one day, then is a library book on seven-day loan overdue if you borrowed it on the twentieth but did not return it until the twenty-eighth? Even if the legal system under which you live has the twenty-ninth as the leap-year day, still the same form of question applies.

We habitually think of the twenty-ninth as being the day added to a leap year, and many dictionaries define bissextus as being February 29, which is clearly an abomination. Rather than people born on that day bemoaning the infrequency of their birthday celebrations, they could with some justification claim the twenty-eighth in every common year, while those born on the twenty-fourth in a leap year are the ones to miss out according to the above statute.

Such an idea might seem bizarre to the ingenuous, but a cursory study of most church almanacs for a leap year will show that common usage has not done away with the rigorously correct dating scheme passed down through the ages from the ancient Romans. For example, the Roman Catholic Church liturgy uses February 24 as the day to be redoubled, with the feast days celebrated on the twenty-fourth through the twenty-eighth in common years being marked on the twenty-fifth through the twenty-ninth in a leap year (the feast day of St. Matthias is moved from the twenty-fourth to the twenty-fifth, for example). Note that this is contradictory to the statute quoted above: in this ecclesiastical calendar the *twenty-fourth* is the day not associated with Saint Matthias in a leap year, whereas the act of Henry III incorrectly implies that the *twenty-fifth* is the leap-year day. Julius Caesar's bissextile year lives on, in various confused forms.

The Puzzling Insertion of Ninety Days

The final question to ask is this: what exactly was Julius Caesar aiming for with his calendar reform? This is again a deceptively simple query. Let us take the year and month lengths as read, because they easily follow from the logic of having an annual cycle based only on the Sun, ignoring the Moon. That basis was inherited from earlier epochs in Roman and Egyptian history. What we really want to know is why he set the *phase* of the calendar such that the vernal equinox, and hence the start of spring, arrived about a week before the end of March.

In essence Julius Caesar defined the phase by adding 90 days to 46 B.C., making it last 445 days and in consequence pulling the equinox back from the end of June, to which it had meandered over the preceding couple of decades. But if he were going to make such a radical change to 46 B.C., why *90* days? For example, why not add 172, which would have made January 1 the date of the equinox, bringing the start of the civil year in line with the beginning of the seasonal year? (One contrary argument is obvious: then the inauguration date of the consuls would have needed to be pushed back by another couple of months, in line with military considerations as discussed earlier.)

Was Julius Caesar trying to put the equinox specifically on March 25? This is often invoked as the traditional date for the equinox (and as we'll see later, the date on which Christmas is celebrated hinges upon this fact). Pliny the Elder (around A.D. 77) wrote that the traditional Roman dates for the equinoxes and solstices were the eighth days (seventh in exclusive counting) before the kalends of Aprilis, Quintilis, October, and Januarius, corresponding to March 25, June 24, September 24, and December 25 on the post-Julian calendar (but, on the pre-Julian calendar, skipping around due to the intercalary months). It is believed that the calendar of Numa Pompilius had the equinox on March 25. If Julius Caesar wanted the vernal equinox to occur on that date, then he failed, because in 45 B.C. this occurred near midnight on March 22-23, and for a century or so thereafter it would swap between those dates in the quadrennial leap-year cycle. If he had wanted the equinox on the twenty-fifth, he should have added only eighty-eight days to 46 B.C. On the other hand, the only Roman festival during the republican and early empire eras which was definitely associated with either the equinoxes or the solstices was *Fors Fortuna,* at the time of the summer solstice, and the calendar reform Julius Caesar enacted *did* put that on June 24.

Was *this* his major aim? A contraindication of any intention on Julius Caesar's part to regularize the occurrence of the equinoxes and solstices on the stated traditional dates stems from the fact that the numbers of days between them are not the same. It is shown in appendix B that currently spring in the Northern Hemisphere is 92.8 days long, while summer, autumn, and winter last 93.7, 89.9, and 88.8 days respectively. Two millennia ago the durations of the seasons were slightly different, but only in the decimal places. Julius Caesar, however, legislated for month lengths leading to the numbers of days between the appropriate twenty-fourth or twenty-fifth days of the months in question being spaced by 91, 92, 92, and 90 (91 in a bissextile year) days. He *could* have put the equinoxes and solstices perennially onto the eighth day (one nundinum) before the kalends simply by choosing different month lengths, and he had plenty of slack to play with. So it looks like he was *not* attempting to regularize the calendar dates of all four season markers. If he was, then he was poorly advised, by Sosigenes or whomever.

We simply don't know why Julius Caesar inserted ninety days into 46 B.C., rather than some greater or lesser number, leading to the current phasing of our calendar against the seasonal year. His reason does seem to be connected with some earlier Roman tradition. Certainly, over the preceding few centuries, the Romans had tried to keep the vernal equinox in late March (although without conspicuous, long-lasting success). If Julius Caesar was attempting to do the same, then this just pushes the question back further: why should the equinox occur *then* on the calendar? As outlined earlier, I believe that a possible solution is that at some stage the early Roman calendar was set up so as to have the start of the year one nundinum after the equinox, in the same way as our New Year is one week after Christmas (which is itself based on the traditional date of the winter solstice), and so have the civil year begin with April (the traditional month of the founding of Rome), with the months of September through December being enumerated according to the Roman inclusive counting system, and *not* as most modern accounts maintain (seventh through tenth months for a year starting with March 1). Ultimately, though, it is unlikely that we'll ever be sure just why Julius Caesar phased the calendar as he did.

It is the astronomical cycles which give us our year, month, and day, and hence the second. To the contrary, the time division we call a *week* is a human invention, which might appear to be of arbitrary length. We have mentioned the week several times above, and also the Roman eight-day equivalent, the nundinum. In the next chapter we will discover how our week got seven days, and in doing so we will branch into religion and astrology, rather than astronomy.

Chapter 7

❧

The year, month, and day all have clear astronomical bases. Our 7-day week, however, does not have an obvious link with any heavenly cycle. Various authors have mused upon its length, usually with a celestial perspective. For example, if one were to divide the lunar month of 29.53 days by four, the nearest integer to the result is seven. But . . . Well, but so what? In the previous chapter I showed that the Roman 8-day market cycle, the nundinum, has a possible linkage with the phases of the Moon; when the Beatles sang "Eight days a week, I 1-uh-uh-uh-uh-ove you," they were clearly mooning over a Roman lass. But our 7-day week has quite different origins. We will see that it is an artificial length of time, but that it evolved from a mixture of astrological and religious beliefs. And astronomy? Well, yes, in the very beginning.

The ascendancy of the 7-day week is quite remarkable, given that before the world became a global community, in relatively recent times, different cultures used disparate periods to define their "weeks." For example, the Mayans had a 20-day reckoning period, while the Incas used 8. Elsewhere other civilizations used 5-, 6-, or 10-day weeks. The Baha'i use a 19-day period. The ancient Chinese used a 10-day week, at least by the first century B.C., a length which was also used rather earlier by the Egyptians: their 360-day year was divided into 36 *decans*, each of 10 days and each associated with a particular star or stellar group (we only have definite knowledge of two of them—Sirius and Orion—nowadays), with the extraneous 5 days not being part of any week.

Often the local climate and hence food preservation considerations controlled the week length, which for the most part represented a market cycle as did the Roman nundinae. In parts of tropical Africa, for example, an eight-day market cycle was used with a four-day subcycle for lesser market days at which the more perishable comestibles were hawked. Many people seem to believe that civilization began with the invention of indoor plumbing (the Romans had that, and the lead killed many of them; note the etymology of *plumbing*), but I would hold that the major step forward was the invention of refrigeration. Having written that, I note that each Saturday morning I go to the fruit and vegetable market to provision for the week, following with a foray into the supermarket; that generally suffices, except for a midweek purchase of bread and milk, echoing the African system mentioned above.

What Is a Week?

At this stage one may ask: what is a week anyway? It is understood here that *week* and *seven days* are not synonymous. The answer is that a week is a cycle which is longer than a day but shorter than a month and which fits in with various necessities, such as shopping, a work/rest rhythm, the frequency with which the laundry needs to be done, and so on. Obviously different cultures have come up with different solutions to the question of how long a week needs to be. While our seven-day week has been in use for over two millennia, one should recognize that some modern societies have tried to displace it. For example, after the French Revolution a ten-day week was introduced in 1792, while the Soviet Union tried first a five- and then a six-day week in the initial decades of the Communist regime, although reversion to the seven-day cycle soon occurred in both states. There seems to be something special about those seven days.

Week is a word you are going to read often in this chapter. Unless otherwise stipulated, henceforth I mean by it a period of seven days. Actually there are other words for that time duration, but they are seldom used. For example, from Greek we derive *hebdomad,* which may mean either the number seven *or* a week. Nowadays its use is confined largely to the ecclesiastical domain (hebdomadal or hebdomadary events, services, or meetings).

But whence our week? The first thing to be clear about is that it is essentially a human construct, whereas the year, the month, and the day are merely our recognition of cycles imposed from above (by celestial objects, I mean, not a deity). Much has been written about other animal species, and whether they live according to some circa-septan rhythm, but the information is at best ambiguous; if there is any such behavior following seven- to eight-day cycles, it is best interpreted as being due to being a fractional part of the month (or connected with lunar phases and tides) rather than being similar to the week, which we have invented ourselves. Then again, some animal behavior is stimulated by human timekeeping: Norwegian scientists studying salmon movements identified a seven-day rhythm, but found that it was interrupted when local forestry workers went on vacation; the salmon had been responding to the rumbling produced by logging trucks.

Another way in which the week is an intruder or outsider compared to the other units of time which we employ is through the way in which it starts. As the clock comes to the end of December 31, a new second, minute, hour, day, month, and year (and maybe a century and a millennium) all begin at the same instant. But not necessarily a week. *About* one in seven years begins with a new week (but not *exactly* one in seven: why this is so I will leave as a puzzle for you to think about, and give the answer later).

Even then, one could ask: when does a week begin anyway? We call Saturday and Sunday the *weekend,* so does the week begin with Monday? If you look at your printed calendar on the kitchen wall, you may find that the columns of dates are printed in the order SMTWTFS, whereas other calendars may sequence the days as MTWTFSS. Which is correct? Is Sunday the last day of the week, because the Bible exhorts the faithful to rest on the seventh day? Where does that leave a Jew, for whom the Sabbath is Saturday? Is the Muslim week different, given that the Islamic holy day is Friday?

Obviously, then, the week is a bit of an oddity. One we are very familiar with, but an oddity nonetheless. In this chapter I will explore the origin and the dominance of the seven-day week as a unit of time, looking at its bases (there are more than one) and the way in which it gradually permeated various near-Mediterranean civilizations and then got handed on to us.

The Origin of the Sabbath

The place to begin is with the Sabbath; or at least with the origin of the word. Since early times there has been a sanctity or mysticism associated with the number seven, with various plausible explanations as to its origin. In reality it seems that there were several distinct origins, each largely lost in the depths of time, but reinforcing each other through their commonality.

The book of Genesis account of the Creation, with God laboring for six days and resting on the seventh, will be familiar to all. Various tracts in the Bible exhort couples to abstain from sexual congress for "two weeks from the onset of the wife's menstruation." In fact, that statement presupposes that a week lasts for seven days, the biblical passages having originated before a period of seven days was commonly recognized to be the unit we call a week.

The root source of the word *Sabbath* appears to be the Babylonian word *sabattu*, or *shabattu*, or variations like *sappatu*: transliteration between alphabets makes the English spelling negotiable. Thus the Yiddish word for the Sabbath is *Shabbat*. Sabattu was considered by the Babylonians to be the evil day of the moon goddess Ishtar, a time when she was thought to be menstruating, at full moon. This may have come about because during a lunar eclipse—which can only occur at full moon—the lunar disk takes on a blood-red hue. For example, consider the following biblical passage (Isaiah 66:23): "And it shall come to pass, that from one new moon to another, and from one sabbath to another, shall all flesh come to worship before me, sayeth the Lord." The correct interpretation is that here *sabbath* means "full moon" and *not* a weekly holy day (Saturday starting on Friday evening, for a Jew; Sunday starting at midnight for a Christian). Like Alice in Wonderland, we must be aware that words mean what their writers intended them to mean, and not some later altered usage.

On the other hand, a distinct Babylonian word was *sibutu*, meaning "the seventh." In each month (taken to begin at new moon) the seventh day was a special day, when rites of purification and expiation were performed. These were associated with the god Apollo, who seems to have begun as a Babylonian invention, later being taken up by the Greeks (hence the name by which we generally know him), who celebrated Apollo on that day with merrymaking and gaiety.

This switching between atonement and festivity on certain dates has occurred frequently. For example, in the Christian churches *feast* days may actually be times when a *fast* is prescribed, and until quite recently no festivities were supposed to occur on Christmas Day. In the 1640s the English Parliament declared this a day of fasting, troops patrolling the streets to ensure that no one was surreptitiously cooking a sumptuous yuletide dinner, with such seasonal observances having been banned by the Puritans in Massachusetts some years before that and fines on miscreants being imposed as late as 1859. The verb *to celebrate* does not necessarily imply partying or living it up; a celebration may be a solemn occasion.

The Transfer to a Seven-Day Cycle

Getting back to Mesopotamia and environs several millennia ago, the number seven gradually became a taboo, as did its multiples. The Babylonians considered the seventh, fourteenth, nineteenth, twenty-first, and twenty-eighth days of a lunar month to be unlucky. Why the nineteenth? Because $7 \times 7 = 49$ was especially ill-omened, and the nineteenth of one month was the forty-ninth day after the start of the preceding one. (On the other hand, the Persians thought of the first, eighth, fifteenth, and twenty-third days of the month as being ones on which to avoid important activity, based in part on *their* week system: they divided a month into two seven-day and two eight-day weeks.) Under this scheme the fourteenth rather than the fifteenth became regarded as the day of full moon (*sabattu*), and thus the meanings of *sibutu* and *sabattu* became melded. Reinforcement came from the seven-day human duration of menstruation, or at least the presumed duration of uncleanliness, as in the biblical injunctions against sexual contact.

The general meaning of *sabattu* was still in the context of a period within a month, however, and not a seven-day cycle. The idea of such a cycle, with every seventh day being regarded as a day of purification when many activities are disallowed, appears to be a Judaic invention, the word *sabattu* being gradually appropriated (in a context disconnected from its original meaning) so as to render the concept of the Sabbath. There is no evidence that the Jews used such a system prior to the Exile (which followed the destruction of the great Temple in Jerusalem by the Babylonians in 586 B.C.), but they were using it a few centuries thereafter, certainly by 100 B.C., in the context of abstaining from work and various other pursuits every seventh day. The important point is that this cycle was disconnected from the lunar month. Rather than being named for the pagan planetary gods as described below, the ancient Jews associated the days of the week with seven archangels, Gabriel and Michael being the only ones now recalled.

The Seven Planetary Deities

We know, therefore, approximately *when* the shifting seven-day cycle we call a week came into being; the next question we must address is *how* and *why* it attained virtually global acceptance, above all other similar cycles. A particular question one might ask is this: given that much of our timekeeping system derives from the Roman calendar, how did the Jewish seven-day week drive out the Roman eight-day nundinum?

The major factor leading to the ascendancy of the Jewish week came from a different belief system entirely: the power of planetary deities. The word *planets* originally included the Sun and the Moon, being derived from the ancient Greek verb *planasthai,* "to wander," and the noun *planetes,* "the wanderers." In ancient understanding, the planets were the seven familiar objects which wandered over the sky relative to the fixed stars (comets being a different class entirely, causing great consternation with their sporadic appearances). All will be familiar with the various early pagan religions in which each of the seven planets was associated with a particular god: thus Jupiter or Jove was the supreme deity of the ancient Romans, but known as Zeus by the Greeks, Thor by the Norsemen, and by other names still in the Middle Eastern civilizations.

While we are familiar, through relatively modern astronomical knowledge, with the actual arrangement of the planets in terms of distances from Earth and the Sun, one must remember that the ancients had access to no such information. The modern heliocentric model of the solar system is often called the Copernican theory, for the Polish astronomer Nicolaus Copernicus (1473–1543) who promulgated it, Galileo Galilei (1564–1642) getting himself into much trouble with the Catholic Church for supporting the idea. The truth eventually became apparent when Johann Kepler (1571–1630) developed his laws of planetary motion based upon the measurements of Tycho Brahe (1546–1601), and the Church at length had to concede (but only late in the twentieth century) that it had made an error with Galileo.

These advances of the Renaissance are all relatively recent, but as early as 270 B.C. Aristarchus of Samos had suggested that Earth and the other planets circuit the Sun. Nevertheless, the earlier geocentric theory was to hold sway for many centuries yet, and Aristarchus's concept did not gain approval. The great Greek philosopher Thales (636–546 B.C.) had thought that Earth was a flat disk floating on a great ocean, the Moon being another disk high in the air which occasionally passed in front of the Sun, producing an eclipse. Pythagoras (ca. 580–500 B.C.) first suggested that Earth is a sphere, the planets and stars revolving around it, a general concept supported by Aristotle (384–322 B.C.), who visualized the celestial objects as moving on spherical surfaces, with Earth at the center. Eratosthenes (ca. 276–194 B.C.) determined the radius of Earth by measuring the angles of the shadows cast at noon in locations of known separation. The Earth-centered theory for the universe was elaborated by the Greek mathematician Ptolemy (Claudius Ptolemaeus, who flourished around A.D. 127–151), and is usually termed the Ptolemaic system. Other civilizations developed similar concepts, but this is the one which has passed down to us in major form. These people were not stupid: their models were wrong, but logically based on the available information and methods of analysis.

If the planets moved on spheres of increasing distance from Earth at the center, then logic dictated that those that moved fastest across the sky (the planets) must be the closest, while the slower ones (the stars) must be the furthest. Within the seven planets, an ordering was arranged according to their speeds of motion and hence increasing distance. Starting with the closest/fastest under such a (false) belief, this order was: Moon, Mercury, Venus, Sun, Mars, Jupiter, Saturn. Certainly from the second century B.C., and possibly from Pythagoras's time, and through to the arrival of the Copernican system, this was the assumed ordering of the planets. In other cultures the arrangement *may* have been arrived at rather earlier: the Babylonians were the first people to cluster the seven planets together, around 500 B.C., although there is no evidence that they used a seven-day cycle based upon them. Further east, in India, a planetary week *may* have been employed in this early period, but evidence is lacking.

This planetary arrangement—the Ptolemaic—is implicit in many pre-Copernican works. For example, the Italian writer Dante (1265–1321), in his *Paradiso,* ascends through seven celestial zones until he reaches the eighth and ninth, known as the *Primum Mobile* and the *Empyrean,* the highest heaven. To understand the basis of such writings one needs to recognize the context of their times and the contemporary state of scientific knowledge.

Another interesting parallel is between musical scales and the Ptolemaic system, which is sometimes referred to as the *music of the spheres.* The seven-day week is cyclic, starting again on the eighth day, just as a musical scale has seven notes, restarting on the eighth and thus being termed an *octave.* In ecclesiastical language an octave is the eighth day after a Church feast (or, as an alternative meaning, the eight-day period commencing from a certain day). There is another similarity between calendrics and musical physics. Many attempts have been made to develop what might be called *rational music,* but the pitches of different notes do not conform, bearing irrational relationships with each other. How boring much music would be, otherwise. Similarly the day, the week, and the month are not aliquot parts of a year, so that attempts to design a "rational calendar" are also futile.

The Planetary Week

From the sequence of planets in the Ptolemaic system is derived what is known as the *planetary week,* which seems to have been a Hellenistic invention, following the conquest of western Asia by Alexander the Great in 336–323 B.C. Because it grew in popularity in the following centuries around Alexandria, this week is often described as being of Egyptian origin, although at that time the Ptolemies (whom we met in chapter 4: Alexander's dynasty, from Macedonia) ruled the region. Do not confuse the Ptolemaic dynasty of rulers of Egypt with Ptolemy the astronomer of the second century A.D., who gave his name to the Ptolemaic system.

The planetary week stems from an allocation of each hour in the day to a particular planet, which is considered to be the *controller* of that hour, in an astrological belief system. The planet of the first hour of the day is called the *regent,* and the day is named for that regent. The planets are written down in reverse order (that is, in terms of decreasing distance from Earth) and allotted to each hour in turn. There were twenty-four hours counted in a day—an Egyptian practice, remember—and whether they are equal or not (they were not, at this stage of history) is no matter of concern. Let us take Saturday as an example to begin with. The hours are controlled by the planets as follows:

Saturn	1	8	15	22
Jupiter	2	9	16	23
Mars	3	10	17	24
Sun	4	11	18	
Venus	5	12	19	
Mercury	6	13	20	
Moon	7	14	21	

Continuing the hour count, the first hour of the *next* day has the Sun as its regent, two planets (Jupiter and Mars) having been skipped. Thus Sunday follows Saturday. The same general rule applies: skip two planets to get the controller of the following day (the Moon, hence Monday). Thus the names of the days derive from an hour lore, which belief persisted into the Middle Ages; for example, both Chaucer and Roger Bacon wrote about it. Similarly, the *month* and the *year* were assigned the regent of *their* first hour. The word *horoscope* is testament to the significance of the hour in similar belief systems.

The final ordering of the planets on this basis is Sun, Moon, Mars, Mercury, Jupiter, Venus, and Saturn. To get this sequence, just write the seven names tabulated above in a circle and then move around that circle, jumping two planets at a time, as in figure 1, on page 80. You will then arrive at the order of the days in the planetary week. Note that this does not correspond with any real physical ordering, such as brightness or sidereal or synodic period (or the actual heliocentric sequence: the Sun, Mercury, Venus, the Moon, Mars, Jupiter, Saturn). It stems from an ancient cosmological theory, astrological beliefs, the twenty-four hours counted in a day, and the supposed significance of the number seven.

The Number Seven Revisited

Above I alluded to the apparent sanctity of the number seven, and this has its foundation in various happenstances, some of which have already been mentioned. Another one involves numerology, in particular based on the Pythagorean 3, 4, 5 right-angled triangle. The sum of the sides making the right angle is $3 + 4 = 7$, the days in a week. The sum of all three sides is $3 + 4 + 5 = 12$, the number of months in a year. The sum of the squares is $3^2 + 4^2 + 5^2 = 50$, which is an important number of years in the Jewish faith: every half-century all Jewish slaves are supposed to be freed, fields handed back to their owners, and all agricultural labor stopped, this yearlong period being called a *Jubilee*. The Catholic Church also celebrates a Jubilee every fifty years. Note that this fiftieth year follows forty-nine (that is, 7×7) ordinary years. The cessation of labor for one year in seven is another Jewish tradition, being

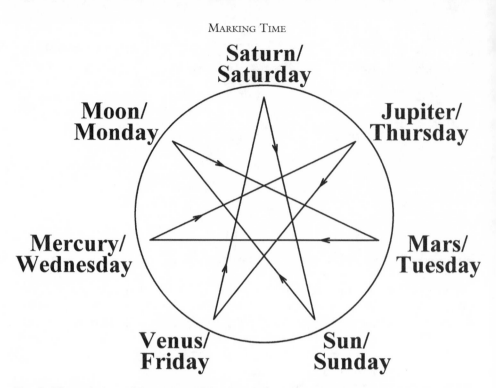

Fig. 1. The ordering of the names of the days derives from the ancient planetary week. The planets were thought to orbit Earth with distances in descending order as in the clockwise sequence shown here, Saturn being the most remote. In each twenty-four-hour day there would be three complete circuits of the seven planets, plus another three steps or two "jumped planets," as shown by the lines within the circle. Thus Saturn's Day was followed by the Sun's Day, which was followed by the Moon's Day, and so on.

called a sabbatical year; this concept will be familiar to all associated with university teaching.

In fact, an alternative explanation for the concept of the Sabbath to that given above, but using the relationship between the numbers seven, fifty, and the annual round, has been suggested by scholars. Pentecost is a Christian festival celebrated on the seventh Sunday after Easter; counting inclusively it is therefore 50 days later, hence the terminology. The Amorites, the Semitic tribes who in the second and third millennia B.C. occupied the biblical land of Canaan (parts of Mesopotamia, Palestine, and Syria) before the arrival of the Israelites, are thought to have used a calendar known as the *pentecontad*. In their year there were seven periods of 50 days ($7 \times 7 + 1$) plus a harvest period of $7 + 7 + 1$ days, making 365 in all.

In common language the number seven or its multiples also appear. Although not much used in the United States, elsewhere in the English-speaking world the term *fortnight* (a diminution of "fourteen nights") is much used. Saying that a magazine is published fortnightly because it appears every second Saturday is more precise than saying that it is bimonthly. When I was a child a popular weekly television program in the United Kingdom was called *Twice a*

Fortnight, which is a circumlocution in more ways than one. Employment of fortnight shows no sign of dying out. On the other hand, despite William Shakespeare using it several times, the term *sennight* ("seven nights," a week) is now little used ("Weary sennights nine times nine, shall he dwindle, peak and pine"—*Macbeth*). In the area of England where I was born, a year is often referred to as a *twelvemonth*, which again Shakespeare employed repeatedly ("I have this twelvemonth been her bedfellow"—*Much Ado About Nothing*). One can have much fun with a concordance of Shakespeare: did you know that he used no words beginning with *x*, and only one starting with a *z* (*zealous*, and that once only)?

The Names of the Weekdays

Until now I may have misled you into believing that the days are named for the planets, but that is not quite true. One needs to differentiate between the gods as the ancients visualized them, and the planets associated with those gods. For example, the planet we call Jupiter the Romans called *stella Jovis*, the star of Jupiter, and his day was called *dies Jovis*, the day of Jupiter. Jupiter was the unseen (except in statuary) god, *stella Jovis* his associated light in the sky. This observation allows a closer understanding of early Christian descriptions of Jesus Christ as being the *True Light* or the *Morning Star* (nowadays we call Venus the morning or evening star dependent upon its phase).

Returning to Jupiter, the equivalent god in Scandinavian mythology is Thor, hence Thursday. Because Thor/Jupiter was the king of the gods, some peoples have held Thursday to be the holiest day of the week: this was the case in Estonia even into the twentieth century. Generally our day names derive through Teutonic languages: Tiw, Tiu, or Tyr was the divine equivalent of Mars, Woden that of Mercury, and Frige or Freya that of Venus.

By now you will likely have jumped to the conclusion that Saturday is just a concatenation of *Saturn's day*. Wrong. For sure the Romans called that day *dies Saturni*, the day of Saturn, but our word *Saturday* does not derive directly from that source: it comes from the god *Saeter* or *Seterne* of Norse legend, an alternative form being *Loki*.

Sunday and Monday we have already met, but an interesting point is why they contrast rather with the distinct Germanic names of the other five days. One explanation put forward is that in northern Europe a five-day week was used by pagan tribes, Sunday and Monday later being added due to the influence of, first, the Roman legions, and later Christian missionaries. The identification of Sunday with the Lord's day in Christian countries, and thus the naming of that day (such as *doménica* in Italian), we will discuss below.

The names of the days in different languages lead to all sorts of clues as to how they came about locally, and also various peculiarities. If you want to use the original Roman deities for days, but don't want to speak Latin, move to Wales: Welsh is the only language which preserves their names in its terms for the days. On the other hand, if you simply want your days to be numbered (ignoring the colloquial meaning of that expression), move to Portugal, with

Russia, Poland, and various Slavic countries also using names based on ordering, while in Germany Wednesday is just *Mittwoch,* "midweek." Anglophones should not start to feel superior: an administrative assistant is often called a *man Friday* based on the character in Daniel Defoe's *Robinson Crusoe,* but the derivative *girl Friday* makes much more sense given that Friday is associated with the (female) goddess of love. Similarly, Tuesday has been used as a girl's given name, despite the fact that the day gets its tag from the (male) god of war, Mars, or Tiw.

But these musings are based on Greco-Roman gods. The Mayan god associated with Venus was a male named Kukulcan, or Quetzalcoatl to the Aztecs and Toltecs. For all I know this tradition could have extended through other Central and South American civilizations down to the Incas and beyond. The figurative location for Defoe's story is a Pacific group known as the Islas Juan Fernández, several hundred miles off the coast of Chile. Nowadays one of the islands is called Isla Robinson Crusoe, and another Isla Alejandro Selkirk for the real-life marooned Scottish sailor on whom the tale was modeled. If Man Friday was from a culture associating the planet Venus with a male god, he would have regarded his epithet as being very masculine indeed.

But I've not finished with Saturn yet. Pop groups may sing lyrics such as "Saturday Night Fever" (the Bee Gees) and "Saturday Night's Alright For Fighting" (Elton John), and those may echo the idea of the meaning of the word *Saturnalia* as unrestrained merrymaking with associated chaos, but again there is no link with Saturday, as such. Saturnalia was originally a Roman festival celebrated over several days in December prior to the winter solstice, followed by the more somber marking of the solstice itself which, we will see, eventually evolved into Christmas.

Saturday (*dies Saturni*) was actually quite the opposite; an unlucky day regarded as being under the control of a maleficent star, leading to our adjective *saturnine,* which can be taken to mean gloomy or taciturn, or may refer explicitly to lead poisoning. The Romans associated not only a god with each planet, but also various entities. The best known would be the link between quicksilver and the planet Mercury, the first term there being synonymous with the chemical element mercury (the symbol for which is Hg, derived from the Latin *hydrargyrum,* literally "liquid silver"). Items linked with Saturn include onions, donkeys (my apologies to the Democratic Party), and lead, demonstrating considerable distaste for that planet and hence the ill-starred day.

If any reader is still wondering about the connection between lead and plumbing, I should point out that the chemical symbol for the element is Pb, from the Latin *plumbum.* The malleable nature of lead, and its low melting point, made it the first material to be used for piping water into houses, two millennia before the advent of extruded plastic. I was born on the edge of the Mendip Hills in Somerset, England, where the Romans mined lead, using some to equip their bathing pools in the settlement which became known as Bath Spa (*Aquae Sulis,* to a Roman colonist, the baths dedicated to the goddess Sulis Minerva).

The Planets and Calendar Reform

In all of the above you will have noticed that the outermost planet involved has been Saturn. In many books it is said that no other planet is visible with the naked eye, making Saturn the most distant planet which *could* have been spotted prior to the invention of the telescope. In fact this is not true: Uranus *can* be seen with the unaided eye, provided you have excellent visual acuity, a dark location from which to observe, and you know where to look. Despite this it was not discovered until 1781, when William Herschel found it from that city of Bath.

Neptune was discovered in 1846, although there is evidence that Galileo saw it through his rudimentary telescope in the early seventeenth century but failed to recognize that it was not a star. Pluto followed in 1930, and there are good reasons to state that there is no massive planet beyond that thousand-mile minnow, although since 1992 several dozen large bodies (up to a few hundred miles in diameter) have been found in a belt at the fringe of the solar system.

One oft-made suggestion regarding calendar reform is that it would be good to have an exact number of weeks in the year, so that all annual holidays could occur on the same day of the week. This could be achieved, say, by having January 1 (plus February 29 in a leap year) counted as being no day of the week, like the 5 epagomenal days of ancient Egypt. Some countries have used calendars predicated upon the need to have a complete number of weeks in a year: even into the twentieth century the people of Iceland employed 364 days over 52 weeks in an ordinary year, and 371 days over 53 weeks in a leap year.

If one allows instead a year which is not filled with complete weeks, one has the problem of what to call these extraneous blank days. Given that the regular days have names based on planets, maybe Uranus, Neptune, and Pluto could get some belated recognition. Perhaps January 1/February 29 could be called UNPday, representing the three outermost planets and United Nations Proclamation day, as well; no PUN intended, of course.

Combining the Jewish and Planetary Weeks

Having meandered into a consideration of the names of the days, we now need to return to our deliberations on how the seven-day week conquered all others. Our story so far has shown that in the Middle East, and in particular the areas we now call Egypt, Israel, and Palestine, around the second century B.C. two distinct seven-day cycles gained popularity: one was the Jewish week, the Sabbath being a day of atonement and purification on which activities of the Judaic people were limited, and the other was the planetary week, whereby each day was considered by the adherents to be reigned over by one of the seven gods with whom the visible planets were associated.

Our septenary week seems to have gained ascendancy due to the reinforcement between these two cycles, which have quite independent origins. But that ascendancy was in the Middle East: how was it transferred to Rome and thence, with time, its empire? Certain occurrences would certainly have made a profound impression upon the Romans: for example, Pompey easily captured Jerusalem in 63 B.C. because he happened to attack on the Sabbath, the Judaic reverence for that day leading to a less than optimal defense being mounted by the Jewish population. Ever since then military tacticians have dictated that offensives begin at times made propitious by their opponent's routines—attack on a Sunday morning while half the people are at church and the other half hungover from the night before, for example.

By the time of Julius Caesar's dalliances in Egypt in the mid–first century B.C. (he fathered a son by Cleopatra and was besotted by her, she only taking up with Mark Antony after her former lover's assassination), the Roman nundinum system was well established. In fact Rome had inherited that system from the Etruscans around 700 B.C., and it was used throughout the Italian peninsula. But just as Julius Caesar and his troops returned from Egypt with calendrical concepts beyond the old Roman system, so they brought back strong impressions of an alternative form of week.

Their attitude would have been especially swayed by the planetary week, because it was based upon a religious system fundamentally similar to their own. From Persia the cult of Mithraism, involving sun worship, had infiltrated Roman life, most especially that of the legions. This should not be confused with straightforward horoscopic astrology, as such, which was officially frowned upon in Rome (in 139 B.C. all astrologers had been expelled).

In this context the planetary week, with days dedicated to the Sun and the Moon in particular, had considerable appeal. By chance it seems the Jewish week was phased such that the Sabbath fell on the day dedicated to Saturn in the planetary week, and since the Romans and others considered this an unlucky planet, the fact that the Jews took this to be a day of rest and solemnization seems to have been misinterpreted as being a time of superstitious dread on their part. One could wonder what might have occurred if the phasing had been one day different, with the Sabbath occurring on *dies Solis* (our Sunday). As it happened the correlation seems to have produced a strong impression on both sides, confirming to the adherents of the planetary week that their system was truthful, while the Jews actually named the planet Saturn (as *Shabtai*) because of its connection with their holy day through the planetary week.

The Ascendancy of the Seven-Day Week

It is impossible to expropriate a specific date for when the seven-day week effectively took over from the eight-day nundinum. A single calendar from around the turn of the era shows both the nundinal days labeled from *a* to *h*, and also a column of days marked *a* through *g* and hence a seven-day cycle; the former were termed the *nundiales* and the latter the *hebdomadales*. A wall inscription at Pompeii, which was destroyed and buried by the eruption of Mount Vesuvius in A.D. 79, shows the *Days of the Gods* in week order, starting with Saturn and ending with Venus. Horace, in his *Satires*, which were written in about 30 B.C., appears to be describing the observance of a seven-day week in Rome.

The picture we see, then, is of a gradual rise in the use of a seven-day cycle from when Julius Caesar returned from Egypt through the following century or two.

By A.D. 200 it was no longer necessary to explain this seven-day cycle, because it had become accepted through common practice. The situation then was similar to the calendar in use in the United States now: there is no U.S. legislation stipulating the calendar to be used, or that March will follow February next year, bar the inheritance of Lord Chesterfield's act of 1751 in the British Parliament through its enforcement in the American colonies. Eventually the week *was* legislated for in the Roman Empire, however, and this allows me at last to introduce the character whose epithet appears at the head of this chapter: Constantine the Great.

The Conversion of Constantine

In the previous chapter I wrote that Constantine lived from A.D. 280. In fact no one is certain when he was born. The *latest* which seems feasible is A.D. 288, but some accounts have his birth as early as 272, and there are reasons to favor that earlier date: when his father, Constantius, died in 306, Constantine was declared Caesar by his troops, and efforts were expended to embellish his youth but apparent maturity. The later reunification of the empire was one of Constantine's epic achievements—this was the time of the tetrarchy, with four capitals, two in the east and two in the west—leading to the bolstered epithet by which we know him.

For most of his early life Constantine was a sun worshiper, but he had been much impressed when, after his predecessor Diocletian passed an edict persecuting Christians in 303, the royal palace was struck by lightning. To Constantine this looked like divine intervention, but Christianity and sun worship were still much intertwined. Much later one finds priests and popes alike being exacerbated by the tendency of their flocks to pray to the Sun, and then turn their backs on it to enter the church building. Churches were often literally *oriented;* that is, arranged so as to face east. In medieval times a church dedicated to a particular saint would be directed toward the place of sunrise on the festival day of that saint.

In A.D. 303 Constantine was not yet ready to desert the Sun, which he hailed as his tutelary god in 311 (he had its representation on his coins, for he thought of it as being his invincible companion). The phrases used were *Soli Invicto Comiti* and *Sol Deus Invictus,* echoing the winter solstice festival of the unconquered sun god: that orb reaches its lowest point in the south at that time, and then ascends again, and this was the basis of the pagan midwinter celebrations which eventually became the modern Christmas. The emperor's beliefs at that time may be identified with this cult of Sol Invictus, which was similar to Mithraism, but distinct (in the same way that the Roman Catholic, Orthodox, and Protestant churches in all their varieties, and the Jewish sects, are distinct facets of the Christian faith).

Constantine's beliefs were ripe for changing, however. In 312, while marching south through Italy in civil war against the other claimants to the emperorship, he saw the image of a cross near the Sun. His description sounds like a parhelion, or sun dog, or maybe a sun pillar, visual phenomena caused by sunlight scattered by ice crystals in the air. If the story as we have received it is based on fact, then this is a possibility, because the vision is associated with the battle of the Milvian Bridge (across the Tiber), which occurred at the end of October, when an ice-laden atmosphere would be feasible. It was claimed that the whole army agreed that they saw it (but they would say that, wouldn't they?).

More suspect, or perhaps hallucinatory, was Constantine's claim that this cross, which he interpreted to be a crucifix, had above it the words *hoc signo victor eris* ("conquer by this" in English). That night Constantine had a dream in which he said that Jesus Christ appeared to him, leading to the salvation of any Christians who were still suffering under the persecution legislated by previous tetrarchs such as Diocletian and Galerius, although Constantine did not himself convert until on his deathbed in 337.

Early in 313, with the Edict of Milan, Constantine and his fellow Augustus, Licinius, decreed that Christian worship should be tolerated throughout the empire, as should other religious practices. Licinius was a worshiper of what would now be called pagan gods; as pontifex maximus, Constantine was the chief priest of the pagan Roman sect, the state religion in his time, so that he could hardly be an overt convert earlier.

Constantine and Licinius developed a long-term antagonism, the latter being treacherous in the extreme. This did not cease until, after a bloody war, Constantine defeated his foe in 324 and thus became sole emperor; but the ascendancy of Christianity had been set in train a decade before.

Constantine's Adoption of the Week

What has this to do with the week? The answer is that it is one of several threads which we need to pull together. Constantine legislated in A.D. 321 for unnecessary work to be forbidden on a Sunday. That is, in fine weather essential agricultural tasks could be conducted, but otherwise the good Roman (who was effectively being instructed to adopt Christian practices) was supposed to set aside one day in seven as a day of rest. Upon this law stands the legality of the week which we have inherited.

But why a Sunday? Why did Sunday become the Lord's day for Christians? It would be hard to dismiss the link with the Sun (that is, the pagan god whose star was the Sun), given Constantine's previous conviction and of course his conversion came about from seeing an apparition in the sky near the Sun. The solar god had supplanted the much earlier Jupiter, and so an identification with *dies Solis* was natural. In the Roman army, in particular, the Mithraists used the planetary week, celebrating the day of the Sun. A Gentile converted from sun worship would find it difficult to throw off that association altogether.

In A.D. 386 Theodosius I, the emperor of the Eastern Roman Empire, decreed that Sunday, the Lord's day, was holy; he abolished the old Roman festival calendar, marking the final overthrow of the pagan state religion and the establishment of Christianity as the official creed of the empire. Thus we see that the year is timed by nature, the unequal lengths of the months by tradition, and the week by authority.

Other Reasons for Adopting Sunday

But there is more to consider. First, the way in which a movement establishes itself is by differentiating itself from other groups. Nowadays, a youth cult identifies itself by dressing differently from the mass of people, cutting and dyeing their hair distinctively, driving specific types of cars or motorbikes, and thus adopting a certain uniformity of behavior. Muhammad adopted the seven-day cycle for Islam, but chose Friday as the holy day, largely to distinguish Muslims. Catholics tend to eat fish and not meat on Fridays. To differentiate themselves from the Jews, early Christians seem to have adopted Sunday rather than Saturday as their Sabbath, leading to yet another meaning for that word, which started out simply as implying the full moon.

Are there any other reasons why Sunday was chosen by the Christians, rather than (say) Wednesday? The answer seems to lie with the biblical account of Christ's Resurrection, which is said to have occurred on the day after the Jewish Sabbath, hence on a Sunday. All of the above considerations seem to have contributed to what was put into law by Constantine the Great in the year 321, our week continuing unbroken ever since. For example, when September 3–13 inclusive of 1752 were skipped in Britain and her colonies, as described in the opening chapter, Wednesday, September 2, was followed by Thursday, September 14.

Which Was the First Day to Be Allocated to a Seven-Day Week?

This opens the following question: which was the first day to be thought of as being a particular day of the week (DOW)? Prior to Constantine's edict of 321, the week had de facto been in use as a cycle for some centuries, although the nundinum had continued in parallel. Is there, for instance, some earlier document labeling a certain date as a Thursday, say? In fact there is nothing definitive, but the first day in history to be allocated a label in our SMTWTFS system can be argued to be August 30 in 30 B.C. The reasoning is as follows.

In the second century A.D. the author Vettius Valens published a description of the way to determine the DOW for any particular date (such as that of one's birth); he used as his basis what was termed the "first year of Augustus," and assumed this to begin with a Sunday. In this context the year is not of Augustus's consulship, but the time when the Augustan era began in Egypt the year after his victory at the battle of Actium over Mark Antony and Cleopatra. Note that at this stage Augustus was still named Octavian: for the counting of the years of our era this fact will at length be seen to be critical.

Augustus entered Alexandria in early August of 30 B.C., Cleopatra famously committing suicide by clutching an asp to her breast rather than adorn his triumph, her not inconsiderable feminine wiles having had less success with Augustus than they had earlier with Julius Caesar and Mark Antony. A new dating era there was established from the start of the next Egyptian year, and their year began with the month of Thoth, which virtually coincided with the Roman September.

The Augustan era in Egypt thus began with the first of Thoth in 30 B.C., which was either August 30 or 31 that year: we cannot be sure. Nor can we be sure of the DOWs for those dates, due to our lack of knowledge of the leap years used in this period, Augustus having needed to delete all leap-year days between about 8 B.C. and A.D. 4. I also note that if the story about Augustus extending the month bearing his name so as to match July were true—this was discussed in chapter 6 and stated to be a myth—then in 30 B.C. there would have been no thirty-first day in August. But with nominal assumptions, January 1 of A.D. 1 was a Saturday, and counting back with triennial leap years it seems that August 30 of 30 B.C. was a Sunday. It is from there that we may count our weeks.

The Significance of the Week to the Rise of Christianity

There seems little doubt that the rise of primitive Christianity owes much to the planetary week and sun worship, with Constantine's vision and dream being pivotal in changing the direction of Roman religion from the dominant Mithraism to Christianity. On the other hand, although the Jewish week was coincidental and assisted the infiltration of the seven-day cycle into the Roman psyche, the Jewish Sabbath occurring on the unfortunate day of Saturn led to a strong rejection of that day by A.D. 360. Sunday (and indeed Friday) were viewed as being beneficent, Saturday maleficent. At least part of the antipathy (to understate the case) of other peoples towards the Jews must surely have come about in this way, which seems especially incongruent given that Saturday is, in contemporary society, the favorite day of the week for many.

Despite the origin of the week, the organized Christian Church soon started to oppose the planetary names for the days. I have previously noted that the days are numbered in Portuguese, but that is not the complete story: the Church especially objected to *Saturday* and *Sunday,* which became known as *sabado* and *domingo.* The ecclesiastical origin of the latter is clear, and the former is obviously based on the (Jewish) Sabbath. In Spanish, just add *el* before those words; in Italian, the names are *sàbato* and *doménica;* in French, *samedi* and *dimanche.* But in those three languages the five weekdays kept their planetary appellations.

Returning to Portugal, Monday through Friday are named as the second through sixth days of the week; in modern Greek the naming is similar. The German *Mittwoch* implies that the week starts with Sunday. Many oriental languages (such as the various Chinese dialects) have names for Monday implying that it is the first day, but others have day names predominantly following the planets. Japanese, for example, use names which in translation are the same as a Roman would have used, but those were not fully adopted until 1873, a solar calendar coincidental with the Gregorian supplanting the previous lunar calendar and the oriental sexagenary system being dropped in favor of the septenary cycle, which had been known since A.D. 807 but was not dominant.

The many languages throughout the Indian subcontinent also use planetary names, but that does not imply that these were copied from European languages: the planetary astrological week arose quite early in India, and it may well be that the Babylonians derived that week not as their own invention, but from further east. If you want to be really confused, go to Malaysia: the formally correct words for the days are derived from Arabic, with Sunday through Thursday being related to the terms for the numbers one through five, but in colloquial Malay Monday through Saturday are numbered one through six!

Which Is the First Day of the Week?

This all provokes the question of when the week should be considered to begin, this having been hinted at earlier in this chapter, when I asked whether, given that Saturday and Sunday comprise the weekend, the week begins with Monday. Portuguese and Greek say otherwise. In Russian, Tuesday is labeled as the second day, Thursday and Friday the fourth and fifth, with Wednesday (*sreda*) being the "middle" of the week; the name for Monday (*ponedel'nik*) implies the day after the preceding week, although originally it meant just "the day after no activity." Saturday is *subbota*, the meaning of which is obvious, and Sunday is *voskresen'e*, meaning "resurrection" or "revival." Thus Russian demonstrates, as might be anticipated, a partial evolution from a religious to a secular system, with Jewish/Christian words for Saturday and Sunday but a week apparently beginning with Monday.

If we take the week as having a Jewish origin, then the Bible should provide an authoritative statement. Biblically, the week begins with Sunday, the day after the Sabbath. Exodus speaks of God laboring for six days to create the world and then resting on the seventh, but that seventh day is the *Jewish* Sabbath, Saturday. All four Gospels say that the Resurrection occurred on the first day of the week, the day after the Sabbath, and hence on Sunday. Other biblical allusions to Sunday as the Gentile Sabbath abound; for example, "Upon the first day of the week, the disciples came together to break bread" (Acts 20:7).

Ecclesiastical usage always has Sunday being the first day of the week. The *Dominical Letter* for each year, which we will see is important regarding the computation of the date of Easter, is the letter assigned to the first Sunday given that January 1 is A.

It is from this lettering that the term *leap year* comes about: March 1 is always a D, so that the inserted day is leaped over in this alphabetical cycle from A to G. (Another way to look at it is that the Dominical Letter regresses through this cycle from G to A one step at a time for common years, but by two steps for a leap year.) As to which day is actually inserted was alluded to in the previous chapter, when I discussed the origin of the term *bissextile year*.

To put out of their misery those who are still puzzling over which day is the leap-year day, look at it this way. If, dear reader, you live in the United Kingdom, or the United States of America, or one of the nations of the British Commonwealth which effectively inherited their civil calendar from the British act of Parliament of 1751 (which we will discuss in chapter 16), then the day inserted into a leap year is February 29, because that act says so, notwithstanding my earlier discussion. If you live elsewhere, then you will need to check the local legislation to find which date is considered to be the additional day. Ecclesiastically, however, and depending upon which church or other religion you adhere to, the intercalary day may or may not be counted as being February 24. I cannot be more definite than this (that statement being meant literally, expressing an inability). Despite what some dictionaries say, if one uses the term *bissextile*, then because of its very specific meaning it must refer to a combination of February 24 and 25, although one would normally just think of the twenty-fourth as being a separate inserted day.

Monday as the First Day

But back to the days of the week. Most modern technological usage makes Monday the first day of the week. For example, check an international airline schedule: the days are numbered from 1 to 7, starting with Monday. In fact this is the global standard, stipulated in ISO-8601, which was gazetted in 1988. (ISO is the International Standards Organization.) Many of the developed countries had already adopted this system by then, most legally making Monday the start of the week in the 1970s.

For international travel this is especially important: it is confusing enough to leave London on a jumbo jet and arrive in Sydney two days later after a twenty-two hour flight, departing after a short vacation for Los Angeles at lunchtime one day and arriving there in time for breakfast on the same date. Modern jet travel has made the confusion of the characters in *Around the World in Eighty Days* (they thought they'd lost their bet by getting home too late, but had forgotten that they'd gained a day in crossing the date line) seem less amusing than it was when Jules Verne wrote the book.

When does the week begin? For most civil purposes, with Monday; for the purposes of the Christian churches, with Sunday. We have seen already that the Jewish *day* begins at sunset, so that it would not be correct to think of the Jewish week as coinciding with the Gentile week, but the lesson is clear enough.

The Distribution of New Year's Days among the Weekdays

It might appear that Constantine the Great was only of peripheral importance with regard to the week, but his influences upon the calender were much greater than merely those mentioned heretofore. Soon I will write more about how his actions were pivotal in delivering the dating system which we use.

But to close, we have some unfinished business with which to deal. Near the start of this chapter I noted that *about* one in seven years begins with a new week, but not *precisely* one in seven, and I said that I'd let you puzzle over that and give an answer later. It's time to put you out of your misery.

The Gregorian calendar has a great cycle of 400 years: in that period there are 303 common years of 365 days and 97 leap years of 366 days, producing a total of 146,097 days. That number is divisible by seven, so that there are exactly 20,871 weeks in four centuries, and thus during that great cycle there are 20,871 of each of the 7 days of the week.

However, 400 is *not* divisible by 7, so that there cannot be an equable distribution of days of the week between the 400 New Year's Days. In fact the distribution is such that January 1 occurs on Monday and Saturday 56 times each, Wednesday and Thursday 57 times each, and Tuesday, Friday, and Sunday 58 times each. Of course, during some period of interest shorter than 400 years (say, during your lifetime) the distribution will be different. New Year's Day is a Saturday in 2000, and a Monday in 2001 (that pattern recurring in 2400 and 2401), even though those are the least likely days for New Year. If the week is taken to begin with Monday, then only one in 7.143 years (in fact, $7\frac{1}{7}$ because $400/56 = 50/7$) starts with a new week. If you prefer Sunday to start the week, then one in 6.897 years kicks off with a new week, slightly more than would be expected if the distribution were random.

One can do similar sums for different dates. Many people superstitiously believe that days which are Friday the thirteenth are unlucky. (Triskaidekaphobia: fear of the number 13.) Some years have only one such date; others have 2 or 3. What is the overall occurrence rate? In the 400-year cycle there are 4,800 days which are the thirteenth of the month. If things were random, then $4,800/7$, or 685.7, would be Fridays. In fact there are 688, so they do occur more often—just—than one might naively expect.

Chapter 8

❧

DIONYSIUS EXIGUUS (A.D. 525)

The single factor which has caused most controversy and division in the Christian religions, and which is pivotal in calendar definitions, is the calculation of the date of Easter. This is called the *computus*. In principle its statement is easy: in any year Easter is the first Sunday after the full moon next occurring after the vernal equinox. In that single sentence is embodied disputes and wrangling almost unimaginable to anyone who is not familiar with Easter's history. From quite soon after the Passion (the Crucifixion of Jesus Christ) through to the present, and doubtless continuing into the future, the debates and vociferous arguments over this matter seem entirely preposterous, even hilarious, to someone like myself not burdened by strong religious beliefs.

The sentence of definition as I have given it above is misleading, in that you might imagine that "full moon" and "vernal equinox" are astronomical phenomena. For Easter, they are not. In appendix A I discuss the concept of the mean sun, the average position of the Sun in the sky, or the place it would be if Earth's orbit were circular and the tilt of our spin axis were zero. The mean sun is related to the real sun by the equation of time, a simple relation easy to program and plot (as in figure A6, on page 375). Our clocks run at a rate based upon this imaginary orb. If we used the real sun, our days would vary in length and our whole timekeeping system would go haywire, and so it makes a lot of sense.

Based on that knowledge you might think that astronomers could define a mean moon, an imaginary construct which represents where the real moon would be *if* the Moon circled Earth at a constant rate, keeping to the same plane. And you would be correct: one can define a mean moon in that way. So is this mean moon used to stipulate when Easter should be celebrated?

Not at all. In determining Easter neither the real or the mean moon are used, but rather an *ecclesiastical moon* (or *calendrical moon*) is defined; the ecclesiastical moon may have a phase which is up to a few days different from the real, astronomical, moon. In effect the ecclesiastical moon may be differently defined by distinct Christian sects, and so I will leave its description aside for the time being. The important thing to note at this juncture is that the Moon as one sees it in the sky is not necessarily a good indicator of when Easter should occur, because an imaginary moon is used in the calculations. Strange, but true.

Neither is the vernal equinox defined cosmically; rather, the *ecclesiastical/ calendrical equinox* is stipulated to be March 21. Note that the ecclesiastical equinox is defined merely as a date, a whole day, whereas to an astronomer the time of the equinox (when Earth crosses the celestial equator) can be specified to a fraction of a second. As we will see, even that date is wrong, even from its basis in the history of the Christian churches. Alice would certainly find the story of Easter to be curiouser and curiouser, a Wonderland domain.

The Importance of the Full Moon

It was previously noted that one of the original motivations for fixing Easter as being soon after full moon was to provide sufficient light to allow early Christians making a pilgrimage to Jerusalem to journey by night. In these well-lit days it is easy to overlook the need for natural light to assist with nocturnal travel.

For example, one of the most influential groups in England in the late eighteenth century, whose interest in applying scientific and technological discoveries in part drove the Industrial Revolution, was the *Lunar Society of Birmingham*. The founding of that society was largely a result of the efforts of William Small, a Scot who returned to Britain in 1764 from Virginia, where he had been a professor at the College of William and Mary but had suffered from ill-health. At the time, Benjamin Franklin was in London, representing the opposition of the American colonies to various taxation measures being imposed on them, and he gave Small an introduction to various manufacturers around Birmingham, the Midlands city which became the heartland of English industry. One of these was Matthew Boulton, with whom Franklin had earlier collaborated on some electrical experiments. The group which Small and Boulton organized adopted Lunar Society as their title because they met for dinner on the Mondays nearest full moon, so as to provide light for their journey home. Small's influence was not limited to the Industrial Revolution, however: he also had an effect upon the American Revolution. One of his students at William and Mary had been the young Thomas Jefferson, who wrote that Small "fixed the destinies of my life." Jefferson's thought, and thus Small's teaching, lives on in the Constitution of the United States, a political realization of the Newtonian schema of the universe, a system of governance in which action and reaction are equal and opposite.

But let me get back to the organization of Easter. While it is true that the needs of pilgrims to Jerusalem was one of the considerations in the early *computus*—or maybe *computi* or *computuses* (pick your plural), because, as we will see, there were several rival methods for calculating the date of Easter, producing various results—there were other matters involved. In chapter 21 the actual date of the Crucifixion will be discussed, the identification being made possible by modern astronomical knowledge coupled with the biblical description of the Moon turning the color of blood. This is clearly a description of a lunar eclipse, during which the moon is darkened as it passes into Earth's shadow, but appears a reddish-brown color due to sunlight which reaches it after passing obliquely through the atmosphere at the edge of Earth (as seen from the Moon), in the same way as a sunset is red and the sky is blue (due to blue light being better scattered by the air molecules, the red end of the spectrum being preferentially transmitted). There is only one time that a lunar eclipse can occur: at full moon. This provides a vital pointer as to when Easter should be celebrated.

We again come back to the duplicious meaning of that word *celebrated*, however. While people often believe it to imply joyous festivities, that is not necessarily the case. It is clear that one celebrates one's birthday, but you can also celebrate a death without it being an Irish wake, with whiskey and Guinness flowing: a funeral can be *celebrated* as a somber affair, with handkerchiefs all around. The distinction is especially moot with regard to Easter: is one celebrating the death of Jesus on the Cross, or his subsequent Resurrection? The answer is that different sects within the broad Christian faith take either one, or the other, or both, of those events as being the appropriate event to be commemorated, leading to the possibility (or perhaps I should write *certainty*) of argument over what should be celebrated, and when.

Biblically, the Crucifixion is said to have occurred on the day before the Jewish Sabbath, hence on a Friday, and we have noted that this was a day of full moon, astronomically speaking (the ecclesiastical moon debates had yet to commence). Which full moon? The Bible makes this the Passover moon, the Jewish festival occurring after the vernal equinox. Similarly, the Resurrection occurred "on the third day," meaning Sunday. That all sounds consistent, but beware: there is a difference of one day between the account given in the Gospel of John (who seems to place the Crucifixion on the fourteenth day of the Hebrew month of Nisan) and those of Matthew, Mark, and Luke (Nisan 15). For present purposes, though, we take it that Jesus was crucified on the day of the first full moon after the vernal equinox, which was a Friday, and he rose from the dead on the following Sunday.

The Easter Controversy

From the above one might imagine that the Easter *computus* would be quite straightforward, but that is far from the case. The "Easter Controversy" could be the subject of an entire book (as it has been in the past, and surely will be in the future: the matter is far from settled yet), so that I can only give a brief sketch here; but I am sure that it will be enough for most readers. Let us start by examining the disputes as they occurred chronologically.

There is no conclusive evidence of Easter being celebrated as a major annual feast in the first century A.D.; that is, in the several decades after the actual Crucifixion (the traditional year of which, A.D. 33, is supported by the astronomical evidence described in chapter 21). By A.D. 120, however, there is evidence of a variety of Easter practices being employed by different Christian sects. One must remember that with communications slow and no printing which would allow the mass dissemination of ideas, and the Christians and Jews in any case being subject to sporadic persecution, it was to be expected that groups separated by even a few hundred miles would soon diverge in their practices.

The first phase of the Easter Controversy was largely concerned with whether Easter might be celebrated on a weekday. The Jewish month of Nisan is the first month of spring, its occurrence being specified by various rules which, it appears, were not always obeyed in antiquity (the Hebrew calendar definition as now applied dates only from A.D. 359). The Passover festival starts on the fourteenth day of Nisan, and a basic rule was that this should occur after the vernal equinox, although more important to the Jews was the start of their year with the month of Tishri, which straddles the autumnal equinox.

In reality, the reckoning of the Hebrew calendar is quite complicated, with six different year lengths being possible, but for present purposes our concern is only that Nisan 14 should postdate the equinox. Remember that the Jewish day begins at dusk (specifically, when three stars are visible) and not at midnight. Traditionally, the Paschal Lamb is sacrificed (to commemorate the Exodus, when Moses led the Jews from Egypt) in the late afternoon on Nisan 14, and eaten after moonrise a few hours later, thus on Nisan 15 (although there is room for quibbling here on an astronomical basis: if the new day starts when three stars are visible, then on average the time when that occurs drifts later around full moon because the greater lunar brightness floods the sky, making it more difficult to spot the requisite number of stars).

It is significant that the fourteenth day may or may not be the date of full moon: on the average, the full moon occurs fourteen days and eighteen hours after conjunction, with a variation in actual times by some hours; also, the appearance of the new moon is typically thirty hours after conjunction (and thus about thirteen days and twelve hours before the full moon), but will be offset from that value by varying amounts dependent upon the latitude and longitude in question. You might care to cast your mind back to chapter 6, where we discussed the origin of the ides in each month, the last six hours or so of which (for a Roman day running midnight to midnight) would often coincide with the first six hours of Nisan 14, which is when the full moon occurs, just after sunset. It is therefore possible, but not definite, that the full moon occurs on Nisan 14, if the start of the month is defined astronomically (by the visibility of the new moon), the following day being about equally likely.

Stepping back to the second century A.D., when the Easter Controversy began, in the Judaic religion Nisan 14 was therefore the important day from much earlier history (the Exodus), it happened to be the apparent time of the Crucifixion, and it might occur on any day of the week. Other Christians (than the Jews) seized upon the fourteenth day of the moon as being the appropriate time to commemorate the Crucifixion, and these people—largely from the East rather than Rome and Romanized Egypt and Palestine—became known as the *Quartodecimans:* they celebrated their Easter, identified with the Crucifixion, on the fourteenth day regardless of the day of the week. In this regard the Quartodecimans coincided with the Jewish Passover, which for the leaders of a fledgling religion (Gentile Christianity being much more recent than the

Judaic faith) was not a desirable factor. This was certainly part of the reason why the Western Church wanted to drop the Quartodeciman practice.

In the Western Church the Resurrection rather than the Crucifixion was regarded as the event to be celebrated, and it was held that the associated feast must occur on a Sunday, the Gentile Sabbath (as it became). By A.D. 150 the matter was being hotly debated, and Saint Polycarp traveled to Rome as a representative of the Eastern Church (note that here I am using the terms *Eastern Church* and *Western Church* as broad geographical divisions, rather than in the better-defined, present-day sense into which they have evolved), to discuss the matter with Pope Anicetus. In the end they had to agree to differ, although communication channels were kept open. This changed toward the end of the century when Pope Victor displayed rather less patience than his predecessors, and excommunicated the Quartodecimans.

The Quartodeciman practice, under a hail of Roman Church invective, largely dwindled away during the following decades, although it survived in places until the fifth century. In essence, the Quartodecimans were fundamentalists who refused to give up on the fourteenth day of the Moon merely because it was also used by the Jews; likewise, the modern Seventh-Day Adventist Church insists upon the primacy of the Sabbath (which for them coincides with the Jewish Sabbath, dusk Friday to dusk Saturday).

The subsequent events show that the aphorism "possession is nine points of the law" is sometimes incorrect; or maybe I should say that the one-tenth sometimes wins out. The term *Passover* derives from the Hebrew *pesach*. The adjective *paschal* is similarly derived. The Quartodecimans therefore might lay claim to the use of paschal in connection with *their* Easter on the fourteenth day of the Moon, since it coincided with the Jewish Passover, but the word *paschal* used in connection with Easter as celebrated later by the Gentile churches is essentially a hijacking, in disregard of its originators and owners. In many European languages the word for Easter is directly sourced through this route. In English our word is thought to derive from the Teutonic mythological moon goddess *Eastre* or *Eostre* (the counterpart of the Babylonian Ishtar) because the feast occurred during her month, known as *Easturmonath* (the Roman calendar not being adopted in England until the eighth century). On the other hand, it may simply originate from the direction *east*, where both the Sun and the full moon rise near the equinox.

It might be useful here to pause to take stock of the various possibilities for commemorating the Crucifixion and the Resurrection. If the fourteenth day of the Moon, and hence the Passover, always fell on a Friday, then all would be well. But it doesn't. Suppose that it falls on a Tuesday. In that circumstance there would ab initio be three alternatives. One would be to celebrate the Crucifixion on the Tuesday (the fourteenth day of the Moon) and the Resurrection on the Sunday, but that would abandon the "on the third day" facet. The second possibility would be to keep to the third day, so that the Crucifixion is celebrated on the Tuesday and the Resurrection on the Thursday, but then the Christian Sabbath has been deserted. The third alternative is what was actually adopted: the lunar phase (the fourteenth day) was abandoned, the Resurrection being recalled always on a Sunday, the Crucifixion on the preceding Friday (called *Good Friday*).

And here is a remarkable point: although the week is of Judaic origin, it was the Gentiles who insisted upon the day of the week for Easter (Sunday), and the Jews who insisted on the day of the month for the Passover (Nisan 14).

The Easter Controversy: Act Two

The second phase of the Easter Controversy grew over the following hundred years or so, the third century A.D., and erupted thereafter. Given that Easter should be celebrated on a Sunday, commemorating the Resurrection and not being associated with a particular lunar phase, how should the date of that Sunday be reckoned? It is here that Constantine the Great reappears in our story.

At about the time of Constantine's uptake of Christianity in 312 there were considerable disputes within the Church regarding when Easter should be scheduled. Again the major argument was between the Eastern parts (now specifically the Christians in Syria and Mesopotamia under the Church of Antioch, the Seleucid capital) and Christians elsewhere, although there were also local variations.

In general the Eastern churches marked Easter as being the Sunday following the Passover, thus devolving responsibility for the astronomical observations and calculations onto the Jews. There were several perceived problems with this. One was that the Jews were regarded by the Gentiles as having committed an enormous sin through involvement with Christ's trial and execution, necessitating (in the Gentile view) that Easter be divorced from Passover, even if the word itself were appropriated for their own use. Another drawback was with the Jewish lunar calendar, which led to irregular embolismic months being intercalated before spring, such that occasionally Nisan 14 preceded the actual equinox, meaning that the Passover festival might occur twice within one year as counted from equinox to equinox. For the Jews this was not a problem, since their year is counted from Rosh Hashanah (a Hebrew phrase literally meaning "beginning of the year"), which is at the start of the month of Tishri, Nisan being six or seven months later, so that only one Passover could occur in the Hebrew year; for the Gentiles, who wanted a year figuratively beginning with the vernal equinox, the Hebrew calendar with leap years containing an extra month, rather than just an extra day, posed a real problem. Or maybe one should say that it was the other way around: the Gentiles invented a reason to disagree with the Jewish system.

In the same way as the days of the week held holy in the Muslim, Jewish, and Gentile Christian monotheistic faiths accentuate the clear differences between them, so it was held that the Christian Easter should be separated from the Jewish Passover. Further, Easter should be celebrated on the same day by all Christians. Even within the other (Western) churches which calculated their own dates for Easter rather than relying on the Jews, there were still differences because of alternative ways for reckoning the dates of the equinox and lunar phase.

The desideratum that all Christians around the world should celebrate Easter on the same day was a resolution of the Council of Arles (in southern France, once a great Roman town), in A.D. 314. With Constantine now emperor and Christianity legalized, even encouraged, communication and progress were now more feasible. The members of that council suggested that the head of the Roman Church should write to all churches in an attempt to ensure such uniformity. But there was still a problem: given that all Christians might agree on a single rule for all, what should that rule be?

The Council of Nicaea

In an attempt to end the disputes Constantine called another Church council, in A.D. 325. We have previously seen that Constantine had become sole emperor in 324 after defeating his rival Licinius in a series of battles around the Sea of Marmara, including a protracted siege of Byzantium. Constantine thereupon decided to make that ancient (dating from ca. 660 B.C.) town rather than Rome his capital, renaming it Constantinople (as it remained until 1930, when it was altered to Istanbul), and turning it into a magnificent city: Clothed in Greek garb, Constantinople was destined to flourish until 1204, when it was conquered by the Crusaders, eventually falling to the Turks in 1453 as they swept through Asia Minor.

Constantine convened his ecumenical council nearby, in Nicaea, about 60 miles southeast across the Bosphorus. In the past Nicaea was a major settlement, the birthplace of Hipparchus in 190 B.C., for example; nowadays it is merely a small town with ancient walls, and named Iznik (or Isnik). The *Nicene Creed* is the formal statement of the underlying tenets of the Christian faith, resulting from the deliberations of the 319 bishops comprising that Council of Nicaea in A.D. 325 (a subsequent council was held there in 787). The changed appellation of that town, and the size and proximity to Arles of the city of Nice in maritime France, have sometimes led people to assume mistakenly that the Nicene Creed derives from Nice.

It has often been written that the Nicene Council decided upon the rules for the calculation of Easter. Even the 1582 papal bull of Gregory XIII states this to be the case. This is incorrect; so much for papal infallibility. The epistle and canons of the Council of Nicaea made no mention of astronomical methods for reckoning the date of Easter, such as the use of the Metonic cycle. All that was said about Easter, in a synodal letter addressed to the Alexandrines, upon whom was to fall the responsibility for the *computus*, was (in translation of course):

> We also send you the good news concerning the unanimous consent
> of all in reference to the celebration of the most solemn feast of
> Easter, for this difference also has been made up by the assistance of
> your prayers: so that all the bretheren in the East, who formerly

celebrated this festival at the same time as the Jews, will in future conform to the Romans and to us, and to all who have of old observed our manner of celebrating Easter.

To the contrary of most reports, the outcome of the Nicene Council was therefore doctrinal rather than chronological. A battle was engaged, between the Gentile and Judaic spheres of the Christian world, and the date of Easter was just an obvious fighting ground. The effect of the agreement reached by the council was that the Gentile side was united, at least in terms of the perceived need to divorce Easter from the Passover as a feast, even if the two were inescapably connected through the heavenly cycles. The subtle way in which this was done, as seen in the passage above, was conducive to harmony throughout the various Gentile churches rather than schism, postponing the eventual split between East and West for many centuries, although there was political trouble ahead, as discussed below.

Augustus De Morgan, Polymath, and Philomath

It seems remarkable that, so far as I am aware, the first person to have recognized the real implication of the Nicene missive was a mid-nineteenth-century astronomer, Augustus De Morgan, who started his inquiry due to contemporary arguments over the date of Easter put forward by people who were ignorant of the fact that the astronomical moon and the ecclesiastical moon are two quite different entities, one real and one imaginary. "One conclusion which I arrived at was, that the Nicene Fathers had a knack of sticking to the question which many later councils could not acquire," wrote De Morgan. He recognized that what they did *not* say was as important as what they *did*.

Would not De Morgan have been biased? you might ask. As a matter of fact he was professor of mathematics at the newly founded University College, London, which had been set up largely on the basis that the universities of Oxford and Cambridge did not admit non-Anglicans. De Morgan's Jewish friend Benjamin Gompertz, a pioneer of statistical theory, had been excluded in this way, due to his religion, and the new University of London had rules prohibiting such discrimination. De Morgan twice resigned his chair when he found that academic appointments were being proposed with preference given on the basis of religion.

With his mathematical and astronomical expertise, plus religious tolerance and lack of dogmatic beliefs, De Morgan was perhaps uniquely qualified to examine the question. But don't think that he was perfect: a perusal of his writings show him to be startlingly sexist. In his view the University of London was the place for Catholics, Jews, Muhammadans, and all others, so long as they were not female.

Getting back to the Nicene Council, the fundamental action was the splitting of Easter from the Passover for basically anti-Semitic reasons, as noted earlier. This is apparent in the rules for calculating Easter today. For example, consider the *Book of Common Prayer* of the Church of England; this has authority both in civil and ecclesiastical law for fixing dates of holidays and feasts. A table is given for finding Easter, and the accompanying explanation states that "if the Full Moon happens upon a Sunday, then . . . the next Sunday after is Easter Day." The reason for this rule is pure and simple. If the "Full Moon" were on a Sunday, then that would likely (but not necessarily) be Nisan 14, and so the day on which Jews observe the Passover. To avoid coincidence between Passover and Easter, the Church rules call for the latter to be postponed by a week.

I may further note that in their determination to avoid any semblence of agreement with Jewish custom, the Gentile churches got themselves into a bit of a tangle. In the sentence quoted above the meaning of "Full Moon" should be recognized to be both (1) the ecclesiastical rather than the astronomical full moon, and (2) actually the fourteenth day of the Moon, the term "Full Moon" being used just to avoid saying so, because that might look like "Nisan 14."

Because of this it happens that a few times each century Easter is celebrated, in what we now call the Western Church in contradistinction to the Eastern Orthodox Church, on a Sunday which is indeed the day of an astronomical full moon: it was ill-informed debate over such events in 1818 and 1845 which provoked De Morgan's investigation. All these attempts to regularize the computation of Easter in defiance of the real motions of the Sun and the Moon (the details of which we will meet in chapter 13) have not been successful in that at times Easter *has* coincided with Passover, most recently in 1954 and 1981, and it is due to happen again in 2123. The converse can also happen, that Easter is on the *second* Sunday after the astronomical full moon, as occurred in 1876: in trying to avoid Passover, Easter was put later than it should be if one were trying to adhere to the Bible.

To anyone who does not understand the actual meaning of "Full Moon" in that sentence from the *Book of Common Prayer* quoted above, these quirks may seem to be contraventions of the Easter rules, and whenever such things happen a flurry of argument and letters to newspaper editors ensues.

The Power Moves East and Italy Is Invaded

Our knowledge of events around the time of the Council of Nicaea stem largely from the accounts of Eusebius, bishop of Caesarea (the Roman capital of Palestine), who was Constantine's chief adviser and is often referenced as the "Father of Church History." The major accomplishment of the Council of Nicaea, judging from soon-following events such as the Council of Antioch in 341, was the unification of the Gentile Christian churches into the Universal Church, from which one may trace the eventual Catholic Church. The foregoing struggles between the Roman state and the Christian Church were largely ameliorated by developments in these decades: Constantine's conversion led to Christianity becoming the state religion, while his eastern translocation of the capital meant that the bishops of Rome (it is a moot point whether those heads of the Church should be called pope prior to Innocent I in A.D. 401) became more important, in the absence of a resident emperor.

This may have been good for the advance of Christianity, but it was bad for later occupants of the Italian peninsula and nearby lands. The military strength was now transferred to the East, and in the late fourth century appeals for the protection of Rome were ignored when invaders threatened, with the result that the Goths (from the northern shores of the Black Sea) under Alaric marched down through Italy and sacked the Eternal City in A.D. 410. This was the first in a series of barbarian invasions over the next 150 years, of which that by Attila the Hun is the best known (contrary to popular belief the Huns were not from the area we call Germany; they were a Mongol race from further east, in the Urals). Attila was forced out of Italy in 452 by the plague, not military might.

The various barbarian incursions led to the complete downfall of the Western Empire by A.D. 476. Without a powerful emperor to subdue them, the religious leaders were certain to bolster their own influence, and within two decades Pope Gelasius I had declared the independence of papal power from both the Church and the state. Until then the emperor had supreme powers over the state religion, wielded in various ways, but those powers were now eroding.

Italy was eventually reconquered by A.D. 554 by Emperor Justinian (of the East, or what is now called the Byzantine Empire), but lost again within two decades to the Lombards, originally a Germanic tribe. One must recall that the federated republic of Italy in its present form is quite recent, with a division into city-states having been the norm throughout most of the last millennium or so: the Lombards ruled much of the Italian peninsula from their capital in Pavia (just south of modern-day Milan) until their final downfall in A.D. 747, while the Byzantine capital in the west, once the resurgent Goths had withdrawn under Justinian's pressure, was Ravenna (near the northern Adriatic coast). Further south Imperial Rome continued as the headquarters of the Western Church.

For calendrical matters it is the final point there which is most pertinent: through war, famine, plague, and the ebbs and flows of strife and torment, the Church in Rome survived and even prospered in the succeeding centuries. After the end of the Lombard reign there was civil strife in Rome, which was rectified from without, through the comparative might of Charlemagne, king of the Franks. In A.D. 800 he was crowned emperor of the West by Pope Leo III, providing some stability. In many ways that year may be regarded as the beginning of the Holy Roman Empire, although formally that is usually pegged against the ascendancy to the emperorship of the German king Otto the Great in 962. Through these years the powers of the papal see were gradually advanced, reaching a zenith when Gregory VII became pope in 1073 and proclaimed his authority over all other rulers, emperors, and kings alike.

The perceived abandonment of the Western Church after Constantinople became the seat of power led to a schism between it and the Eastern Church, which in this context now encompassed the whole eastern end of the Mediterranean. Almost two centuries were to elapse before this rift was healed in A.D.

519, while Italy was still occupied by pagans, the visit of Pope John I to Constantinople soon thereafter sealing the rapprochement which helped to stimulate Justinian's forays west in the subsequent decades.

The point here is that the splitting of the churches at this stage was because of political causes, rather than doctrinal differences. This was not the case in their later disputes. The head of the Eastern Church, the patriarch of Constantinople, never obtained the same level of autonomy as did the Roman pope, but claimed superior authority through the use of the title *ecumenical patriarch:* this basic argument has continued to this day.

This and numerous doctrinal disagreements led to a series of worsening fractures appearing between A.D. 680 and 725, culminating in a final bifurcation in 1054 when the Roman/Western Church excommunicated all those of the East. The hostilities having been declared, the wounds festered for over a century until 1182, when the Greeks of Constantinople, good Christians all, massacred the Latins who lived there. This drew another fine Christian response from the West, the Venetian-backed Fourth Crusade taking and sacking Constantinople in 1204, as previously noted.

The Easter Controversy: Act Three

Constantine's decision to move his capital led to the Western Empire lying undefended, belly up to the barbarians. But at the same time his calling of the Council of Nicaea led to the unification of the Gentile churches and the discrete and discreet separation from the Judaic faith. The latter act brought the Church together, at least theoretically, while the former resulted, after a few decades, in its dismemberment. Christianity later reconciled its far-flung parts, then argued afresh, and managed, despite the tension, to get along for some centuries, before finally blowing asunder with bloody reprisals on all sides. This story puts me in mind of a dysfunctional family, the members held together by the ties of common origin and experience, until one day something finally triggers an explosion of emotion and resentment.

These developments meant that despite the Nicene communication quoted earlier, which implied homogeneity in the *computus,* a variety of methods for determining the date of Easter were developed, with just one thing in common: they were predicated upon a requirement that the Jewish Passover be avoided. This comment applies both to the early history, in the fourth and fifth centuries, and the later history through to the terminal split between Western (Roman Catholic) and Eastern (Orthodox) churches in the eleventh century, their calendrical differences continuing to this day. At present, we are concerned with the early phases.

How, then, was Easter reckoned after Nicaea? In effect the council passed the problem on (in modern parlance we would say that it was "not part of its terms of reference"), apparently suggesting, as noted earlier, that the Church of Alexandria was likely the most proficient in making such computations. This should not be a surprise: we have already seen that most of our calendar is based on Egyptian know-how.

As a result of the Alexandrine deliberations, the Eastern Church adopted a date between Nisan 15 and 21 as being appropriate for Easter. In order to accentuate the fact that Easter was now divorced from any connection with the Jewish Passover, instead of using the Hebrew month of Nisan to express the lunar phase, I will follow other writers in referring to this range of dates as being *luna* 15–21 (and in any case the Gentiles had accused the Jews of inconstancy in ensuring that Nisan 14 occurred after the vernal equinox, which we have seen is largely inconsequential for the Hebrew calendar). In this representation one would say that the Quartodecimans took Easter to be *luna* 14, with no range permissible. Differing from the East, and for reasons which are obscure, the Western (or Roman) Church calculated Easter on a date constrained to be *luna* 16–22. It may simply have followed from the reasoning that if the Crucifixion occurred on Nisan 14/*luna* 14, then the Resurrection must have been on *luna* 16 and thus Easter should occur no earlier than this. Alternatively, the explanation might be connected with the contradictory accounts in the Gospels, pertaining to whether the Crucifixion occurred on Nisan 14 or 15. The effect of the inequality is the significant thing here: it meant that in six years out of seven East and West might agree, but in the other year there was a one-month gap between the times that Easter was celebrated by each. This disagreement continued for two centuries.

I wrote above that East and West *might* agree for six years out of seven, but that would be the case only if their *computuses* tallied in all other respects. That is, if their dates for the equinox and new moon were identical: remember that these are ecclesiastical equinoxes and moons, to be calculated in advance, and not real observed phenomena. In fact it was by no means the case that the dates for the equinoxes and the moons assumed by East and West were the same. Let us see why.

In chapter 5 we met the Metonic cycle of nineteen years, and I noted that the Babylonians appear to have anticipated this discovery some time before Meton. It seems that the Jews calculated their lunisolar calendar using this nineteen-year cycle inherited from the Babylonians: they brought it back with them when they returned to Judaea and Egypt after the Exile. Thus, although this cycle may have been a simple and effective way to predict the times of forthcoming new moons and hence dates for Easter, there would have been some disinclination on the part of Gentiles to use it because of its Judaic connections.

Nevertheless, the Metonic cycle *was* used by some sects. Even fifty years before Nicaea, Anatolius, bishop of Laodicea (now Latakia, on the coast of Syria), had been using this nineteen-year round as a convenient instrument. In the Church of Alexandria it was decided to adopt this method, although March 21 was assumed as the ecclesiastical equinox, rather than the March 19 used by Anatolius, leading to about one Easter in fifteen being on widely different dates even though they were using the same cycle and the same *luna* 15–21 rule.

Elsewhere other cycles were used. The Metonic cycle renders an average of 12.3684 months per year (235/19), compared to a requirement based on the mean lengths of the vernal equinox year and the synodic month of 12.3683. But the ancients did not know that requirement to such precision, so that it was not possible to assess properly the relative merits of other cycles. A cycle much used, including by the Jews for their Passover *computus,* was one of eighty-four years, which may be thought of as being four Metonic cycles plus one *octaeteris.* This renders 1,039 months, or 12.3690 months per year. Not too bad, but it led to a drift in the lunar phase of about thirty minutes per year compared to the Metonic cycle's six or seven minutes, and thus a full one-day discrepancy within a half-century or so. As we will see in chapter 9, use of this eighty-four-year cycle persisted in the British Isles for several centuries more.

In Rome this cycle had been employed, apparently picked up from the long-standing Hebrew eighty-four-year reckoning, but superseded by a sixteen-year cycle due to Hippolytus in A.D. 222. The latter was soon found to be wanting, and the Church resumed the eighty-four-year cycle. A problem was that the Roman Church assumed the equinox to be March 18, somewhat too early. In A.D. 312 the date for the ecclesiastical equinox was moved to March 25, that being the traditional date and also the approximate actual date of the astronomical equinox in Christ's lifetime; but precession had moved it in the intervening centuries. This system—eighty-four-year cycle for the Moon, March 25 for the equinox, Easter on *luna* 16–22—was called the *supputatio romana vetus;* "supputation" is a lovely word which we will meet again in chapter 16.

In A.D. 343 another change was made in Rome, pushing the equinox back to March 21 so as to agree with the Alexandrines, but the differences in the *computuses* still resulted in the dates of Easter differing in the years 346, 349, 350, 357, 360, 387, and 417. The quick succession early in that sequence led to the Roman and Alexandrian churches agreeing to yield in turn, but neither would shift from their dates in 387 and 417, provoking great disharmony. When there was a discrepancy again in 444, Pope Leo the Great tried mightily to persuade the Alexandrines to shift to the Roman *computus,* but failed. While the major aim of the Council of Nicaea had been achieved, with the church distinguished from the synagogue, the sentiment expressed by the Council of Arles, that all Christians should celebrate Easter at the same time, was under threat.

Even before he became pope in A.D. 461, Saint Hilarius had asked Bishop Victorius of Aquitaine—the ancient Gallic province bordering the eastern coast of the Bay of Biscay—to investigate alternative cycles, and the latter devised in 457 what is called the *annus magnus,* a cycle lasting 532 years. This is variously called the *Canon Paschalis,* or the *Victorian cycle* (a title which is somewhat confusing given much later British history). The 19-year cycle of Meton we may term the *lunar cycle* in this context. There is also a *solar cycle,* consisting of 28 years: the quadrennial leap-year cycle under the Julian calendar, multiplied by the seven days of the week: a few minutes with pencil and paper should satisfy you that the sequence of days of the week for January 1 perennially repeats on that cycle under the Julian (but not the Gregorian) calendar. Multiplying 19 by 28 one derives the 532-year great cycle. Given that this is a multiple of the Metonic cycle used in Alexandria, its adoption by Rome would have been promising, the remaining barrier to harmony being the Roman utilization of *luna* 16-22.

Dionysius Exiguus

Again, like the Seventh Cavalry, the eponymous hero of this chapter arrives late on the scene. Dionysius Exiguus was a monk who lived and worked in Rome. Considering his calendrical importance, it is disappointing that comparatively little is known about him. We known that he was from Scythia, a large, nomad-inhabited region in southwestern Russia, partly in Europe but partly Asian, too, but we do not know when he was born. Apparently he arrived in Rome in the last few years of the fifth century. He was certainly dead by 556, and our best guess at his expiry puts it in the year 546.

His name has been much misrepresented in modern accounts, the impression being given that he was small in stature. The modern version of Dionysius is Denis, and Exiguus may be translated in various ways dependent upon the context. Many writers have called him "Denis the Little" or "Denis the Diminutive" (for alliteration's sake), making him sound like one of Robin Hood's Merry Men, the second part of his appellation being interpreted as being a nickname, like you might call someone Shorty (even if he were a star basketball player). Actually, the evidence is stronger that the Exiguus part was a self-applied addendum, meaning "the lesser" and reflecting humility on his part. Whatever the truth may be, let me refer to him simply as Dionysius.

Many accounts also make it sound like Dionysius was either a preeminent Church figure, perhaps the abbot of a monastery, or else a lowly cleric who, unwittingly, did only one significant thing in his life, catapulting him to fame and immortality, like a suburb-domiciled insurance salesman who happens to catch the ball knocked into the left-field bleachers, producing the winning home run, bases loaded in the bottom of the ninth of the deciding game in the World Series. Dionysius was neither.

In fact Dionysius was a preeminent scholar of the records of the Church. In the first decades of the sixth century he painstakingly codified all extant decretals (papal letters and bulls largely responding to doctrinal questions), by A.D. 520 adding to this corpus all available decrees of the various councils and canons of the provincial synods. In passing I note that this included the documents resulting from the Nicene Council. Dionysius was an expert, then, on Church records and their implications, and he was living on the pope's doorstep. He was the obvious person for the pope (Saint John I) to ask to report on when Easter should henceforth be celebrated throughout the Christian Church.

Dionysius's considered report was an excellent example of diplomatic accomplishment, although it is not clear that it was necessarily meant to be. That is, his method of Easter computation has in essence survived to the present day above all others through a process of natural selection: it survived and propagated because it was acceptable to all, although it rejected facets of the systems advocated by different churches.

Taking authority from the Council of Nicaea, which had delegated responsibility for the *computus* to the Alexandrine Church, he adopted the *luna* 15–21 rule, rejecting the one in use in Rome, and the March 21 equinox previously accepted by Rome, and calculated a ninety-five-year table (five times the nineteen-year lunar cycle) as had previous patriarchs of Alexandria, in particular Saint Cyril. The table of the latter, who had died in 444, was coming to an end by Dionysius's time, and the heads of that Church by then were more agreeable than Cyril, who had been a great lover of conflict, leading to prolonged differences with Rome. On the other hand, Dionysius objected to the era (year numbers) used by the Church of Alexandria, which counted from when Diocletian became emperor in 284 (the *anno Diocletiani* system), on the quite-justifiable grounds that Diocletian was a persecutor of Christians.

Not everyone immediately began using Dionysius's table. For example, in Gaul the Church insisted upon continuing the use of Victorius's table and thus celebrating Easter independently until Charlemagne's time, several centuries later. Regarding the evolution of the calendar, the notable holdout was Britain, and the implications of this we will consider in the next chapter. Let us just note that all later Easter tables calculated using Dionysius's *computus* are generally termed *Dionysiac tables*, even though he himself did not compute them.

Before looking at how our era's reference point (the start of A.D. 1) was established, we should be clear that neither the 95-year nor the 532-year cycle is perfect, because the Metonic cycle is not perfect. It is not really a cycle at all, but an ephemeral rather than perpetual solution. In its 19-year duration the shift in the mean phase of the Moon is more than a couple of hours, and so almost half a day in 95 years, and several days over the longer cycle. It is this, in the main, that leads to the discrepancy between the ecclesiastical moon (an imaginary orb based on an erroneous presumption that the Metonic cycle is perfect) and the astronomical moon, resulting in the requirement that every so often (every couple of centuries or so) a one-day adjustment to the ecclesiastical moon is introduced. I quite deliberately wrote that last sentence in the present tense: I am describing what is required nowadays, not just in the past.

Our Era Is Defined

It was necessary that computations of the date prescribed for Easter be made some years ahead of time, to allow for the information to be disseminated throughout the Church. Dionysius completed the task set him by the pope in 525, producing an Easter table for the years we now label as A.D. 532–626. As aforesaid, the earlier table of Cyril of Alexandria used the Diocletian era, which Dionysius wished to avoid. The 532-year table of Victorius, drawn up some seven decades earlier, used as its origin the era of the Passion (years counted from the traditional date of A.D. 33), which is known as the *annus Passionis*. Another rival dating system used by some, especially for civil matters, was the Spanish Era, having an arbitrary start date retrospectively set as being January 1, 38 B.C., which was adopted as a representative time for the beginning of Roman rule in Spain. In fact the Spanish Era system was used in Spain and Portugal until the fifteenth century, so one might wonder which year Christopher Columbus thought it was when he set sail for the Americas in 1492. In the Middle East the ancient civilizations used other systems, such as the Chaldean Era of Nabonassar, which began in 747 B.C.

One thing sticks out here: the numerical agreement between the cycle length of Victorius and the first year of Dionysius's Easter table. Now Dionysius did not call that year *anno Domini* 532, although many later accounts give that impression. In fact it was another two centuries before *anyone* started using the familiar A.D. system. Nevertheless, it is indeed at Dionysius's feet that we need to lay the responsibility for defining the system which we use, and so we need to consider what he actually accomplished.

All that Dionysius wanted to do was to calculate the dates of Easter back to the Incarnation of Jesus Christ (according to his *computus* as opposed to when the events were actually celebrated, of course on different dates in different churches). Previously we have met the concept of a proleptic calendar; that is, a calendar calculated later and projected back in time. Of course this does not change the dates which the people of the time employed any more than one might hope to find a coin with the inscription "45 B.C." minted on it. A proleptic calendar is just a later familiar framework applied for convenience after the fact.

What Dionysius did was to calculate a forward Easter table lasting for 95 years, but also to consider a proleptic Easter table going backward over the great-cycle duration of 532 years (although not using the *computus* of Victorius, which he had rejected in favor of the Alexandrine system). This took Dionysius back beyond the Crucifixion, which Victorius had used as his anchoring date, and one may wonder what meaning to ascribe to a computation of Easter for the years when Christ was alive on Earth.

The essential point here is that Dionysius was interested in figurative Easter dates within the years only, and he was *not* trying to define an era. His concern was not with chronology. It's just that the first year of the 532-year cycle preceding his own was close to the actual birth year of Christ, giving him the opportunity of a new framework for labeling the years for his own reference purposes.

Many books will tell you that Dionysius made A.D. 1 the year of Christ's birth. This is simply untrue, on several counts. A few paragraphs back I mentioned the Incarnation of Jesus. This does not refer to his birth, but rather his conception (in blunt language). Most will be familiar with the biblical account of the Annunciation, when the angel Gabriel announced to the Virgin Mary that she would bear the child of God. This is now celebrated as Annunciation Day, or the Feast of the Incarnation, or several variants of those terms, alternatively Lady Day or the Feast Day of Saint Mary. Whichever name one wants to know it by, it is the date which is important here: the familiar March 25. But this was arrived at by backward reckoning.

In the preceding chapter we saw that the pagan midwinter festival celebrating the unconquered sun, at the winter solstice, was held on December 25; under Constantine, this became accepted as the appropriate date for Christmas, celebrating Christ's birth. In fact, that is the traditional date for the winter solstice on the Julian calendar, and it had moved by several days by the fourth century, but no matter: the actual birth of Christ according to all the evidence was at a totally different time of year, anyway.

The early Church had favored January 6 as the date to celebrate Christmas as an alternative to the pagan midwinter festival which followed the debauched week of Saturnalia, this later date being variously regarded as the anniversary of the Coming of the Magi, and also the baptism of Christ soon after his thirtieth birthday. On the liturgical calendar January 6 is Epiphany, a word derived from the Greek term for manifestation; in civil parlance it is better known as Twelfth Night. Clement of Alexandria (ca. 150–ca. 215) mentioned celebrations being held at that time. It has been suggested that the date was already a Roman festival, and well regarded by the Church because of its peaceful connection: it was the date in 27 B.C. when Augustus closed the gates of the Temple of Janus, signifying harmony throughout the Roman domain.

The important point here is that, Church opposition or not, by A.D. 354 a Roman civil almanac was stating that December 25 was the anniversary of Christ being born at Bethlehem in Judaea, and from 375 at least this date was pretty much universally recognized as being the date for Christmas observation. It is what was later accepted which is essential here. Although the Roman civil year after the Julian reform began with January 1, the rise of the recognition of Christmas as December 25 after Christianity became the state religion meant that this date was generally regarded as heralding the New Year. This practice was to persist for many centuries throughout the Christian world.

Counting back the gestation period of nine months from Christmas, one comes to the appropriate date for the Incarnation: March 25, the vernal equinox (to within the accuracy of the people's knowledge) around the start of the era, the traditional date for that astronomical event at the onset of the Julian calendar.

On the other hand, some accounts state that the dates for the Incarnation and the Nativity were settled in the opposite way. It is said that Hippolytus (ca. 170–ca. 236) had decided that the Crucifixion had occurred on March 25 in A.D. 29, and that the Annunciation had occurred on the same date, all gravitating to the equinox. In this schema the Nativity then occurs nine months later, on the winter solstice.

Things are, of course, as they are, but it is interesting to speculate how history would have been affected if the human gestation period were eight or ten months rather than equal to the time between the equinox and the solstice.

Being in receipt of the above accepted dates within the year, Dionysius calculated which year was involved in the framework of the era of Rome; that is, the year *ab urbe condita* (A.U.C). His assigned dates were:

Annunciation: March 25, 753 A.U.C.

Nativity: December 25, 753 A.U.C.

The year 753 A.U.C. is that which we nowadays label 1 B.C. When reading that at first it may not seem to make sense: according to this chronology Christ was born in the first year *before Christ*. This quandary we will settle below. But the essential point here is that according to Dionysius's calculations, the traditional date of the Nativity was December 25 in 753 A.U.C., which we call 1 B.C., and the instant we use as the starting point of our era came a week later, when the figurative clock ticked into January 1 in A.D. 1.

The Feast of the Circumcision

The problem one may have with the fact that Dionysius's framework has Christ's appearance on Earth occurring during 1 B.C., whether one wants to use the Incarnation or the Nativity as the appropriate event, arises in this context because most readers would count a person's life as beginning from birth, or maybe conception. In the Orthodox Jewish faith, however, life does not properly begin until a boy is named and circumcised. For example:

> *And he that is eight days old shall be circumcised among you, every man child in your generations.* (Genesis 17:12)

> *And when eight days were accomplished for the circumcising of the child, his name was called Jesus . . .* (Luke 2:21)

Now is the time to count on your fingers, taking into account the fact that this is all based upon traditional acceptances of the dates involved. If Jesus's life is taken to begin on the day he was circumcised and named, which is the eighth day postparturition (the same day of the week as that when he was born), then this puts it on January 1, A.D. 1. If you look up a liturgical calendar, you will find that January 1 is termed the Feast of the Circumcision. Many histories state that the early Christian churches opposed the use of January 1 as New Year for some centuries, claiming that it was a pagan festival, but in hindsight it seems more likely that this opposition was anti-Semitic in nature, forced by a desire to differentiate the church from the synagogue. But by the sixth century the Feast of the Circumcision was beginning to be accepted as a holy festival, making Dionysius's reckoning a possibility, although it was not until 1090 that it took its proper place in the liturgical calendar.

According to Jewish tradition, then, the description of our era is correct, *anno Domini* meaning "the year of our Lord," which should indeed start from January 1, A.D. 1, for an accepted traditional Nativity on December 25, 1 B.C. Whether that traditional date is factually correct (and some would not even accept the reality of a person named Jesus Christ) is another matter. We have seen already that a variety of methods of reckoning eras and dates for the start of the year have been employed, and we will meet more later. Right now we are interested in the format whereby January 1 begins the year, in accord (by chance or not) with the Julian reform. The correct term for this is *stilo circumcisionis;* the style of the circumcision. Like it or not, that's what we use, although as the method spread throughout Christianity it became known instead as *stilo communis,* perhaps again demonstrating anti-Semitic tendencies.

Regarding the Feast of the Circumcision occurring on the eighth day after Christmas due to the Jewish rite being scheduled for that time (unless the baby is weak or significantly premature), let me relate a personal anecdote. Both of my sons are carriers of Von Willebrand's syndrome, inherited through their mother's line. Almost all that I knew about it, until I started reading about the calendar, was that it represents a low level of blood clotting factor VIII, which may lead to prolonged bleeding from relatively minor wounds and easy bruising. I imagined that there were also factors I to VII, and maybe IX onward, too, and the boys were just low in one factor, like someone suffering from scurvy because they are low in vitamin C although their other vitamin and mineral levels are fine. How dumb I was. Factor VIII gets its name because it has only reached an appreciable level in the bloodstream of a newborn by the eighth day postpartum, and factors I–VII were figments of my imagination. Jewish male babies are circumcised on the eighth day after birth because they could well bleed to death if the procedure were carried out earlier. Perhaps in this we see another part of the puzzle of the origin of the seven-day week, the eighth day, or the octave, being the same day of week as that of the birth.

The Error in the Dating of Our Era

Throughout this book I have been making implicit use of the terms A.D. (*anno Domini*) and B.C. (before Christ). Later I will note upon when these terms originated. Some will object and prefer the chronological systems C.E. and B.C.E. (Christian/common era and before the common era) as being secular. Despite being an atheist myself, I do not object to the B.C./A.D. system, because it is that which is best understood, in general terms. To me that is no different from using the metric system in all the scientific research which I do, and yet thinking of myself as being five feet nine inches tall. If one is going to argue about using the B.C./A.D. convention, then one should also reject all other aspects of our dating system, because as we have seen our time of day, days of the week, and months of the year all have origins tied in with pagan, astrological, Jewish, and Christian beliefs. The Quaker system of doing away with all such references I can appreciate (even if I do not wish to share); the selective argument against B.C./A.D. I have no time for.

Nevertheless, one must recognize that the B.C./A.D. system is *wrong* in that its origin was incorrectly arrived at. Dionysius made an error. It is appropriate that we consider that error here.

If you read most any history book on the topic, you will see that it states that Herod the Great, the ruler of Judaea, died in 4 B.C. (750 A.U.C.). The problem with that is that if Christ was born during Herod's reign, as the biblical account demands, then the Nativity must have been several years earlier than indicated by the B.C./A.D. framework which we use. His actual birth year, and perhaps even the date to within a few weeks, may be arrived at from astronomical records, and we will discuss this in chapter 21, but right now we are interested in how Dionysius's mistake occurred.

A first step would be to question whether the accepted date for Herod's death, in late March or early April of 4 B.C., is correct. It is surprising how flimsy the evidence is for this, being based on sparse records indeed. In terms of material things, there is only the lack of coins bearing his name and dates after that time. Historically, the only account of the place and epoch which survives is that of the writer (one hesitates to call him an historian) later known as Flavius Josephus, an acknowledged falsifier of records who had apparently been a Jewish commander in the defense of Judaea but who later went over to the Roman side—in short, a liar, a turncoat, and a traitor to his faith and people. One could imagine his character and credentials as a witness being torn apart by a defense lawyer in a modern-day trial. In any case Josephus was writing almost a century after the fact. On the basis of various pieces of evidence it has been suggested by some that Herod may have lived through to the start of 1 B.C. If this were the case, then in the biblical context Jesus might have been born as late as the middle of 3 B.C., only thirty months prior to his traditional birth a week before the end of 1 B.C. The error in the era is still present, but of lesser magnitude.

It is conventional, though, to assume that Dionysius's error is by a matter of about four years, and for that there is a simple explanation. Apparently Dionysius fixed on the year of the Nativity by accepting a statement by Clement of Alexandria that Jesus was born in the twenty-eighth year of the reign of Augustus. The problem is how to interpret that statement. Augustus became undisputed ruler after he defeated Antony and Cleopatra at the battle of Actium in 31 B.C., this marking the de facto establishment of the Roman Empire. Trouble is, at that time he was still called Octavian. It was not until January of 27 B.C. that the name "Augustus" was bestowed upon him by the Senate (the term he used for himself was "Princeps," or first-among-equals, not wanting to make the same mistake of self-aggrandizement as had his great-uncle Julius Caesar). Dionysius appears to have interpreted the twenty-eighth year as referring to that of Augustus the name, rather than Augustus the person, and in any case a report from Alexandria would be confusing to a resident of Rome because the Egyptians used a different era, starting from when Augustus later arrived there in triumph and Cleopatra committed suicide. Clement's statement, whether factually correct or not, would seem to refer to 4 B.C. (and one could argue about when the regnal years began—with the Actium date of September 2, or the calendar year?), but Dionysius interpreted it as meaning the end of 1 B.C.

An alternative explanation which has been put forward is that Dionysius used Saint Luke's Gospel as the major basis for his chronology, accepting the statement that John the Baptist began to preach in the fifteenth year of the reign of Tiberius Caesar (A.D. 28-29), and that shortly thereafter Jesus began to teach, when he was about thirty years old. An objection to that interpretation stems from Dionysius adopting the Alexandrine *computus* based on post-dating *luna* 14: if he had been so impressed with the Gospel of Luke, would he not have used Luke's account of the Crucifixion occurring on Nisan 15 rather than the Nisan 14/*luna* 14 which only Saint John presents?

As a matter of fact it seems that under any interpretation one would be pushed to accept that Christ was born in the twenty-eighth year of Augustus's rule, but our calendar is as it is. It results from a series of evolutionary steps, some of which seem ludicrous, and most are definitely suboptimal. Evolution has left humans with various useless and potentially dangerous physiological features, like our vermiform appendix, which may rupture and cause death, or the blind spot in each eye caused by our retinas being back to front, unlike the better form (I cannot say design) in the octopus. The zero point for our B.C./A.D. crossover (the end of the year 753 A.U.C., or any number of other possible reference points on alternative era-counting systems) appears to be some years after the birth of Jesus Christ, but that arbitrary instant happens to be the time reference which we use.

Which Came First?

Very deliberately I have not mentioned various matters until near the close of this chapter. The first is the concept of the year zero. In the B.C./A.D. scheme there is no year zero. After 31 December 1 B.C. came A.D. 1 January 1. In the past I have found that trying to explain that there is no year zero just confuses people more than letting the matter lie, so I will not say much more. To give an example of the confusion, let me quote *Webster's Dictionary*. In trying to illustrate clearly (!) the usage of A.D., *Webster's* gives this example: "From 20 B.C. to A.D. 50 is 70 years." In fact the elapsed time is only seventy complete years if one counts from the very beginning of 20 B.C. to the very end of A.D. 50. If you did the count the other way around—from December 31, 20 B.C., to January 1, A.D. 50—then the gap in time is only sixty-eight years. If you chose a specific date in both 20 B.C. and A.D. 50, say July 4, then the time between is sixty-nine years.

If you object to that no-year-zero scheme, then don't use it: use the astronomers' counting scheme, with negative year numbers. In that scheme the year before A.D. 1 was the year 0 (or 1 B.C.), the year before that was -1 (2 B.C.), and so on. In tables of cometary apparitions you will find that comet Swift-Tuttle, although formally discovered in A.D. 1862, was recently identified in Chinese observations in the year -68 (69 B.C.). This scheme of negative year numbers was invented by Jacques Cassini in 1740, one of a family of astronomers of whom four were successively directors of the Paris Observatory.

I wrote above that "after 31 December 1 BC came AD 1 January 1," and I think that writing the dates that way has much to recommend it, since it emphasizes the switch in counting directions. People often argue about whether the month or the day should be cited first, and it is general practice in the United States to write month/day, whereas in most of the rest of the world the form day/month is more usual: United States government forms have changed noticeably over the past decade so as to stipulate that the date format be day/month/year, which seems bizarre because the international standard

ISO-8601 of 1988, which may be viewed as being particularly significant given the current furor over the Year 2000 computer bug, warrants for the date to be given in the form YYYY-MM-DD. As an astronomer, I agree with that: astronomers write dates starting with the year. If you think about it and lay aside previous prejudice, you'll see that makes a great deal of sense. Put the most significant digit first. If you ask a storekeeper how much a certain item costs, he or she doesn't reply, "Ninety-nine cents and two dollars." The same reasoning should apply to dates. I was born on 1955 June 11. For dates B.C., in astronomical research I use the zero and negative year numbers prior to that, but for historical matters I would choose to reverse the sequence of descriptors: the battle of Actium occurred on 2 September 31 B.C. (Julian calendar).

Next, one could ask when the terms A.D. and B.C. started being used. They seem obvious to us, through frequent usage, but it was not always so. As a schoolboy I enjoyed reading a spoof on British history entitled *1066 and All That*, by W. C. Sellar and R. J. Yeatman. Every so often a pop quiz appears in the text. One question asks: "Has it never occurred to you that the Romans counted backwards? (Be honest.)" The reader can see that this apparently silly query has truth on two levels. In our later B.C. convention *we* count backwards through Roman history from 753 B.C., although they of course did not; but our perusal of their calendar shows that indeed they did tend to count the days backwards to the kalends, nones, and ides.

The previous question in the quiz was: "Which came first, A.D. or B.C.? (Be careful.)" And the answer to that? Well, of course the years we label B.C. preceded the years we label A.D., but as regards the terminology, A.D. came before B.C. We will see in chapter 10 that *anno Domini* was an eighth-century invention, but acceptance of the idea of labeling earlier years as *before Christ* came much later still, as recently as the seventeenth century. The person responsible was another Dionysius: Dionysius Petavius, a Jesuit priest, who in 1627 argued that this was the best system to use (although naturally he was advocating the Latin *ante Christum* rather than the modern English *before Christ;* the alternative *ante Domini* renders an ambiguous acronym). Of course, those earlier years had not been ignored or unnumbered, it is just that they were labeled according to other chronologies such as the Roman A.U.C., the Olympiad cycle, or the Jewish *anno mundi* ("year of the world") which counts from 3761 B.C., while only a few had made use of year counts prior to the Incarnation. The appended B.C. for year numbers has only been generally employed over the past few centuries since Petavius insisted upon its superiority.

Dionysius Exiguus Has the Final Word

Much has been made here and elsewhere of Dionysius Exiguus being responsible for getting our era numbering wrong. I have mentioned above that this was not his intention; that is, I am saying that it was not even his aim to provide a chronology for later people to use, he was just trying to compute dates for Easter and develop a framework for referring to past years which avoided the persecutor Diocletian.

Nevertheless, we should be aware of what Dionysius did write with regard to defining years. In fact, he distinguished between three different years: civil, liturgical, and ecclesiastical. The civil year in his time was that of the Julian calendar, beginning with January 1 (we are going to see that over the following millennium this date was deserted as Church authority overwhelmed that of the state, although we still back-reckon dates largely according to this date for New Year). Dionysius stipulated a liturgical year beginning with the paschal full moon (*luna* 14), the implication of which we have already studied.

The way in which his work developed such later importance, however, is through his ecclesiastical year beginning with March 25, the traditional date of the Incarnation. The first year in Dionysius's sequence of ecclesiastical years ran from March 25 of 753 A.U.C. (1 B.C.) to March 24 of 754 A.U.C. (A.D. 1). He could claim divine authority in this respect, because March is the prescribed start of the year in the Bible: "This month shall be unto you the beginning of months; it shall be the first month of the year to you" (Exodus 12:2).

In this sequence Dionysius labeled years with the term *anno ab Incarnatione* ("the year of the Incarnation"), and much later these have been referred to as the *year of Grace*. Sometimes this is written as *annus ab incarnatione domini nostri Jesu Christi,* the meaning of which is clear, the critical point being that these are years counted from the Incarnation, which Dionysius reckoned as March 25, 1 B.C. It was only rather later tinkering with the date used for the start of the year which led to the *anno Domini* system, which Dionysius did *not* originate.

Chapter 9

∽

Although people nowadays have reason to recall Dionysius Exiguus, making fun of his imagined nickname (Denis the Little) and blaming him for our dating system being in error, in his own time both he and his work faded into obscurity. We have earlier seen how the Western Roman Empire was overrun by barbarians throughout much of the fifth to eighth centuries, with no great upturn in civil fortunes until Charlemagne restored order.

To some extent the Church was insulated against this chaos, as evidenced, for example, by the account of Saint Augustine (in his famous history *The City of God*) of the sacking of Rome by Alaric and the Visigoths in 410. Augustine, writing only three years after the event, gave a vivid account of how the invaders set aside large churches and temples, allowing Christians to gather there and thus escape the slaughter outside their sanctuary, and he argued that this was proof of the protecting hand of God. Similar incursions by the Huns, the Vandals, the Lombards, and other waves of illiterate tribesmen over the next century or so all but put a stop to scholarship. The only haven was, to a limited extent, within the Church. Dionysius was an example of a scholar who managed to survive, if not thrive. The thread of learning was thin and stretched taut, but it held.

The Easter table of Dionysius survived only for paradoxical reasons. The Ostrogoths ruled their part of the Italian peninsula from Ravenna, in the northeast, and their servant and public spokesman was named Cassiodorus. He was not a Goth: he was a Roman statesman and author, with an eye for the main chance. In modern times we would call him a spin doctor, or a PR man.

Cassiodorus was not all bad, though, and he used his money to establish monasteries. One in particular was called the Vivarium, and it was located on his ancestral lands bordering the Gulf of Squillace in what is now Calabria. If Italy is shaped like a boot, this gulf is the small dip in the bottom of your foot just behind your big toe. Splendid this monastery was, largely insulated from the perennial land-going ravages of the barbarians throughout the main peninsula to the north (the leg of the boot), until it was devastated by the seafaring Saracens in the ninth century.

Here is the paradox. When Italy was temporarily reconquered in the years leading up to 554 by Emperor (of the East) Justinian, the great Roman

Cassiodorus was out of a job because his paymasters, the Goths, were put to flight, and he retreated south to Squillace: thus the Roman was forced to flee from Ravenna and Rome because the Roman Empire reasserted itself. If this were not the case, then our calendar history would be different, because Cassiodorus took with him a copy of Dionysius's Easter tables for A.D. 532–626 and deposited it in the library at the Vivarium, where it was safe from the swings and arrows of misfortune as Italy was battled over in subsequent decades.

Most of Christendom was using the Dionysiac *computus* for Easter by this stage, except that Gaul was persisting with the original Victorian tabulation, and a few pockets of Quartodecimans and other miscreants were still insisting on their own rules. The table of Dionysius was not perennial, however, since he had calculated only a ninety-five-year sequence and, as we have seen, the Metonic cycle is not perfect. A decade before the original set of tables was due to expire, a monk at Squillace named Felix Gillitanus was delegated to prepare a new set for 627–721. But the fortunes of the Vivarium monastery were on a downturn, its original sponsor being long since deceased (Cassiodorus died in 585). In about 636 its library was dispersed, and the documents containing Felix's Easter table found their way to Rome.

This timing is important, yet another chance event in the calendar's evolution. To see why, we need to travel northwest a good deal, and consider what was happening in the British Isles. By the *British Isles* I mean all of those islands off continental Europe which now comprise the United Kingdom and the Republic of Ireland. By *England* I mean vaguely the political area now known by that name, but let's not worry too much about the actual borders. This area covers, to first order, the area settled (some would say *civilized*) by the Romans. *Wales* and *Scotland* likewise are modern political divisions, but areas where the Romans had trouble. *Ireland* is across the water, and we are not worried here about its later political division. At the time in question it was usually called *Hibernia*.

The Christianization of the British Isles

Most of the early history of the British Isles is obscure. Although we have written records from the Middle Eastern civilizations dating back many centuries B.C., and of course relatively well-preserved structures there and also from the Greek and Roman cultures, from Britain comparatively little is available to tell us about its early inhabitants even though the evidence of Stonehenge and its like, and later Bronze and Iron Age sites, indicate that these were not the savages that many accounts would like to suggest.

One thing which is clear is that the tendency of modern Britons to think of the local population as being largely homogeneous, with the major distinguishing feature being regional accents in the spoken word, is incorrect; in fact, the peoples who have contributed their genes to the British population over the past five millennia or so have been many and various. I am not just thinking of the Vikings marauding and practicing rapine; rather, I refer to wholesale waves of immigration from distant parts. For example, recent arguments in which disgruntled Britons have complained about archaeological discoveries indicating many megalithic monuments to have been built by incoming peoples from France and northern Europe rather than "real" locals make no sense at all, especially when one considers later arrivals such as the Celts, in around 900 B.C., and the Normans in 1066. British and Irish blood is nothing if not heterogeneous. Barely a few centuries have ever passed without some new tribes arriving from across the seas and making their mark on the future culture and population.

In chapter 6 I wrote of the visits of Julius Caesar to southeastern England in 55 and 54 B.C. This is an episode which is much misunderstood. Ask the average Englishman when the Romans invaded and he is likely to credit Julius Caesar with the accomplishment, believing that he defeated a fierce, woad-painted Boadicea and then proclaimed *veni, vidi, vici* ("I came, I saw, I conquered"). In fact Caesar's incursion was temporary and unsatisfactory, and it was the much-maligned Emperor Claudius who actually invaded Britain almost a century later. Julius Caesar conquered much, but did not have an appetite for primitive Britain; he made his famous proclamation after campaigning in eastern Europe. And Boadicea came even after Claudius, leading an unsuccessful revolt against the already ensconced Romans before her death in A.D. 60.

Boadicea and her contemporaries were just the start of centuries of problems the Romans faced with the inhabitants of the British Isles, who were pushed back to the peripheries. The Celts may have come originally from the Continent, but, apart from a remnant population in the Breton region of France, the Romans penned them back into Ireland, Scotland, Wales, and Cornwall. Hadrian's Wall is one example of the expense to which the Romans went in trying to pacify the parts of Britain which they wanted to settle and exploit, trying to keep out those who would wreak havoc, and it was never an easy task.

After about A.D. 350 order began to disintegrate, and the problem of the Gothic hordes marching on central Europe from the east, coupled with the Jutes, the Angles, and the Saxons trying to invade Britain, led to a gradual Roman withdrawal completed around 410, the year that Rome itself was first sacked.

The Western Empire was collapsing on all fronts, and in Britain anarchy reigned as the Picts and the Scotti rampaged down from the north and the west. Famine, disease, and the concomitant collapse of town life led to the disintegration of the previously organized society. The period beginning from the final downfall of the Western Empire in A.D. 476 is called the Dark Ages for a good reason: little is known of what happened without scholars to record events, and Christianity had hitherto made little progress in the British Isles, although there were pockets of believers here and there. Significantly, one place was in Ireland, where missionaries had arrived from the Continent in the fourth century.

At the start of this chapter I referred to a Saint Augustine, the Church historian and philosopher who lived from 354 to 430. There is another Saint Augustine of a later age, about two centuries after, and he is pivotal in the present stage of our story.

This later Augustine was a monk who arrived in Kent, the southeastern corner of England just across the narrow English channel from France, in 597. In that decade the Roman Church, with Gregory the Great as pope, had started a concerted effort to spread Christianity through Germany and the countries along the Danube, and Augustine was sent west in an attempt to convert the heathens then occupying Britain. He established the cathedral at Canterbury and set about his task. Augustine was not to get far, as he died in 604, but the changes he initiated were important.

At that time England was divided into seven kingdoms: the Saxons ruled Essex, Wessex, and Sussex; the Angles dominated the more northerly Mercia, Northumberland, and East Anglia; while the Jutes' kingdom was Kent. Augustine converted the Jutes of Kent to the Roman Church, and served as the first archbishop of Canterbury. One of his problems was that elsewhere in the British Isles were other Christian sects using Easter *computuses* they had inherited from previous contacts with the distant churches of both Rome and the East. This is why the monks who had much earlier made their way to Ireland were significant: they had been bypassed by the subsequent debates on the Easter question. To spread the influence of the Roman Church, Augustine and his successors needed to promulgate observance of Easter on the dates decreed by Rome, and eradicate the anachronistic *computus* which had been used in Ireland for 250 years, and was now being spread down through Britain as the Irish Church expanded its influence.

The Easter Controversy in Britain

Along Augustine's supply line from Rome came the Dionysiac Easter table, and by 604 this was being copied by scribes in Canterbury. There is no evidence for any adoption of the Church calendar prior to this, and little is known of the calendrical system the various peoples employed in England in the vacuum left after the Roman legions and many colonists departed two centuries before, apart from the fact that they had a year beginning at midwinter, taken as the original December 25. In this southeastern part of England, the celebration of Easter according to the Roman *computus* as laid down by Dionysius became well established over the next few decades. One presumes that when the specific ninety-five-year table of Dionysius ran out in 626, an extrapolation was calculated. By about that time, and certainly within a decade (due to the known dissolution of the library at Squillace), the table for 627–721 due to Felix was brought to Rome; from there it could be copied and sent to the outposts of the Roman Church, such as Canterbury.

Kent was only one of the seven kingdoms of England, however, and the far-flung Celtic lands were a different matter again. As aforementioned, by the early fifth century Christianity had established a toehold in these more distant parts, and a sect known as the Columban (or Celtic, or Irish) Church had later evolved, founded in Scotland by Saint Columba (521–597) and his Irish missionaries. The significant point here is that the Columban Church was not in easy contact with Rome due to the large distance, the previous antipathy (to put it politely) towards the Roman state in Ireland and Scotland, and the downfall of the Roman era in Britain leaving all in chaos.

The calendar of this Church was calculated according to the old eighty-four-year cycle which we met earlier, and also differed from the *computus* now followed (post-Dionysius) in most of Christendom in that when *luna* 14 fell on a Sunday it was permitted to be Easter Day, leading to possible coincidence with Passover. This meant that they were out of step with the practices of the contemporary Roman Church, which was now encroaching upon the British Isles, and this was sure to lead to an eventual showdown. If all Christians were to mark Easter on the same day and avoid the Jewish Passover, then something had to give. The Easter controversy had been long since settled around the Mediterranean, but in Britain the battle lines were just being drawn.

Bordering on Scotland, Northumbria under King (later Saint) Oswald had become a vociferous adherent to the Columban Church from about 633 onward. Northumberland had earlier come under the influence of Augustinian missionaries, but it was then conquered by the Mercians, who remained heathen until their king Penda died in 655. Penda defeated Oswald in battle in 642, and promptly dismembered him; Oswald's followers collected various parts of his body and distributed them to several churches (his head went to Lindisfarne and is now in Durham), leading to a cult of Oswald and eventually his sainthood. After the death of Penda, Northumbria reverted to Christianity—and in particular the Columban Church—under King Oswy, who had been exiled to the north in the interim.

Through such times one sees a north-south schism fomenting with the opponent being the Roman Church stronghold in Canterbury. In between was Mercia, which maintained its pagan beliefs. With that kingdom converting to Christianity after Penda's death, the buffer zone was gone.

Heeding the Nicene Creed, the Columban Church made some attempt to understand the current doctrine of the Roman Church, and sent at least two delegates to Rome to obtain information and instruction: Benedict Biscop went in 653, accompanied by Wilfrid, who, after a protracted stay in France, returned in 658. These were both rich Northumbrian nobles who, after much discussion with the pope on doctrinal questions, returned with many valuable ecclesiastical items, such as silken cassocks; but the most important were extensive collections of Church documents, including copies of Felix's Easter tables. Dictionaries of the saints recall these men as Saint Benedict Biscop (628–689) and Saint Wilfrid (633–709). The common man has another reason to remember Biscop: he introduced to Britain the concept of glass windows.

Much argument ensued on the Easter question, although there were other grounds of debate. One was the supposed role of the bishops in the hierarchy of the fledgling Church in England. Another—one must say ludicrously, in an incredulous outsider's view—was vociferous disagreements over the appropriate form of the tonsure: the style of head shaving for monks.

These matters were brought to a time of decision in the 660s, apparently due to the fact that Oswy was married to a Kentish lady, Queen Eanfleda, who was heavily under the influence of her personal priest Romanus. The latter's name says it all. In one year it happened that the Columban Church had scheduled Easter Sunday on a *luna* 14, whereas the Roman *computus* stipulated that Easter must be a week later. This meant that the Columban Easter Day feasting coincided with the Roman Palm Sunday, a day of atonement, resulting in King Oswy attending the celebrations while his consort was fasting and so unable to join him.

The upshot of this was that Oswy decided that the matter must be brought to a resolution, and so he called the famous Synod of Whitby in that fishing town in northern England in 664, where the Abbess Hilda presided over a newly founded convent; the ruins of Whitby Abbey remain a great tourist attraction today. Most accounts present what happened at the synod as being a triumph of reason over an inferior *computus* (but then the winner always gets to write the history); I have a few discordant comments to make below. It seems that Wilfrid championed the cause of the Dionysiac cycle (of 19 years, compoundable into the 532-year great cycle), and the argument was finally swung that way when he claimed the authority of Saint Peter, which much impressed Oswy, since he did not want to offend the keeper of the keys to the gates of Heaven.

The rest, as they say, is history, with the Dionysiac *computus* being accepted throughout much of the previous Columban domain, although the monks on the tiny island of Iona in the Western Isles of Scotland held out on the eighty-four-year cycle until 715. Between 669 and 690 Theodore of Tarsus, archbishop of Canterbury, was instrumental in bringing together the seven different kingdoms, united now in religion, to form what became the English nation. The Synod of Whitby was a pivotal event, then, in both the history of the British Isles and the evolution of the calendar, because it was the acceptance and preservation of the Dionysiac *computus* in Northumberland which led to our dating system, as we will see in the next chapter.

The Subterfuge of Whitby

But things are not quite as simple as the above account would suggest. I said that I had a few discordant points to make, and here they are.

First, the events leading up to the Synod of Whitby, and the proceedings there, are usually described in terms which make it sound like a gentle family dispute, with King Oswy just being a bit miffed by his queen not being with him at the Easter feast. As a matter of fact, Oswy was a blood-stained monarch who had carried out many unchristian acts, including the murder of his cousin Oswin. Oswy had founded various monasteries not so much out of goodwill, but more as an act of expiation.

There is an extensive account of the actual debate at Whitby which has been handed down to us, but again we need to remember that the victor writes the history, and we might perhaps look for other definite evidence of the circumstances.

One thing which is not known for sure is the date of the synod. Some writers have argued that it was held in 663 rather than 664, but apparently on the basis of misunderstandings of the annalistic years used by later recorders of events; that is, different start dates for a year were used by different writers, and often one needs to resort to specific verifiable events in order to calibrate the time frames. An example is the great comet of 676, which we know appeared late in that year from other consistent European and Chinese records, but at least one English annalist allotted it to 677 because he was using a year starting on September 1. It seems certain that the Synod of Whitby was held at some time toward the middle of 664, but on an indeterminate date. In fact, the date not being recorded seems to have been part of a subterfuge designed to fool potential opponents of the Roman Church.

Another great event occurred in the British Isles in 664, a total eclipse of the Sun. Seen first across the north of Ireland, this is the earliest such eclipse to have been definitely recorded in England. As a recent research paper by Dublin academics Daniel McCarthy and Aidan Breen points out, it seems remarkable that the possible link between the synod and the eclipse has not previously been examined, because they occurred close in time and both involved the Moon. Perhaps the records were fudged deliberately to obscure the connection.

We know when the eclipse occurred, because we can compute such things with utmost precision, given other eclipse records which allow the deceleration in Earth's spin rate over the past few millennia to be ascertained. It was on May 1. We know the track which the eclipse took across northern England (see figure 2, on page 124); Whitby is close to the center, but most of the major monasteries in the north of England were also within the path of totality, and many of these had recently been founded by Oswy as part of the resurgent Columban Church. History also records another major event in England in 664: an outbreak of the bubonic plague, which seems to have come soon after the eclipse.

The apparent sequence of terrifying eclipse, pestilence, and then the synod seems too unlikely to have occurred by chance. That is, in the atmosphere of dread following the eclipse, which may have been interpreted as a sign of God's wrath over the Easter question, and the following plague, which would have been linked to the eclipse in the minds of cleric and peasant alike (does not the Bible refer to such demonstrations of God's anger?), it appears arguable that Oswy hurriedly called the Synod of Whitby in an effort to assuage the vengeance of God which had been provoked by his own establishment of several new monasteries under the "false doctrine" of the Columban Church. Under such circumstances the outcome of the synod would have been a foregone conclusion, and the accounts of the proceedings largely an invention to provide a covering story. I've been to employment interviews run like this.

The role of the eclipse in this connection, and the fact that it must have predated the synod, is confirmed by a letter to King Oswy from Pope Vitalian in 665, in which the latter wrote that "we know how you have been converted to the true and Apostolic Faith by the shielding right hand of God." The conversion in question was not from heathenism to Christianity, but from the Columban to the Roman Church. The "right hand of God" is here the Moon,

Fig. 2. The ground track of the total solar eclipse of May 1, A.D. 664, across northern England. (Adapted from D. McCarthy and A. Breen, "An Evaluation of Astronomical Observations in the Irish Annals," *Vistas in Astronomy* 41 (1997): pp. 117–138; based upon computations by F. Richard Stephenson of the University of Durham.) Also shown are several early ecclesiastical centers mentioned in the text.

which had obscured the Sun in a swath passing across most of Oswy's new monasteries.

The Falsified Record

There is a last fly in the ointment, which points to a subterfuge having been carried out in order to convince Oswy to convert. Obviously Wilfrid, the champion of the Roman stance regarding the *computus,* could not have arranged the eclipse (or the plague) in order to impress Oswy, but he may well have been opportunistic. While the connection between the events described above has only recently been revealed by McCarthy and Breen, it has long been known that the record of the date of the eclipse is in error. Under the circumstances it might not be too strong a word to say that the record seems to have been falsified. The discrepancy has long been a puzzle to chronologists, having been pointed out at least as early as 1590.

We are sure that the eclipse occurred on May 1, both from modern astronomical back-calculations and also accounts of it preserved in Irish and Continental annals. But the English account has it on May 3. This could not have been a simple slip of the quill because Roman dating was used, the fifth of the nones being stated rather than the kalends, as would be appropriate for the first day of the month. The explanation seems to be that the Dionysiac tables (actually the extrapolation prepared by Felix) had a date for the new moon of May 3, and that is what was recorded after the fact as the date of the eclipse, despite it having actually occurred on May 1.

Let us examine what *must* have happened, astronomically, and then what *seems* to have happened on the human front. The solar eclipse occurred at conjunction, as it must, and the new moon (when the thin crescent appears in the west soon after sunset) follows about thirty hours after conjunction, on average, but the actual time depends critically upon the geographical location of the observation point and the season. For example, at Athens the time between conjunction and apparent new moon varies between about twenty-three hours in March and sixty-nine hours in September, whereas at Babylon, at a lower latitude, the corresponding range is between sixteen and forty-two hours. In northern England in early May, a time gap of around twenty-six hours would have been feasible.

The eclipse of 664 occurred in late afternoon in England, and that in itself allows various calibrations to be done. Sunset we know occurred just after 7:30 P.M. Whitby time on May 2, the day after the eclipse, and calculations of the position of the Moon right then show that it would have been barely visible as a crescent very near—within three or four degrees of—the limiting visibility position. Similar computations for Rome, which is rather further east so that sunset would have occurred earlier, show that again the Moon would have been just high enough in the sky to be observable, but its position on the horizon would have been rather closer to the setting sun, which could confound things. Of course, all visibility assessments depend upon an assumption that the sky is clear so that clouds or haze do not interfere.

The above is based upon modern-day computations of the relative positions of the Sun and the Moon. The Dionysiac tables which were at hand in England would firstly have been based upon conditions for Rome or, even worse as regards disagreement with observations in Whitby, for Jerusalem, and for each location new moon spotting on May 2 would have been possible but marginal. On that basis alone one might expect the tables to have indicated that the new moon was due on May 3.

The Dionysiac *computus,* however, was based on the imprecise Metonic cycle. The year 664 was the final year of a 19-year cycle (i.e., 664 was 133 years, seven cycles of 19, after 532). Even if the Sun-Moon relation had been recalibrated for the start of that cycle, by the end of it the real Moon in the sky would be almost ninety minutes behind its computed position, and so would have been predicted to have dropped below the horizon on May 2 before being seen, delaying the new moon until May 3. That is what the imperfect human knowledge would have shown in the tables.

Wilfrid was arguing for the perfection of the Dionysiac tables he had brought from Rome. In retrospect, using modern knowledge, we can calculate that a new moon might have been visible, clouds permitting, on May 2, but that the limitation of the computation techniques would have led to May 3 being the predicted date. Wilfrid and his cohort were not in our knowledgeable situation, though. The eclipse was startling and they wanted to play upon it, and so at the synod they bluffed that it had happened on May 3 and pointed to that date having been predicted for the new moon in their boasted tables. The synod was held at least several weeks after the eclipse, and few people would be able to recall for certain when a specific event occurred: in a court case, when a lawyer asks a witness what they were doing on some specified date several weeks previous, an immediate definite response usually represents foreknowledge that the question was going to be asked and a checking of diaries.

The Dionysiac tables employed were generally accurate enough for their desired purpose. At the Synod of Whitby, however, they were used as part of a deliberate subterfuge, a double-bluff deceit which led to the date of an important eclipse being incorrectly recorded and still causing puzzlement over 1,300 years later. May 3 was the date recorded for the eclipse, two days later than it actually occurred.

It seems unfair to leave this shadow hanging over Wilfrid and his party, for one cannot be sure that a deliberate misrepresentation of the facts occurred. On the other hand, the evidence seems strong. As is well known, ignorance of the law is no defense, whether it be the law of the land or the laws of celestial mechanics. An understanding of eclipse phenomena (such as how the Moon orbits Earth while Earth orbits the Sun) was still almost a millennium off, and some early writers displayed almost unbelievable stupidity in this connection, making statements which are clearly at variance with simple observations. I began this chapter with a mention of the earlier Saint Augustine's book *The City of God*, a history of Rome. Referring to a legend about an eclipse occurring at the time of the death of Romulus, the founder of the city, Augustine showed his ignorance in trying to ascribe the eclipse which seems to have occurred at the Crucifixion as being through the direct intervention of God rather than a natural event:

> . . . *this latter obscuration of the sun did not occur by the natural laws of the heavenly bodies, because it was then the Jewish Passover, which is held only at full moon, whereas natural eclipses of the sun happen only at the last quarter of the moon.*

First, the Crucifixion eclipse apparently described in the Bible was a lunar, rather than a solar, eclipse (see chapter 21). But in any case, eclipses do not occur at either waxing or waning quarters: lunar eclipses are constrained to the full moon, and solar eclipses can happen only when the Moon is at conjunction, a day or two prior to the new moon. It follows that a claim that Wilfrid's party did not comprehend the distinction between conjunction and new moon is no defense. If they had understood the difference, then they could have argued for the veracity of the Dionysiac tables because the lunar visibility was marginal, even with clear skies, on the second evening after the eclipse, but pronounced on May 3, as predicted. If they did not understand the difference, then it seems that they willfully misrepresented the date on which the eclipse had occurred. Either way, the success of this subterfuge led to the conversion of the entire British Isles to the Roman Church and eventually to our calendar.

In several places above I have referred to "one English annalist" explicitly or, it turns out, implicitly through such statements as "history also records another major event in England." There was but one major chronicler of the history of Britain through to the end of the seventh century, through the eye of whom we view what may or may not have transpired in those poorly remembered ages, and by the actions of whom the *anno Domini* dating system eventually became dominant. In the next chapter we meet this man.

Chapter 10

❦

THE VENERABLE BEDE (A.D. 725)

It may be that I have used the word *pivotal* too often in my account of the evolution of the calendar. The problem, however, is that each of the twists and turns in its history have indeed been pivotal. If the asteroid or comet which slammed into Earth 65 million years ago had been delayed by just a few minutes on its course about the Sun, then the subsequent history of life on Earth would have been different, and although *Homo sapiens* may have evolved sooner, or later, April would not be called April, and you would not be reading this book now. Neither would you be reading it if Genghis Khan, Napoléon, and Hitler had not lived, if Newton had not discovered gravity, and if the members of the Lewis and Clark expedition had all succumbed to typhoid before reaching the Rockies. These are all incidental events which have contributed to this volume resting in your hands at this instant, but they are not central to the evolution of the calendar as has been Julius Caesar, or Constantine the Great, or Cassiodorus taking a copy of the Dionysiac Easter tables to Squillace, or various other characters we have yet to meet. The next great figure on our visiting list awaits his introduction now: the Venerable Bede.

Bede was a Northumbrian monk, formally regarded as a saint but with a name almost universally associated with the term *Venerable*. Clearly he did something special. As earlier foreshadowed, his importance lies in his writing the only history of Britain through to the end of the seventh century prepared at the time and which, through various copies, has survived to give us a reasonable picture of the era. In writing his annals Bede began the usage of the counting of time we call *anno Domini,* providing us with the framework, incorrect though it is, which we use to number our years. As Kenneth Harrison (an historian at the University of Cambridge) wrote, "Although the Anglo-Saxons did not invent the Christian era, still less the Julian calendar on which it is founded, they promoted it to become a standard reckoning for historians, and in everyday use, over many parts of the world."

We have already met Bede, although you will not have recognized it. Almost all that we know of Britain after the Romans departed, and the rise of Christianity, stems from Bede's analysis of the vast quantities of fragmentary information he distilled over his decades of work. Like Dionysius, there are many things we do not know about Bede. He may have been born in 672, or 673, or perhaps some other year around then; we are quite sure that he died in 735. To understand more about him we need to look at the context of his times.

Outposts of Learning

While we are familiar with thinking of Rome as being the Eternal City, and of various other western European cities as having long histories of culture and learning, the Dark Ages were gloomy indeed from the perspective of the advance of civilization. Scholarship, which in itself does not produce things of immediate use (such as food, arms, or shelter), depends upon the labors of others, and when the Western Roman Empire tumbled the organizational system which produced the excess supporting scholars fell with it.

The only continuing strand was within the Church, whose conservative stance and power over the peasantry has enabled it to weather two millennia through thick and thin, and one must say that the average churchman tends to have done rather better than the common people. The invasion by the hordes from eastern Europe during the fifth and sixth centuries of Italy, plus much of central and western Europe as far as Spain, led to a fragmentation and isolation even of the Church, and by the year 700 learning in western Europe was confined largely to the great monasteries of Ireland and Northumberland. The tyranny of distance provides some benefits.

Fifty miles up the coast of northeastern England from Whitby is the mouth of the Tyne, the great modern-day industrial city of Newcastle straddling that river somewhat inland. Near where the Tyne empties into the North Sea stands the town of Jarrow. In the year 682 Benedict Biscop founded a new monastery here—as part of the Roman Church, of course—having eight years before established a similar institution at nearby Wearmouth. Many great religious centers were scattered along this coastline, the most famous nowadays being that on the ancient fabled island of Lindisfarne. From about the age of seven Bede was educated by Biscop at Wearmouth, within a few years moving to Biscop's relatively new establishment at Jarrow, where he lived and worked for the rest of his life. This move is important: at Jarrow the abbot, and Bede's mentor, was named Ceolfrith, who himself was ordained by Wilfrid in 669 and followed the latter's position over Easter. Accessible to Bede at Jarrow was a large collection of manuscripts which had been brought back to England by Biscop from his five trips to Rome, the last being a couple of years before his death in 689.

Bede, either of self-volition or as an assigned task, spent much of his twenties compiling a history or chronicle of early England, distilling from different sources the events which occurred in each year, and also descriptions of natural phenomena such as the motions of the planets, the occurrences of eclipses, the timing of the tides, and so on. He then moved on to an explanation of the Easter *computus,* for which he was uniquely prepared and equipped, in terms of the ready availability of the Church documents from Rome. Of course, these were times long before books were printed, so that one cannot think of a Bedan (one can tell how prominent Bede is in history from the fact that *Bedan* is an oft-used adjective in medieval studies) manuscript as having been published in a particular year with a particular title, but the following indicates his major works. His first exposition, titled *De Temporibus,* was completed in 703, but was felt to be too compressed, and so he at length wrote a much longer account, *De Temporum Ratione* ("on the reckoning of time"), which was finished in 725. The title indicates that this was an important step in the history of the calendar; in this work he computed the dates for Easter over a complete 532-year cycle, from Dionysius's beginning in A.D. 532 through 1063. This account was amplified over the next few years as Bede compiled his *Historia Ecclesiastica Gentis Anglorum* (*Ecclesiastical History of the English People*), completed in 731, in which the dating system employed was the familiar *anno Domini,* although with a twist, as we will see.

The Eclipse of 664 Revisited

In the previous chapter I wrote that the English records had the eclipse of 664 occurring on May 3, whereas we know that it actually happened on May 1. Those records, as such, are five separate accounts given by Bede in his various works. This has been the matter of some debate, as I foreshadowed. Some authors assume that Bede made no errors (intended or otherwise), while others are more realistic, recognizing that he was hugely careful, but human. Others again have accused Bede of deliberately misrepresenting the circumstances of the eclipse.

My own interpretation is much more charitable. In his chronicles Bede seems to have very deliberately drawn attention to this eclipse, saying that it was an event "remembered well" in England, and yet he mentions no other eclipses before or after. In my view, Bede recognized that the date given for the eclipse was counterfeit, as I discussed under the heading "The Subterfuge of Whitby," but only a few decades after the event he did not feel able to draw explicit attention to the fiddle, the Roman Church's dominance in Britain still being shaky—the Columban Church was still a rival, and the Synod of Whitby was within living memory—and thus likely to be undermined by any demonstration that the Dionysiac *computus* was far from perfect, especially if coupled with shady dealings by the Roman party headed by Wilfrid. My interpretation is that Bede therefore schemed to highlight the erroneous eclipse record without correcting it, in the confident expectation that long after his death and with the Roman Church being immovably ensconced someone would recognize the discrepancy and identify what actually happened. Maybe Bede's anticipated posthumous justification could now be completed.

Bede's Calendrical Significance

The Venerable Bede has a significance in the history of the calendar far beyond quibbling over whether the eclipse of 664 was misrepresented, however. In a short while we will come to the year-numbering aspect, but first let us consider the way in which a date within a month was given in that era.

Earlier noted was the fact that in dating the eclipse Bede wrote that it occurred on the fifth of the nones (May 3) rather than the kalends (May 1), which you will recognize as being the Roman dating method. Bede, though, later began to employ the concurrent method of dating favored by the Greeks and others in the East. For example, he might write *die XXIIII mensis Septembris* (literally "day twenty-four of the month of September"), which to us is much more familiar than the Roman system. The significance of this specific date—the traditional date of the autumnal equinox—we will meet below, but for present purposes we should note that the old system of Roman dating was being eroded.

It is in connection with year numbers that Bede is most significant. We have already seen that there was a panoply of numbering systems in use, such as the Roman *ab urbe condita,* the Jewish *anno mundi,* the Spanish Era tallying from 38 B.C., the *anno Passionis* counting from the Crucifixion, the *anno Diocletiani* which Dionysius Exiguus wanted to avoid, and so on. In chapter 8 I noted that although the abbreviation B.C. (the Latin equivalent being *ante Christum*) does not seem to have been used prior to the seventeenth century, a few scholars had much earlier actually made use of year counts numbering backward from the Incarnation date according to Dionysius. One of these, perhaps the first, was Bede; for example, in his history of England he gave the time of the arrival of Julius Caesar in Britain as a certain number of years earlier than the Incarnation. The familiar B.C. system, and similarly the A.D. convention, were therefore pioneered by the Venerable Bede, even if he used other specific terms to label them (because a rose by any other name is still a rose).

The Indiction

The most-used numbering system, however, we have yet to meet, and again Bede is central to the story. Let us introduce it through a letter sent by Pope Gregory the Great to King Aethelberht of Kent; we would simply date the letter June 22, A.D. 601, but Gregory needed to write: "Given the 10th day of the Kalends of July in the 19th year of the reign of our most devout lord Mauricius Tiberius Augustus, in the 18th year after the consulship of the same lord, the 4th Indiction." Given regnal years which may (or may not) start on the actual date of accession to a seat of power, rather than a standard start-of-year date, the florid language was necessary. The single thing which stipulates the year in this context is the final clause: the year number in what was known as the *indiction.*

In the previous chapter I mentioned some confusion over the dating of the great comet of 676, writing that "at least one English annalist allotted it to 677 because he was using a year starting on September 1." As you might presume, I was referring to Bede, but do not imagine that *he* was confused: Bede knew which year he meant, and it is later readers who did not understand the indiction who got confused (although Bede contributed to the mess because he altered the start date for his annalistic years to different days in September). So let me tell you about the indiction.

You may have thought that Constantine the Great had done enough calendarwise, but the long-term use of the indiction was the result of his influence, although not really his invention, it having been introduced a few years previously by Diocletian. It is a taxation cycle, lasting fifteen years (three quinquennia), and given that taxes need to be levied (to raise funds to pay scholars among other things) it was a good idea. Lasting in all for a decade and a half, a third of a working lifetime nowadays and a larger proportion in the past, it allowed for forward planning with confident expectation of what economic cut the government would expect from your profits. In my view the few-year parliamentary terms of modern democracies with tax laws altered willy-nilly provide a great discouragement to entrepreneurs, regardless of the actual tax rates. With the indiction, people knew where they stood.

Constantine legislated for the indiction in 312, with a start date of September 1. Thus the year we post hoc call A.D. 327 was another first year of the indiction, being fifteen years after 312, and 342 was the start of a new cycle again. Just think how useful the indiction is to chronologists wanting to ascertain the dates of historical events: if the year of the indiction is given, the relatively short cycle time allows no doubt to be entertained about which year (in the A.D. system) was involved. You might think of the indiction as being a bizarre construct because you are unfamiliar with it, but in fact it continued to be used sparingly throughout the remnants of the Roman domain until Napoléon Bonaparte formally abolished the Holy Roman Empire in 1806.

In constructing his tabulation of earlier dates of Easter, Dionysius had used the indiction cycle as an unambiguous label: there is no reason one cannot extrapolate it back before A.D. 312 in the same way that we call the years prior to A.D. 1 "before Christ." Dionysius had consistently used September 1 as the start date, and Bede, picking up on Dionysius's work, at first did likewise. But what happened thereafter has caused great trouble to later generations, due to inconsistency and the duplicity of the names used. At various times and in various places the following start dates for the indiction have been employed:

September 1: As introduced by Constantine in 312, this is called the *Greek* or *Byzantine* or *Constantinopolitan indiction,* and it was largely used by the Romans.

September 23: Marking the birthdate of Augustus Caesar, sometimes called the *imperial* or *Caesarean indiction,* and often confused with the following.

September 24: Called the *indiction of Constantine,* or *Bedan indiction,* it seems to have been invented by Bede likely to accord with the equinox (which was not on this date, but actually near the seventeenth, at the time), and was used by him in his later chronicles.

September 29: This was the date of Michaelmas, which some thought to be more appropriate for an ecclesiastical calendar (even though the indiction was a civil taxation cycle).

December 25 (or January 1): This was called the *Roman* or *pontifical indiction,* although that should not be taken to mean that either the Romans or the popes used it.

Obviously the situation is most confusing, if one wants to interpret medieval manuscripts. One monastic order might have employed a certain indiction date for their chronicles, while another just ten miles away had decided upon an alternative. In Rome the popes continued to use the September 1 date until the end of the eleventh century, but then followed Bede and switched to September 24 by the time of Pope Alexander III (who died in 1181). Other Christian sects used Easter as the start of the year, even though it could vary by a month, especially if they numbered their era according to the *anno Passionis* system beginning with A.D. 33.

Calling December 25 or January 1 an *indiction* date is doubly confusing, because in reality this wasn't a taxation year reference point, but rather the start of the calendar year: in Rome January 1 had been maintained since the introduction of the Julian calendar, but elsewhere (such as in Britain) the people of Bede's time tended to think of midwinter or Christmas as being the end of one year and the beginning of the next, due to the festivities at the time. This usage continued for some centuries more; for example, a prominent chronicler of the thirteenth century, Matthew Paris, employed December 25 for the start of his annalistic years.

The Start of Bede's Years

It was from this jumble of dating systems that Bede was trying to make sense. He was interested in Dionysius's tables of Easter dates not just for future calculations of the appropriate dates to use, but also because he realized that the complicated cycle of Easter dates allowed past years to be reckoned so long as they were given in the records at his disposal. Knowing that Easter had been celebrated on, say, April 12 according to the Dionysiac *computus,* he could derive the year in terms of Dionysius's *anno ab Incarnatione,* especially if the indiction year were given, too.

The problem is that Bede did not stop there. Dionysius's years began consistently with March 25. In his earlier immature writings Bede had used December 25, then abandoned it in favor of September, initially the indiction date of the first, but later the twenty-fourth. No one is quite sure why Bede did this, but it seems to have been a matter of fashion. Taking the last of Bede's works as being definitive, one would say that Dionysius's years started with the presumed vernal equinox, and Bede's with the autumnal equinox, a gap of six months.

This six-month jump (as we will see in the next chapter) is an essential deviation in the history of our year-numbering system. Although it simply meant that Bede's year labels counted from September of 1 B.C. rather than March of 1 B.C. (as did Dionysius's system), the repositioning of the origin had a knock-on effect.

Of more concern here is this question: given that there were many year-numbering systems in use, and that the framework of Dionysius had been ignored for two centuries, why did Bede's uptake of that erroneous calculation lead to its eventual acceptance throughout the world? The answer seems to be connected with a form of millennial madness.

Millenarianism and the *Anni Mundi*

The excitement over the year A.D. 2000 is one breed of millenarianism, but there is a much stronger one associated with the lapsing of six thousand years. The Bible speaks of a thousand years being as one day in the sight of the Lord (Psalm 90), so that as one thinks of a seven-day week with the last in that cycle being the Sabbath, so the end of the sixth millennium brings about a sabbatical period.

To some believers this promises great things, a following thousand years of peace and plenty, such a philosophy being termed *chiliasm* (from the Greek word *chilia*, meaning "thousand"). The more pessimistic think of it as being the time of dread when the awful day of divine judgment dawns; this contrary outlook is *apocalypticism*, its study being termed *eschatology*.

Many disparate civilizations have, for unknown reasons, assumed beginnings of time occurring a few thousand years B.C. For example, the Mayans date the last creation of the world to 3114 B.C.; in chapter 15 we will meet Archbishop Ussher's 4004 B.C.; the Jewish count from 3761 B.C. we have already seen. Alternative *anni mundi* ("years of the world," counting from some assumed origin) which were argued for within the Christian Church in its early centuries had beginnings near 5509 and 5200 B.C. Let us refer to years counted from these suggested originations AM1 and AM2, both of which are different from the Jewish *anno mundi*.

Starting with the earlier of these, and keeping in mind the fact that the B.C./A.D. system to which we are habituated came much later, as the fifth century A.D. dawned it was actually the century starting AM1 5900. Since doomsayers rather than optimists almost always hold sway (just read your daily newspaper), anyone regarding the AM1 system as truthful would have been filled with dread: the Judgment Day at the end of the year 6000 was at hand. That belief would have been reinforced by matters on Earth rather than imagined signs from Heaven, in that the invaders from eastern Europe were starting their conquest of the western Roman domain, sacking Rome in 410, 455, and finally in 476.

A solution which the Church has employed many times was introduced around 420–440, largely through the writings of Saint Augustine: change the ground rules. It was decided, despite the apparently contradictory evidence (the barbarian Goths murdering, raping, and looting, which must have felt much like the Apocalypse to the plebs), that a mistake had been made and the clock was reset on the AM2 scale, postponing Judgment Day to a comfortable three hundred years into the future, actually in the year we call A.D. 801.

The implications of this for people like Bede are obvious. With all the Church documents in hand, Bede could see that his era beginning around

A.D. 700 was counted as the AM2 5900s, meaning that the Apocalypse was due in the foreseeable future.

Bede chose the same solution to that problem as had Saint Augustine, circumventing the public and clerical hysteria which might be provoked: renumber the years. First he studied the Old Testament and decided that the correct *annus mundi* should count from 3952 B.C.; one may imagine that in A.D. 2048 a few people will be getting itchy feet on that basis.

There is a problem with this, however, in that it put the prognosticated Apocalypse too far into the future for Church purposes, threats of impending eternal damnation being useful for controlling the masses. It was on that basis that Bede picked up on Dionysius's chronological framework, which Bede saw would put the new millennium (not the sabbatical millennium, but a useful device anyway) only a few centuries hence.

It was into this environment, in which a new year numbering based firmly on the Christian faith was much desired, that Bede's system was spawned. The Carolingians, the new rulers of France and Germany, seized upon this dating convention, which we now call *anno Domini*. Within a few decades Charlemagne was expanding the Carolingian domain, taking the new year-numbering system with him as he conquered much of Europe, sweeping the barbarians from northern Italy and freeing Rome by 800. Charlemagne was crowned on Christmas Day, the first day of the sabbatical millennium on the AM2 scale (December 25 being the start of the year 801). One may wonder whether the clerics in Rome viewed this as being a good thing (a chiliastic attitude) or a bad one (an apocalyptic view): although Europe had been in chaos from the civil perspective, the Church had thrived and widely spread its influence while there was no emperor to limit its power. Outside of the Church, however, the people recognized the year to be A.D. 801, rather than any *anno mundi* 6000, and the Apocalypse had been postponed.

By this route the Bedan year-numbering system, derived from his writings based on precious copies of documents taken to England from Rome over a century before, was brought to the Eternal City. Although the popes did not finally accept the A.D. system until 962, when the Saxon Otto I was crowned as Holy Roman Emperor, and then with some hiccups (the impending millennium change on the A.D. scale caused some problems, as we will learn later, but the Church in Rome would have seen that it afforded a useful instrument of control), the dominance of the joint production of Dionysius and Bede was now practically assured. There was only one outstanding question left to be settled: when should New Year be marked?

Chapter 11

❧

After the internecine battles of the seventh century between the various Anglo-Saxon tribes, the period when Bede worked in the first decades of the eighth century was largely quiescent. Peace was not to reign for long in the northeast of England where he had lived, however. In 787 the Danes invaded, followed by the great Viking assaults of 856–875. England was still far from homogeneous, awaiting leaders such as Alfred the Great (king of the West Saxons between 871 and 899) to form the nation gradually into one.

Despite these invaders being largely heathen, worshiping pagan gods such as the planets (hence our day names), in the monasteries and churches Christianity survived and indeed flourished.

The Role of Saint Oswald

In chapter 9 we found that there were two Saint Augustines, separated in time by a couple of centuries, and we also met a Saint Oswald, who was martyred in the year 642. Skipping forward another three centuries we come across a second Saint Oswald. Born of Danish parents, in the 950s this Oswald was inducted as a monk at Fleury on the Loire in France, returning to Britain and being made bishop of Worcester in 961. He was not a popular appointment due in part to his background, and there was much friction with the canons of Worcester. As a consequence he began to build a new church with the intention of superseding the extant cathedral. While in France, Oswald had been inspired by a cult surrounding the Virgin Mary, who had hitherto been a peripheral figure in Christian worship, and he dedicated his church to Saint Mary.

Although Oswald went on to become archbishop of York in 972, his new cathedral for Worcester had an important effect in that it introduced the cult of Saint Mary to England, and over the next few decades this grew in prominence, with many more churches being dedicated to her. As a consequence the feast day of Saint Mary—Lady Day, or Annunciation Day, our old friend March 25—started to become one of the major marker points in the year.

Even in Bede's era the usage of Christmas Day as the start and end of the year had begun to die out, while his date (or dates) in September had never been widely used, being restricted mainly to Church chroniclers. While other

New Year dates were possible, such as the January 1 of the persistent Julian calendar, among the common people it seemed obvious that the major celebration of the year, around which the date of Easter hinged anyway, should become recognized as the time from which year numbers should be counted.

The Ascendancy of March 25

Previously I noted that Bede used the traditional date of September 24 as the autumnal equinox, even though the discrepancy between the mean Julian year and the astronomical year had led to this moving to a date almost a week earlier by his time, and he knew this very well. In the same way, March 25 had not been abandoned as the date for the vernal equinox, at least for ecclesiastical purposes. That is, the original reason for putting the assumed Annunciation Day on March 25 was maintained. For example, one of the earliest Church foundations in western England was at Glastonbury, near where I was born in Somerset. The *Leofric Missal* is a calendar for A.D. 970 drawn up at Glastonbury Abbey and now preserved in the Bodleian Library in Oxford. It shows the equinoxes and the solstices on their *traditional* dates (March 25, June 24, September 24, December 25), even though the real astronomical phenomena associated with those terms occurred about a week before those dates. The use of March 25 for New Year and the ecclesiastical vernal equinox, resurgent around the end of the tenth century, was therefore consistent with what one might expect, given the biblical injunction quoted at the end of chapter 8, Christians being exhorted to count March as the first month of the year.

There was another reason for Lady Day assuming great popular significance at this time, which has been largely overlooked by modern writers. With the present furor over the turn of the millennium, one could imagine that at the end of the first millennium on the *anno Domini* scheme there would have been great concern over what was to transpire. Some, the apocalyptically inclined, thought that Armageddon would come after the completion of the year 999, howsoever the start and end of the year was determined, while others looked forward to times of milk and honey, their reward for their perennial earthly tribulations.

There was also a significant alternative belief pertaining to the timetable defining Judgment Day. Many were convinced that the Second Coming of Christ would occur when Good Friday coincided with the Feast of the Annunciation, providing another reason for people to regard March 25 quite literally as the day to be reckoned with. By chance this occurred in 970, 981, and 992, but not again until 1065. On each occasion the Church leaders were provided with a chance to whip up hysteria within the populace, each time arguing afterward that only their prayers had managed to postpone the Apocalypse (tough luck on those pious few confidently anticipating Paradise). No wonder March 25 became central to the public and Church consciousness.

Within those few decades it became common in England to count the year as beginning on March 25, and this is termed the *stilo Annunciationis*. But wait. A mistake was made.

Ordinals and Cardinals

It seems to me that this is as good a place as any to clarify what the year numbers imply in the B.C./A.D. era system.

Mathematicians distinguish between *cardinal* and *ordinal* numbers. Cardinal numbers are simply labels for quantities, like one, two, three. Ordinal numbers refer to a position in a sequence, such as first, second, third. Our year-counting scheme employs ordinal numbers. Thus A.D. 1 implies the first year of the life of Jesus Christ (as reckoned from the traditional date of his naming and circumcision, on January 1 in that year). The previous year is termed 1 B.C. because it was the first year before A.D. 1 (i.e., an ordering is implied). In such a system, clearly the concept of a year zero has no meaning: one cannot have a year which was zero years before or after an event. The astronomers' system of numbering years $(-3, -2, -1, 0, 1, 2, 3)$ only makes sense because *cardinal* numbers are used.

Perhaps using dates closer to the present might make things clearer. Regnal years for monarchs generally use ordinal numbers: some act of Parliament might say that it was passed in the thirty-seventh year of the reign of Queen Elizabeth II, and that implies that she has been on the throne for thirty-six years plus part of a year. If you were born on July 1 in 1970, then on January 1, 2000, you will be twenty-nine years old, and in your thirtieth year of life, but in your thirty-first calendar year (count them, 1970 to 2000 inclusive). As I write, in early 1999, Cliff Richard is fifty-eight years old, but he has had pop music hits in five different decades since his first in the late 1950s, and I imagine that he confidently anticipates continuing into his sixth or even seventh decade as a singing star.

Another example of ordinal as opposed to cardinal numbers which might clarify matters even better is one's height. I am five feet nine inches tall, in my socks. That means that I am *in* my sixth foot. But I am not over six feet high. Similarly in the year A.D. 2000, only 1,999 complete years will have elapsed since the start of our era. The 2,000th year has finished when the clock ticks into 2001.

The final clincher in this regard is to note the way we label centuries. The term *nineteenth century* is recognized to cover the era where the year number began with 18. Correctly, the year 1800 was the last of the seventeenth century, and the year 1900 the final one in the nineteenth. The message should be clear. Our centuries we label with ordinals, at least in English; other languages may use alternative conventions.

1 B.C. or A.D. 1?

We saw that the Venerable Bede had promoted a year-numbering method in which he counted from a date in September, the start of his sequence being in the year which we label 1 B.C. When the cult of St. Mary became ascendant towards the end of the tenth century, the original meaning of the term *anno ab Incarnatione*, referring to Dionysius's count from the Incarnation/Annunciation, was recalled, and things slotted nicely into place: people began to count years from March 25 again.

The problem was that they fell into the wrong slot. According to Dionysius's tally, years should have been counted from March 25 in 1 B.C., but instead an epoch beginning on that date in A.D. 1 was effectively instigated. This might be considered to be closer to the "real" start date of January 1 in that year, being only three months after rather than nine months before the origin of our era, but that ignores the actuality of the reasoning which went into Dionysius's framework.

If you reconsider what we learned in chapter 1 about the alteration of the British calendar in the 1750s, you will see what I mean. The year 1751 had almost three months knocked out of it in order to let 1752 begin on January 1. If the origin of the annual counting scheme then in use had been March 25, 1 B.C., as Dionysius had deployed, then the year numbers in use would have been one larger, and a little over nine months (March 25–December 31) would have needed to have been deleted to achieve the same ends.

All these numbers tend to lead to confusion, and so it is useful to describe each of these systems with a single phrase. We have previously stated that an assumed New Year's Day on March 25 is called the *stilo Annunciationis*. Obviously this style comes in two varieties, one with year numbering beginning in 1 B.C. and the other on that date in A.D. 1. The former is called the *stilo Pisanus* and the latter the *stilo Florentinus*, being associated in later times with the cities of Pisa and Florence even though neither style was invented in those cities. The *stilo Pisanus*, which is more logical and was effectively developed by Dionysius, was employed in Pisa right through to the eighteenth century. The *stilo Florentinus*, let us be clear, was used in Britain and the colonies until the end of 1751. Elsewhere in what we now call Italy, other city-states insisted upon other dates for New Year's Day—such as Venice, which maintained the old Roman republican March 1 (called the *more Venato*, or Venetian style) until 1797, when it was conquered by Napoléon.

If this seems confusing to read about now, just think what it was like in medieval times. Reginald Poole, keeper of the archives at the University of Oxford, was a prominent chronologist early in the twentieth century. Writing in the *English Historical Review* in 1918 he gave the following example:

> If we suppose a traveller to set out from Venice on March 1, 1245, the first day of the Venetian year, he would find himself in 1244 when he reached Florence; and if after a short stay he went to Pisa, the year 1246 would already have begun there. Continuing his journey westward, he would find himself again in 1245 when he entered Provence, and on arriving in France before Easter (April 16) he would be once more in 1244. This seems a bewildering tangle of dates.

One can only agree with Poole's sentiment. Within Europe different states were variously using January 1, March 1, March 25, Easter, September 1, September 24, and December 25 for New Year, and even if they agreed on the date to utilize, it did not follow that their year numbering would be the same.

Chain Links of Differing Length

The fact that the assumed date for New Year's Day has been moved around from time to time will be discomfiting for some readers, who might like to believe fondly that the year is a period of constant duration. Years are quantized, but not necessarily of equal duration, in the same way as photons of lights are quantized, but are not of the precise same energy: blue light is more energetic than red, and even in a laser, about as close as one can get to monochromatism, the photons still have a small range of energies. Of course the discussion in appendix B, where many distinct definitions for the year, each of different length, will have shown that the duration of a year depends on what one means by the term, but still many would imagine that each of those units making up the count on the *anno Domini* system must be of standard length.

That is simply not the case. Under the Julian calendar, three out of four years lasted for 365 days, the other for 366. Since the Gregorian reform, the years 1700 (unless you lived in a holdout Protestant country), 1800, and 1900 have been 365-day years. In the Catholic countries, by and large, A.D. 1582 had only 355 days. In Britain and its colonies, A.D. 1751 had only 282 days and 1752, a leap year in that it had a February 29, lasted for but 355. In the last few decades we have taken to adding leap seconds to the year every now and again; prior to that the day length was inconstant in absolute terms because Earth's spin has slowed, as we will discuss in detail in chapter 22. Those are but tiny variations, however, insignificant compared to the wholesale, almost random changes in the specific calendar years used by the Roman republic, culminating in Julius Caesar's Year of Confusion, 46 B.C., with 445 days.

Against this backdrop we should consider the shenanigans between A.D. 700 and 1000, when the year's start was shifted from March 25 to September 24 and then on again to March 25, along the way effectively losing a year from the count because of the two 6-month years which, one should realize, cannot really be allotted to any particular time, since even adjacent monasteries might be using different marker points for their annalistic years. The A.D. scheme we use I have often termed a *framework,* but more correctly it is like a chain which we have later laid over past proceedings in order to mark off points along the axis of time. A chain we all recognize to be a series of discrete links, and in a chain we may count the number of links, the answer being an integer. But it does not necessarily follow that the links are all of the same size. And that's what our series of year numbers is like: a chain whose links may be of differing length.

One can look back and use a later-defined standard link length as an appropriate measure to determine in an absolute sense when a particular event occurred. For example, we say that the battle of Pydna in 168 B.C. (itself an imposition of a dating scheme far after the fact) actually occurred on June 21, on the proleptic Julian calendar, even though the Romans dated it as being on September 3; its proximity to a lunar eclipse allows us to be quite definite in that. In such calculations one assumes a standard year length, and astronomers often work in mean Julian years (each lasting for precisely 365.25 days), even though there is no astronomical basis for such a sum: it is just a more-or-less round number. The basis for our dating of the event, however, is truly astronomical, in that we have employed precise modern knowledge of the movement of celestial bodies to determine the instant in question with an accuracy far beyond that attainable by the people of the time. In the next chapter we will consider what sorts of phenomena allow us to do such things, permitting the definition of times with arbitrary accuracy and the chronological cross-correlation between accounts recorded by different cultures.

Chapter 12

❧

RETROSPECTIVE DATING

It is clear that the year numbers in our era definition have been retrospectively defined for most of the early phases. For the later period, from which we have printed records, this has led to various pertinent examples. I began this book with George Washington, whose birth date and birth year were revised during his lifetime. One can go back just a little further. History books tell you that Isaac Newton was born on Christmas Day in 1642, but that is only because Britain was still using the Julian calendar; elsewhere on the Continent it was already 1643. The great authors Shakespeare (in England) and Cervantes (in Spain) died on the same *date* in 1616, but not on the same *day*, because of the ten-day offset between the Julian and Gregorian calendars their respective nations were using.

As one goes backward in time it becomes more difficult to unravel Reginald Poole's "bewildering tangle of dates," but it is not impossible. I previously quoted some passages from the comic account of British history by Sellar and Yeatman, *1066 and All That*. The book gets its name from the one date in history known to every British schoolchild: the battle of Hastings in 1066, when the Normans under William the Conqueror defeated King Harold. I was born in a place called Midsomer Norton; the second part of its appellation is a diminutive of "Norman town." Britain's history swung on William's victory, in many ways.

But how do we know that the battle took place in 1066? The actual date seems to have been October 14 and that helps, because it is far from any of the New Year dates used by the English or the French, so that they might agree in their accounts. England had by then begun to use March 25, whereas in France the Christmas style was in force. With the Normans having conquered the Saxons, one might have expected them to have imposed their own calendrical system on England, but instead the converse occurred: by A.D. 1111, the French had adopted the Lady Day system. It did not last long, however, and by 1215 France (or at least part of it) had moved over to an observance of Easter as the start of each year; hence the statement by Poole which was quoted in the previous chapter, with Provence using one dating system but much of the rest of the area we now call France using an alternative year beginning with Easter. But

do not think of France as being backward, calendrically speaking: they introduced January 1 as New Year's Day in 1564, decades ahead of most other
countries.

Above all other considerations there is one phenomenon which allows us
to be sure that the battle of Hastings took place in 1066 rather than 1065 or
1067, and that is Halley's comet, famously depicted in the Bayeux Tapestry.
Bright comets like Halley do not appear often, and records of an apparition
witnessed in 1066 tally with the position and motion we are able to compute
using later observations and our knowledge of its dynamical evolution. In
fact, we have records of Halley's comet dating back as far as 240 B.C., and it
has been a long-term project of various historians to find even earlier mentions of its appearances, making use of Babylonian clay tablets and so on, because we are able to follow its orbit by computer back to at least the fifteenth
century B.C. The Norman Conquest was indeed in 1066. Halley's comet tells
us so.

Comets as Clocks

Other great comets in history similarly provide place-holding functions, allowing us to calibrate events recorded at the time against a definite chronological
framework which we have later stipulated.

Some comets, like Halley's, have regular orbital periods, allowing us to use
them like clocks. In the case of Halley's comet, it returns every 76 years or so,
and we can compute when and in what part of the sky it could have been seen,
allowing us to search historical records for candidate sightings which then enable us to assign years, months, and perhaps even specific days to those records.
There are only a few other comets which are sufficiently bright, and of short
enough orbital period, for us to do this. One is comet Swift-Tuttle, which returns about every 130 years and for which we have records dating back to 69
B.C. We saw this comet in 1992, and expect it again in 2126.

There are many other bright comets which appear in the sky, however.
These mostly have orbital periods of some millennia at least, so that they will
not have been seen previously; at least, not since the dawn of civilization and
written records. But their blazes of glory across the heavens, witnessed and
noted by cultures spread around the globe and so out of contact, enable us to
use them as chronological markers. The great comet of 676 was mentioned already. The comet seen late in 442 allows us to know when Attila withdrew from
Italy in the face of the plague, even if he was less than assiduous in ensuring that
a diary of his campaigns was maintained, because we can easily count off the
ten years after the comet using contemporary accounts of his retreat.

A very famous comet is that which has been associated in error by some with
a foretelling of Julius Caesar's murder. In fact, the comet appeared some
months later. Its story has recently been explored by John T. Ramsey and A.
Lewis Licht of the University of Illinois at Chicago, in a book entitled *The
Comet of 44 B.C. and Caesar's Funeral Games*. My friend Brian Marsden, an expert on comets, begins his foreword to that volume as follows:

When beggars die, there are no comets seen;
The heavens themselves blaze forth the death of princes.

These lines, attributed to Calpurnia on the morning of March 15, 44 B.C., as she implored her husband not to go into the office that fateful day, are arguably the most famous of all literary allusions to comets. In writing them, Shakespeare was not only expressing the traditional association between disasters on the earth and the appearance of portents in the sky, but he was also demonstrating his appreciation of the historical account of the specific comet that allegedly appeared soon after the assassination of Julius Caesar.

The historian Ramsey looked into the history of this fabled comet, and found that the Chinese recorded a bright comet in May and June of 44 B.C., the only one within some years of that date. Since these ancients in Cathay had no reason to invent a comet to enhance the significance of the assassination far away in Rome, it seems clear that the comet was real: some have doubted its reality because the earliest documentary evidence stems from some notes by Augustus a couple of decades after the fact, and one might suggest that he was merely trying to embellish the memory of his adoptive father. The documentation from Asia lays that suspicion to rest.

The Chinese saw it at least two months after the murder, so that it would not have been viewed by the Romans as a celestial portent of things to come, as Shakespeare's setting might be interpreted as suggesting. In fact the comet was recorded only as a bright, daylight apparition from Rome in the last ten days of July—the physicist Licht shows that the two sets of records from Europe and east Asia could be accommodated by a single comet, allowing its orbit to be determined—and so one may wonder whether the linkage of the comet with the murder and the naming of that month for Julius Caesar might be related. It was the month he was born, and seeing the comet appear then may have impressed upon the Senate the link between the man and the month, which they decided to formalize by changing its name.

Before passing on from Caesar's comet, an intriguing feature of our present use of comets might be mentioned. As the quote from Shakespeare's play given above exemplifies, in the past comets and other celestial phenomena such as eclipses were recorded and argued about because the people believed that they were portents of things due to happen in the *future*. My point here is that now we use such sightings as tools for determining when significant events occurred in the *past*. The comet of 44 B.C. was linked with Caesar's murder because it appeared soon afterward, and superstitious belief over the eons may have led to the idea that it had been a prophetic sign. In reality the comet came rather later, as the accounts of Augustus and others show. In this case there was really no doubt about when the assassination occurred, but for many other historical events a reasonable dating has been made feasible through proximity to some cometary apparition.

Almost all ancient literature contains allusions to celestial events, and the Bible is no exception, with more than a dozen distinct reports of comets being included therein along with many eclipses. For example: "And when the Sun was going down . . . great darkness fell upon him" (Genesis 15:12–17). Here the "him" refers to Abraham in Canaan, and it is possible to identify this description with a computed solar eclipse occurring on May 9, 1533 B.C., which would have occurred at about 6:30 P.M. local time (that is, when the Sun was going down). In the biblical account, a great comet was said to have been seen the following year, which can thus be dated to 1532 B.C.

In such a way the original astronomical input (the computed eclipse) has allowed an historical event to be dated with some precision, but there is a feedback effect: perhaps a known comet can be tentatively traced back to 1532 B.C., and the rudimentary observational information thus available from that era used to improve our understanding of its orbital evolution. For example, comet Hale-Bopp, which was discovered in mid-1995, we know last passed through the inner solar system (making it potentially visible) around 2214 B.C., but our uncertainty in that year is by about a decade. Maybe a candidate comet will be found in some ancient record, and its brightness and direction of motion will allow us to identify it with Hale-Bopp. That would be a step forward for cometary science, and history.

Various natural events around 760 B.C., culminating in a major earthquake which damaged King Solomon's Temple in Jerusalem and caused a tsunami in the Sea of Galilee, clearly had a considerable effect upon the people of Judaea. The fact that the rumble was felt over eight hundred miles away allows us to estimate that the earthquake was about magnitude 7.3 on the Richter scale. In the Bible there are eight separate allusions to a solar eclipse around that time (for example, Job 9:6–7), which may be identified as that of June 15, 763 B.C. This was apparently followed closely by a bright comet, and if this was Halley's comet (we cannot be sure due to the sparsity of the information) then it would indeed have been visible in August of 763 B.C., a useful step backward from the earliest definite observations I cited earlier for that particular object. The combination of these pieces of information leaves us pretty sure that the earthquake happened four years later, according to the biblical chronology, in 759 B.C. Comets are useful for chronological purposes, then.

Other Chronological Markers

Although great comets seen around the globe provide excellent markers allowing cross-correlation between written histories, they are relatively infrequent and other widely observed phenomena may provide useful references.

Astronomically, there are several other pointers, such as supernovas like that of A.D. 1006, and that in 1054 which has since produced the well-known Crab Nebula from the debris ejected by the stellar explosion. Conjunctions (comings-together in the sky) of the brighter planets—Mars, Jupiter, and Saturn in particular—are also events which occur infrequently enough to be notable to ancient peoples, but are computable given our knowledge of planetary motions. The possible connection between such conjunctions and the star of Bethlehem mystery we will meet in chapter 21.

As I write, we are awaiting the great Leonid meteor shower of 1999. Such spectacles have occurred every thirty-three years since A.D. 902 and are recorded in Chinese, Arab, and European chronicles. Other meteoric outbursts stretch back even further in time, as do descriptions of major auroral displays which broadly follow the eleven-year sunspot activity cycle.

Sunspots themselves have been recorded for an extensive period, since A.D. 187 in Britain and from 28 B.C. in China. Although not easily charted in pre-telescopic times, the presence of some obscuration in the atmosphere such as dust storms or smoke from major forest fires made it possible for large sunspots (dark, lower-temperature formations on the solar surface) to be viewed and recorded. In China, most ancient sunspot observations were made between December and April, when the air is laden with dust from central Asia blown east by the dry winter monsoon.

Eclipses provide another excellent time marker, especially total solar eclipses which have very narrow (typically one hundred miles wide) tracks across the ground, and we have already mentioned such events as recorded in the Bible. One account which has long puzzled scholars, because it describes the Sun as stopping still during an eclipse, and even moving backward, is in Joshua 10:12–14, which appears to have been the solar eclipse of September 30, 1131 B.C.

Peculiar combinations of eclipses are valuable; for example, this series was seen in the early ninth century A.D.:

Date	Year	Type	Location
February 11	807	solar	France and England
July 16	809	solar	England
June 20	810	lunar	Europe
November 30	810	solar	France
December 14	810	lunar	Europe

This sequence happened to occur near the end of Charlemagne's life, and the cause was explained to him. Also seen around this time was a very prominent sunspot, which persisted between March 17 and 24 in A.D. 807, also being noted by the Chinese.

In historical accounts, many eclipses have been associated with significant events, but not always with veracity. An example is the death of Alexander the Great in 323 B.C., from disease at the early age of thirty-three in Babylon. Some accounts have an eclipse accompanying or foreshadowing his last breaths, but we are pretty certain that he expired in June (based upon calibrating the calendar in use by his people using reports of the positions of the planets), whereas the eclipses visible that year from Babylon (a solar/lunar pair separated by a fortnight) did not occur until October. Why the confusion? Earlier, in 331 B.C., we are quite sure that Alexander had profited from a lunar eclipse just before his decisive battle with King Darius III at Gaugamela in Assyria, victory opening up all of the Persian Empire and the East so far as India for occupation by his Macedonian dynasty. Perhaps history confuses these or, as in the case of Julius Caesar, backward vision swapped the order of events, it seeming right that so stellar a character should depart in the company of a dramatic celestial phenomenon.

It would be easy to dismiss all suggested links between celestial events and human woes, such as the deaths of kings, as being superstitious nonsense, but there is always the possibility of a causal relationship. Primitive societies in the modern world provide examples of expiration apparently being caused by the power of suggestion or belief, as when someone "points the bone" at a detested rival, and such scenes associated with Voodoo or similar religions are often depicted in Hollywood movies. Could an eclipse cause someone's death? Yes, if they believed strongly enough that it was a portent of their demise. The son and successor of Charlemagne was Emperor Louis I, known as "the Pious," and perhaps he was too reverential. He died in A.D. 840 at the age of sixty-two, apparently never having recovered from the fright of seeing the eclipse which occurred about a month before, on May 5.

Much later in time foreknowledge of eclipse schedules has allowed the learned to gain an advantage over others. For example, during his fourth voyage to the New World, Christopher Columbus was marooned on the north coast of Jamaica, where hostile natives refused to provision him with food or water, or assist with repairs to his battered ships. But knowing that a lunar eclipse was due (on February 29 in 1504), Columbus was able to threaten the aborigines with the wrath of God, telling them how His sign would be manifest. The actual eclipse occurring as Columbus said it would led to their speedy adoption of a most amenable nature regarding his future requests.

On the other hand, the failure of sky watchers to predict eclipses can also have dire consequences, since their paymasters like to be forewarned, especially in societies where celestial phenomena are believed to be portents of major change. The earliest eclipse we have on record is that of October 22 in 2137 B.C., which was seen from China. Unfortunately the astronomers royal, named Hsi and Ho, were lax and failed to predict its occurrence, the emperor being so angered that they lost their heads.

Obviously, astronomical phenomena which may be precisely back-calculated allow us to stipulate when certain events must have taken place and thus calibrate the calendars for different sites, although they are not the only tools available. Other quite mundane concerns may allow records to be aligned. Most English people of at least my age will remember the harsh winter of 1963, and medieval chronicles likewise contain synopses of exceptional weather which permit the dating to be rectified, whether winters killing livestock or droughts shriveling the crops. In several places I have mentioned pandemics, such as the plague, which have spread through contiguous regions within a year or so, like the Black Death sweeping through Europe in the fourteenth century. Such outbreaks would obviously figure prominently in accounts of the times; similarly, in many places the death records for 1918 are dominated not by soldiers killed in the Great War, but by victims of the influenza epidemic. Any notable event which was not restricted to a single locality allows chronologies to be correlated, in principle.

The Need for an Overarching Dating System

In my description of various events above I have not been explicit about the calendar used, but the epoch tends to imply the calendar. For example, the eclipse of September 30, 1131 B.C., apparently mentioned in the Bible: my insinuation there is that this is the date on the proleptic Julian calendar (the calendar introduced by Julius Caesar to start from 45 B.C., but projected backward in time by us). Generally it may be presumed that any date prior to the Gregorian reform in A.D. 1582 is given on the Julian calendar, unless some other system is stipulated. Then again, the locality may also be significant, it being understood that Isaac Newton's birth date was Christmas Day 1642 because he was born in England, which persisted on the Julian calendar for another 170 years after the Catholic calendar reform.

Don't think that such matters are all in the distant past. The well-known "October Revolution" of 1917 which resulted in the formation of the Soviet Union actually occurred on November 7 on the calendar used almost everywhere else, but in Russia it was still October 25 because they had not yet converted from the Julian calendar. In fact, the earlier revolution of that year, in which the czars were overthrown, also has a misnomer from the same cause: it is called the "February Revolution" despite having occurred in March (Gregorian).

In calculating that date of September 30, 1131 B.C., the basic way in which the astronomical computations were done was to count *days* rather than *years* backward, and then later convert to a date using a simple program which spits out the date on the Julian (or other) calendar. Indeed, to save confusion over when observations are made, modern-day astronomers normally use an accumulating day-number system called the *Julian Date*. This is a prime example of something which is obvious and useful once you are habituated to it, but confusing when you first meet it, because it is not a date at all, but rather a combination of time and date which allows one to stipulate an instant in time.

The Julian Date is the universal standard by which the flow of time is tallied. If you dig into a computer's time and date programs, odds are that you will find that it is counting the Julian Date or some derivative of it. So we'd better understand the origin, evolution, and definition of this quantity.

The Holocene Era System

The Julian Date (J.D.) effectively encompasses the Universal time system, and from that and my preceding paragraph you might imagine with just cause that the J.D. must be some modern invention pertaining to computers and astronomy which I could and should have detailed earlier. The fact is, though, that the framework on which the J.D. system rests was defined over four hundred years ago, and in this book we have only now covered all of the historical ground which is needed in order to describe it.

Just imagine that you'd like to define some calendrical system in which all the dates are positive. Doubtless you have already realized that would be a useful thing, from such quandaries as are thrown up by our B.C./A.D. system. For example, how long was it between the death of Julius Caesar on March 15, 44 B.C., and that of Augustus on August 19, A.D. 14? Answer in years (plus a fraction), and also as a number of days. Quickly now! Even if you are adept at mental arithmetic, the combined confusion of there being no year zero, years B.C. counting downward, years A.D. upward, while the months consistently run in the same direction, leads to problems. Imagine instead expressing the fraction of a year as a decimal, so that the start of July 1 of 1 B.C. is 1.5 B.C., and July 1, A.D. 1, is A.D. 1.5. At first sight, if presented merely with 1.5 B.C. and A.D. 1.5, one might think that there are three years between those dates, whereas the answer is just one year.

Having set you the ticklish problem above, I guess I'd better solve it. It's easiest to express it in the astronomers' dating system. In that frame, Julius Caesar died at about -42.80 and Augustus at $+14.63$, given that two decimal places is sufficient so long as we're not going to quibble about times of day or leap year matters. So, 57.43 years is the difference. The actual number of days could be problematical for intervening dates, due to the mistake made in enacting the Julian reform, but Augustus rectified that, making it feasible to tally them up without worrying about which years were actually counted as leap years in Rome given that the two dates in question bracket the discrepant era. The answer is 20,970 days, adding them all up. But imagine how much easier it would be if both of those dates were allotted specific numbers in some all-encompassing scheme which could be easily programmed, allowing rapid calculations without needing to count on your fingers. As we will see, this is what the Julian Date system achieves. But let us first consider some background.

The late Cesare Emiliani of the University of Miami, a prominent researcher of the environmental changes during the geological era termed the Holocene, which covers basically the 10,000 years or so since the end of the last Ice Age, once suggested a novel and simple dating revision based on his professional expertise: simply add 10,000 to the *anno Domini* system. Thus A.D. 1 would be the year 10,001, and A.D. 1999 would become the year 11,999. One simply adds 10,000 to all years A.D. For the years B.C., one subtracts the number from 10,001, such that 1 B.C. becomes 10,000. Easy. Another method is to just subtract every digit from 9, and then add 2 to the result. Thus Alexander the Great died in 9678. Julius Caesar died in 9957, Augustus in 10,014, and there were 57 years, 5 months, and 4 days between their deaths. Piece of cake.

This could be called the *Holocene Era* (abbreviated H.E.), and its appeal stems from several considerations:

1. It is religion-independent in that it does not require the use of the B.C./A.D. labels with their associated implications.
2. It is based on real Earth history but with an arbitrary start date (we cannot stipulate a precise time when the Holocene began any more than one could state a precise second when a certain rainstorm started).
3. A transition from ordinal to cardinal numbers could be made, so

that the year H.E. 12,000 would start 12,000 years after the assumed beginning of the era, rather that A.D. 2000 starting 1,999 years after the assumed reference point of the B.C./A.D. system.

4. The H.E. system has a reference point predating all written history, so that all dates in history would be positive, allowing easier comparisons of chronologies.

5. If implemented when Emiliani suggested it a few years back, the software conversion necessitated by the Year 2000 Millennium Bug furor could easily have encompassed this addition of a digit to the front of the year number.

The fact of the matter is, however, that Emiliani's scheme has not been taken up, and the opportunity seems to have been lost.

The Julian Period Defined

There was a much earlier system which included many of the attractions of Emiliani's system, however, and this is the aforementioned Julian Date scheme. Back in the late sixteenth century—in fact in 1582, making it concurrent with the Gregorian reform, although largely unrelated to that calendar change—the French scholar Joseph Justus Scaliger (1540–1609) was considering the general question of uniquely identifying years, and he came up with a system which was later adapted to produce the J.D. reference.

Scaliger's system is based upon three cycles of years we have already described. Two of these are the 19-year lunar/Metonic cycle, and the 28-year solar/dominical cycle over which the weekdays and calendar dates repeat on the Julian calendar. Together these produced the 532-year *annus magnus* of Victorius of Aquitaine which we met in chapter 8. On top of that, the indiction provided another cycle, 15 years in length. Together, these three systems result in a potential cycle of 7,980 years.

For any date in this almost eight-millennial cycle the year may be stipulated by a number in the lunar cycle (called the golden number, G, between 1 and 19 inclusive), in the solar cycle (S, between 1 and 28), and in the indiction (I, between 1 and 15). Thus Scaliger could characterize any year in the previous patchwork of chronologies by (S,G,I). Why not just the year number, you might ask? The point here is that backwards calculations had mostly been based on these quantities, because the Dionysiac Easter cycles went as G, the days of the week and the dates (controlling when Easter Sunday would occur) as S, while the indiction cycle was also central to most annals. Thus Scaliger's (S,G,I) system was of great utility for reckoning past events.

The 7,980-year timescale is also of considerable utility because it allows for a span covering all of history, and then some. Scaliger called this the *Julian Period*. We have seen that Dionysius Exiguus computed tables of Easter for his immediate future starting with A.D. 532, having erroneously arrived at a date for the Incarnation in 1 B.C. That is one complete *annus magnus* earlier, so that it follows that the value of G was 1. The phasing of the solar cycle (which may be thought of as a leap-year cycle) was such that 1 B.C. was characterized by

$S = 9$. Given that $I = 1$ in A.D. 312, when Constantine began his long-lasting indiction cycle, it follows that reckoning it backward the year 1 B.C. would have $I = 3$. Thus 1 B.C. was (9,1,3) to Scaliger, and he realized that the combination (1,1,1) would be appropriate for 4713 B.C. A rather convenient date, because it precedes all written history, and was in the right ballpark for existing concepts of the age of the world in Scaliger's time. Year 1 in Scaliger's Julian Period framework is therefore 4713 B.C. (or −4712 to an astronomer). Since the first Julian Period will not finish until the end of A.D. 3267, when all possible combinations of (S,G,I) are exhausted at last, Scaliger's numbers will not start to recycle for a good time into the future, and so we won't worry about it.

Finally I should tell you how to calculate the values of (S,G,I) for any year, perhaps that of your birth:

> *Solar cycle, S:* Add 8 to the year A.D., then divide by 28. Take the remainder and add 1 to get S.

> *Lunar cycle, G:* Divide the year A.D. by 19, add 1 to the remainder, and that is G.

> *Indiction, I:* Add 2 to the year A.D., divide by 15, add 1 to the remainder to get I.

For example, A.D. 1999 is (20,5,7); A.D. 2000 is (21,6,8); A.D. 2007 is (28,13,15); A.D. 2008 is (1,14,1); A.D. 3267 is (28,19,15); and A.D. 3268 is (1,1,1). For years B.C.? Use the astronomers' years, such that 4713 B.C. is entered as −4712 in the sums. As I have claimed earlier, the astronomers' dating system makes more sense for such calculations, even when calculating parameters like S and G which are for ecclesiastical rather than astronomical purposes.

There is one thing we need to clarify before moving on. I wrote above that the twenty-eight-year solar/dominical cycle is the period over which the days of the week repeat on the same calendar dates throughout the year, *on the Julian calendar*. But the Gregorian calendar is that which is in near-universal use. The same twenty-eight-year cycle applies in this case *except* if the twenty-eight years bracket one of the years divisible by four which is not a leap year (1700, 1800, 1900, 2100 . . .). In that case there is a hiccup, and because of this the length of time over which the calendar effectively repeats in the Gregorian system is the full four hundred years, as discussed right at the end of chapter 7. This means that the solar/dominical cycle is strictly four hundred years for the Gregorian calendar; nevertheless, for the purposes of labeling years within the Julian Period we use the system just described, as defined by Scaliger.

The Julian Date

Scaliger, then, arrived at an arguable and useful initial epoch to employ for a historical dating system in which all the year numbers are positive, and the important cycles are each directly used in producing the year number. For astronomical applications, though, which we have seen are essential for determining dates for important events which may be tied against comets and eclipses and so on, it is better to have day numbers. One may calculate the days in the years on the proleptic Julian calendar all the way back to 4713 B.C. (leap year every fourth year) and apply that form of calculation all the way through to A.D. 1582, each day having a specific number associated in a sequence which began on January 1, 4713 B.C. Such a number is termed a *Julian day number*. After 1582 one has to be careful, and take into account when the calendar was reformed in whichever nation with which you are concerned. Thus Shakespeare and Cervantes may have died on the same date, but the Julian day numbers associated with their deaths differ by ten, that being the difference between the Julian and Gregorian calendars in the seventeenth century.

Sounds quite straightforward. The problem is that astronomers are a contrary bunch. Rather than starting the Julian day numbers with *one* at the midnight starting January 1, 4713 B.C., the count starts with *zero* at Greenwich mean noon on that day. Thus J.D. number one runs from noon on January 2 to noon on January 3. This day number starting at noon might seem bizarre, given that practically all cultures have used day starts at either dusk, midnight, or dawn, but it was used for many purposes right into this century, astronomers being loath to dispense with their days-start-at-noon convention. As mentioned earlier, this system was used to define Greenwich mean time until 1925.

Actually this convention for J.D. is quite logical (said the astronomer). Unlike the ordinal dating system we generally use (for example 2/2/1999 means the 2nd day of the 2nd month of the 1,999th year, *not* that 2 days, 2 months, and 1,999 years have passed), the Julian day number is cardinal, and so counts the number of complete days which have *elapsed* already. Thus the Julian day number is an integer, the number of days already passed. One can then move on from that to the *Julian Date*, which generally is given with several decimal places. Note that I introduced the Julian Date and its abbreviation earlier without defining it. The J.D. is the Julian day number plus the fraction of a day which has elapsed since the preceding noon. For example, Julian day number 2,451,545 begins at noon on January 1, A.D. 2000, making the midnight before (when everyone is celebrating) J.D. 2,451,544.5 precisely. If you count up the days elapsed since 4713 B.C., accounting for the 10 days deleted from 1582 and the 3 leap-year days which did not appear in 1700, 1800, and 1900, you will find that the sums are correct.

Obviously having such a large number to label a day can be inconvenient, as is having days assumed to start at noon, whereas for almost all other timekeeping purposes in the civil world the day starts at midnight. For this reason something called the *Modified Julian Date* (M.J.D.) was introduced. This is the J.D. reduced by 2,400,000.5, the decimal at the end moving the day start back from noon to midnight. On this scale M.J.D. 0 began at the start of November 17, A.D. 1858. The start of January 1 in the year 2000 is at M.J.D. 51,544.0, and the new millennium begins on January 1, A.D. 2001, the instant being M.J.D. 51,910.0. Note that the J.D./M.J.D. may be given with arbitrary precision: waiting for the precise second before I strike the enter button this sentence was stored in my computer at M.J.D. 51,126.32118, to the nearest second. That number stipulates an instant of time independent of time zone, dating system, and so on.

I wrote above that *astronomers* defined this J.D./M.J.D. system based upon the place-marking point at the start of 4713 B.C. advocated by Scaliger over four hundred years ago. Actually it was John Herschel, son of the discoverer of Uranus, who suggested the Julian Date scheme in the nineteenth century, but do not imagine that it is anachronistic: in 1997 the International Astronomical Union decided to fix the various definitions which I have given in the preceding paragraphs, the deliberations being made necessary by technological advances requiring the length of the day to be stipulated more precisely.

In principle the J.D./M.J.D. system is foolproof, but in the same way as the terms *Universal time* and *Greenwich mean time* have obtained multiple meanings due to laxity in usage, so have the phrases *Julian Date* and *Julian day number*. Some people have taken to numbering the days within a year, starting from one, as being Julian day numbers. This is doubly wrong because the correct system is cardinal, telling of the number of days elapsed since the origin of the count, and there are many other better terms which could be used (for example, *civil day number*).

How the Julian Date Got Its Name

As of yet I have not told you how the Julian *Period* system got its name. It is often claimed that Scaliger named it in honor of his father, whose appellation was Julius Caesar Scaliger (why do parents do this to their children?), another prominent scholar and philosopher, who was born in Italy but later moved to France. If this is true—a contrary claim is that the Julian Period gets its name from its derivation based on the solar cycle in the Julian calendar—then our Julian *Date* gets its name from something named for someone who was himself named for Julius Caesar, the dating system's name therefore not coming directly from the great Roman so central to our calendar.

At the risk of belaboring the point, let me give a parallel example. There is a minor planet which orbits the Sun in the asteroid belt, out between Mars and Jupiter; it is numbered 2,309 in the master list maintained by the International Astronomical Union. Look it up in the data banks, and you will find that its name is given as *Mr. Spock,* which will be familiar to most readers. The asteroid was discovered by Jim Gibson, an astronomer who formerly worked at Mount Palomar observatory in California. The citation bestowing this title upon that specific hunk of interplanetary rock reads as follows:

> *Named for the ginger short-haired tabby cat (1967–) who selected the discoverer and his soon-to-be wife at a cat show . . . He was named after the character in the television program* Star Trek *who was also imperturbable, logical, intelligent and had pointed ears.*

Mr. Spock the asteroid was therefore named for Mr. Spock the long-deceased feline who was named for Mr. Spock the Vulcan who was named for . . . I guess we'd need to ask the *Star Trek* scriptwriters what they thought about the famous American writer on child rearing, Dr. Benjamin Spock.

A Great British Invention?

There is no doubt that the Julian Period/Date system stems from its promotion by Scaliger in the late sixteenth century. It does not follow, though, that he invented it. Given the perennial Anglo-French antipathy, one could well understand that if they could, the British would claim priority, above a Frenchman like Scaliger, even if he were of Italian stock.

In his famous *Annals of the World* of the mid–seventeenth century, James Ussher (archbishop of Armagh and Primate of All Ireland) wrote that

> *there arises the period of 7980 years, which was first (if I mistake not) observed by Robert Lotharing, Bishop of Hereford, in our island of Britain, and 500 years after by Joseph Scaliger fitted for chronological uses, and called by the name of the Julian Period, because it contained a cycle of so many Julian years.*

Leaving aside any debate over the origin of the Julian epithet in this connection, one might wonder why Robert Lotharing has been largely forgotten. In fact, my introductory paragraph above is a little misleading, as perhaps is Ussher's statement, which seems to be staking a claim for the invention being British. Because Lotharing was French.

I began this chapter with the battle of Hastings in 1066, and now we need to return there. While all Little Englanders might like to imagine that Britain was a place of peace and light invaded by the dastardly Normans, who deliberately shot Harold through the eye with an arrow (good marksmanship that would be), in fact the islands were a place of huge factionality and war-torn strife. We have already described the significance of the cathedral built at Worcester in the tenth century. Only about twenty-five miles away, on the border between England and Wales, lies Hereford, which had been a bishopric since soon after Saint Augustine arrived, and also had a modest cathedral. In the few years following A.D. 1050 Herefordshire was ravaged by the Welsh invader Gruffydd ab Llywelyn, abetted by Alfgar, king of the East Angles. They sacked the city and stole every valuable item from its cathedral, setting it ablaze.

Into this vacuum came the Normans. A great friend of William the Conqueror was the aforementioned Robert Lotharing, a man of learning who had studied at many continental universities. Having been born in the former French province we now call Lorraine, he was known as Robert de Lotharing: the area had previously been known as Lotharingia, named after Emperor Lothair I, the son of Louis I whom we met previously (he who was scared to death by an eclipse in 840). Robert was appointed bishop of Hereford by William in 1079, and he immediately set about rebuilding the cathedral; the present establishment contains a memorial to him.

Over the next sixteen years, until his death in 1095, Robert de Lotharing was instrumental in turning Hereford into one of the major centers of Christian worship in Britain. But he still found time for his scholarly pursuits, one of which seems to have been the invention of the great 7,980-year cycle which we call the Julian Period.

Calendar Recycling

Yet again there is a little point we need to clear up before moving on. I noted above that the solar cycle of twenty-eight years leads to the dates on the calendar repeating on the same days of the week with that grand periodicity. This is always true for the Julian calendar; for the Gregorian calendar it is true for all years between 1901 and 2099, which I will presume to cover the lifetimes of most of my readers, but disobeyed if the span bridges the non–leap years of 1900 or 2100, requiring a four-hundred-year cycle in all. Leaving that aside, twenty-eight years is the cyclicity we are interested in from my present perspective, which is this. You have a printed calendar on your kitchen wall which you particularly like. How many years before you could use it again? That is, how long before (say) April begins with a Thursday, your birthday is on a Saturday, and so on, all the date/day combinations being identical?

Well, the overall cycle is twenty-eight years, and that's how long it is between leap years repeating themselves (a leap year starting on a Monday does not occur again for another 28 years: the four-year leap cycle multiplied by the seven days in the week). For common years, the wait time for prospective calendar recyclers is much shorter, though. The first year after a leap year repeats six years later, so that the civil calendar for 2007 is the same as that for 2001. It's a longer gap for the second and third years after leap years, however, eleven years being required. Thus the calendar for 2013 is the same as that for 2002, and 2014 the same as that for 2003.

The Gregorian calendar can produce much longer waits, of up to forty years. An example is 1872, a leap year. Because that was twenty-eight years before 1900, its dates were repeated in the latter year on the Julian calendar but not on the Gregorian, in which it was a common year. The next leap year with the same day/date combinations was 1912. For an inveterate calendar recycler, that would have been a long wait. But it's not as long as the suprahuman four hundred years which one might naively have expected.

Chapter 13

❧

POPE GREGORY XIII (A.D. 1582)

onsider two points we have just discussed in chapter 12: (1) the reference point for the Julian Date scheme is on January 1, 4713 B.C.; (2) unless otherwise stipulated, the calendrical scheme used for such an epoch is the Julian: the proleptic Julian calendar. Okay, if I now asked you whether that particular January 1 was a week after the winter solstice, what would you reply? With that question in mind, peruse the following.

Around the start of our common era (around the time of the birth of Jesus Christ), the year started with January 1 and that was indeed near to a week after the winter solstice, which was on or about December 25, because that's the way in which Julius Caesar defined and phased the calendar which bears his name (actually he got it slightly wrong as we saw in chapter 6, putting the solstice a couple of days earlier, but we'll leave that aside). But the Julian calendar is such that the average year lasts for precisely 365.25 days, because there is a leap year every fourth round. In reality we have seen that the year better appropriate for defining a calendar actually lasts for a little less than that. Here we will not worry about whether we should be using the mean tropical year, or the year between vernal equinoxes, and so on, but we'll just jump ahead and say that the Gregorian reform of the calendar gives a close approximation to the mean year duration needed to keep the equinoxes and solstices on more or less the same dates. The Julian calendar lets these drift away rather quickly because it uses an overlong average year. A good guide to the time of the solstice will therefore be given by the Gregorian calendar.

If one wants to calculate a proleptic Gregorian calendar, it is quite simple to work backward. The Gregorian is the same as the Julian calendar except that three leap years are dropped from every four centuries, leaving ninety-seven in every four hundred years. One must now ask *which* leap years are dropped, for years B.C.

It seems to be a fluke—the result of a one-in-four chance—that leap years are those for which the numbers *anno Domini* are divisible by four: A.D. 12, 16, 20 . . . 1560, 1564, 1568, 1572, 1576, 1580, for example. One could suggest that this comes about from the choice of Dionysius Exiguus of that year we call 1 B.C. (753 A.U.C.) as that of the Incarnation, although 5 B.C., a year which would be favored for the Incarnation and/or Nativity by many historians, would have led to the same result (that is, leap years still being in years A.D. divisible by four, because the number would simply be larger by four). Leaving all such puzzling aside, the fact remains that this is the way things are.

Although we have seen that between 45 B.C. and about 9 B.C. the leap days were erroneously inserted in every third year, Augustus thereafter imposing a moratorium for a decade or so, it is conventional in chronology to let every fourth year on the Julian calendar be considered to be a leap year. Thus we count all of 1 B.C., 5 B.C., 9 B.C. . . . 101 B.C., 105 B.C., 109 B.C. . . . 1001 B.C., 1005 B.C., 1009 B.C., and so on as being leap years (and similarly A.D. 4 and A.D. 8, although the former may not have been an actual leap year at the time). All this shows another advantage of the astronomers' year-counting convention: those B.C. years are labeled 0, −4, −8 . . . −100, −104, −108, and so on, clearly divisible by four. To identify B.C. years on the Julian calendar which are leap, one needs to decrease the number by one and then see whether the result is divisible by four.

But what about the proleptic Gregorian calendar? In that case the years which are missed out and counted as being common years with only 365 days are those three in every four hundred whose B.C. numbers end in 01 without the preceding digits being divisible by four. Thus, on that calendar, 101 B.C., 201 B.C., 301 B.C., 501 B.C., 601 B.C., 701 B.C., 901 B.C., 1001 B.C., 1101 B.C., 1301 B.C., . . . were *not* leap years, in this after-the-fact dating scheme.

Because the Gregorian reform deleted ten (and only ten) days from A.D. 1582, the proleptic Gregorian calendar only renders identical dates with the Julian calendar between March 1, A.D. 200, and February 28, A.D. 300. If you work your way backward from that time to 4713 B.C., you will find that thirty-eight leap-year days are missing from the proleptic Gregorian calendar which do appear on the Julian. The effect of this is that for a particular day the Gregorian date in that early era is thirty-eight days prior to that on the Julian calendar. Thus Monday, November 24, 4714 B.C. (Gregorian), describes the same twenty-four-hour period as Monday, January 1, 4713 B.C. (Julian). On the proleptic Gregorian calendar the Julian Date system starts with zero at noon on November 24, 4714 B.C.

The significant thing about the Gregorian calendar in this sense is that it more accurately indicates the time of year with respect to the seasons, and with regard to the equinoxes and solstices. Above I asked you whether that January 1 (in 4713 B.C., Julian) was a week after the winter solstice. The answer is *no:* that date was four or so weeks *before* the winter solstice, which had moved to the end of January, although it remains on more or less the same date in the Gregorian calendar (i.e., about a month after November 24, 4714 B.C., Gregorian). The Gregorian calendar does not *perfectly* track the equinoxes and solstices, but it does a good enough job for me to be able to say that.

The Drifting Julian Calendar

The above shows that going backward in time the Julian calendar moves away from the astronomical reference points within the year which, for various reasons, one might like to hold on a near-consistent date. The example was in terms of one of the solstices, but the same generally applies to the equinoxes, and in particular we have seen that the vernal equinox is of fundamental significance for calendrical definition because the date of Easter hinges upon it. Going backward in time from the era of Julius Caesar, the equinox on the Julian calendar was later and later in terms of the date, occurring toward the end of April if one goes back to about 4700 B.C. And the converse is also true: proceeding forward in time through the first millennium A.D., the date of the vernal equinox was moving steadily backward from the March 22/23 which it occupied at the onset of the Julian calendar. The calendar, then, was allowing the vital reference points to drift away from their original dates. Traditionally the date should have been March 25, but Julius Caesar (or Sosigenes) had made a slight mistake, putting ninety rather than eighty-eight extra days into 46 B.C., so the Julian calendar started out wrong, and then got progressively worse because its mean year is overlong. Things were amiss.

It would be a mistake to believe that the Alexandrians, Greeks, or Romans of 2,000 years ago knew the precise instant of the vernal equinox in the way which we can now define it to a split second. Their knowledge was limited to being able to recognize when it would occur to within a day or so, and we have seen how different churches in the first several centuries A.D. were assuming different dates for the equinox. Part of the reason for this was that they knew that it was moving inexorably forward, although it was not clear to them how quickly, and of course for future calculations of when Easter should be celebrated it was necessary to be able to say when the equinox was *due*, rather than when it had occurred in the past.

By the time of the Council of Nicaea in A.D. 325, the date of the vernal equinox was thought to be March 21, 4 days ahead of the traditional date. In fact, the true average date (shifting back and forth in the quadrennial leap-year cycle) was on March 20, 2 or 3 days earlier than the March 22/23 of Julius Caesar because in the intervening 370 years the equinox had regressed by about 2.8 days as a consequence of the mean Julian year being a little longer than the vernal equinox year. At the time of Nicaea, then, the astronomical equinox was actually on March 20, but the date assumed in the Alexandrine *computus* was 21, and that's all that mattered when it came to fixing Easter.

It was all that mattered except that a one-day error occurring when you can only determine the date of an event like the equinox to a precision of a day or so is acceptable, whereas over the next millennium the error grew (by about a day every 130 years) while the ability of astronomers to determine the time of the equinox improved. By A.D. 525 Dionysius Exiguus knew that the astronomical equinox was actually occurring on about March 18/19, and by A.D. 725 the Venerable Bede knew that it had pushed back even further, to the seventeenth or so. Nevertheless, the Easter *computus* they promulgated continued to employ an assumed ecclesiastical equinox of March 21.

Not Just the Sun, but Also the Moon

To us it may seem obvious that a correction—a reality check—was required if the equinox assumed by the Church was to be kept in line with the celestial actuality, but remember that here we have been looking solely at the location of the Sun in the sky. Easter depends not only on the Sun, through the equinox definition, but also on the Moon.

While the real astronomical moon was circuiting Earth and displaying its changing phases, the Church was following its ecclesiastical moon, which was assumed to follow the Metonic cycle. And we have already seen that such a cycle is not precise. Nineteen Julian years last for an average of 6,839 days and 18 hours precisely, whereas 235 mean lunations last for about 89 minutes less than that. From this perspective one might say that the required year length to make the Metonic cycle exact needs to be 89/19, or about 4.7, minutes shorter than the 365.25-day Julian year. For example, this is indeed the approach taken in defining the Hebrew calendar: the mean lunation is *defined* to last for precisely 29 days, 12 hours, 44 minutes, and 3⅓ seconds, very close to the real value, and 235 of those units is *defined* to be the duration of 19 Jewish years each with a length of 365 days, 5 hours, 55 minutes, 25 seconds plus ²⁵⁄₅₇ of a second. Thus it is fundamentally the Moon rather than the Sun which governs the Jewish year.

On the other hand, if one wants a year length truly defined by the Sun, then one has to take into account the fact that the vernal equinox year is about eleven minutes shorter than the Julian year. Overall, in the sixteenth century the astronomical moon was clearly ahead of the ecclesiastical moon, by more than four whole days, and to some extent one could argue that this was compensating (by about 4.7/11, or over 40 percent) for the astronomical sun being ahead of the ecclesiastical sun as denoted by the time of the equinox.

This point is often missed by people writing about the reform of the calendar, who only address the question of the Julian year being too long according to our real motion about the Sun. The Gregorian reform was an adjustment of the calendar for religious rather than scientific purposes, and the motion of the Moon was also hugely significant in that regard, because it controlled Easter. It is just not as simple as the story often told of Pope Gregory XIII being taken to the Tower of the Four Winds in the Vatican for the astronomer Ignazio Danti to show him the image of the Sun on the meridian line on the floor of the *Sala del Calendrio* (Calendar Room), Gregory being impressed by the Sun being ten days away from where it should be for an equinox on March 20/21 and suddenly deciding that the calendar must be adjusted. As we'll see, Gregory XIII had long since been convinced of the need for change, but the requisite reform was recognized to be more complicated than simplistic accounts can convey. On top of the dual sun/moon complexity, one should recall that in this era it was still not accepted that Earth moved about the Sun rather than the converse, especially within the Catholic Church: the forced recantation by Galileo Galilei in that regard was still several decades away.

The Background to the Gregorian Reform

Leaving such considerations aside, it was clear to all those who looked into the matter that the equinox was drifting earlier in the year, and that sooner or later action would be required. The idea that the Catholic Church suddenly woke up in the late sixteenth century and decided to amend the calendar is utterly wrong, because the need for change had long been recognized. For example, the great English philosopher Roger Bacon had written to Pope Clement IV in the mid–thirteenth century, sending his treatise on the error of the Julian calendar. A couple of decades earlier John Holywood, better known under the flamboyant name of Johannes de Sacrobosco, had also pointed out that the equinox was coming too early in the year.

By the fifteenth century the trickle of memoirs by mathematicians and other learned men had become a deluge, with Rome being inundated with publications showing how the calendar was in error, many of them from the states of Germany and Poland which had become the center of astronomical knowledge. In the same way as John Holywood used a rather florid nom de plume, so did one of these Teutonic astronomers, Johann Müller. He took the latinized name Regiomontanus, from the medieval city of Königsberg ("King's mountain"), where he was born; this was in Franconia, part of western Germany.

In 1472 Regiomontanus became the first person to print an astronomical almanac, and two years later he came to Rome at the invitation of Pope Sixtus IV to attend to the reconstruction of the calendar. But before he had time to make major inroads on his task of revising one of the most conservative constructs in a most conservative organization, he died prematurely, and so calendrical reform was put off yet again.

There was no doubt among sixteenth-century papal advisers that their many communicants were correct about the advance of the equinox in the calendar. Julius II, who was pope between 1503 and 1513, considered various proposals; we should be glad that he did not accept any of them, else the Julian calendar would have been replaced by . . . Well, what could we have called it? The search for a solution was carried on by Leo X, the pontiff from 1513 to 1521, but then the matter was left to rest for a while.

Pope Gregory XIII

In 1545 the Council of Trent began its deliberations, continuing off and on for eighteen years, in the town of Trento in northern Italy. For the Church this was a much-troubled century, with the Reformation movement in northern and western Europe eroding the power of the Roman Church, leading eventually to the creation of the breakaway Protestant churches such as the Lutheran in the north, and the eventual Church of England following King Henry VIII's split with Rome over his demand that his marriage to Catherine of Aragon be annulled. Many and varied were the discussions at Trento aimed at aggressively tackling the upstart Protestants, and a prominent protagonist was an Italian bishop, later cardinal, named Ugo Buoncompagni.

Apart from its anti-Protestant acts, the Council of Trent authorized the pope to take the matter of the calendar in hand and instigate an appropriate revision. In 1572, nine years after the completion of the council's deliberations, this same Ugo Buoncompagni was elected pope and took the name Gregory XIII.

In many ways this was a case of the poacher turning game warden. Although ecclesiastical histories steer rather delicately around the subject, Gregory XIII clearly had an early life which was far from spotless, and one might allude to his pursuits having included wine, women, and song. Hardly the curriculum vitae one might expect for a future pope, but not only was he a reformed man, the Catholic Church needed a strong leader to combat, rather literally, the Reformation.

One of Gregory's first acts was to have the Catholic Church celebrate the Saint Bartholomew's Day Massacre, in which three thousand Huguenots (French Protestants) were slaughtered in Paris: these were bloodthirsty times. He tried to mount insurrections in Ireland and elsewhere against the British under Elizabeth I, who had been excommunicated and worse (a *fatwah* was declared against the queen, with Catholics enjoined to try to assassinate her, as a leader of the heretics). Elsewhere Gregory XIII had other wars to fight, with the Turks approaching from the East. Europe was in turmoil.

The Catholic Church may have thought that it was getting a powerful and efficient leader to pull them through these troubled times, but actually Gregory was largely ineffective, and the various measures he enacted (seizing land and property to fund wars, for example) were hugely unpopular, with the result that he was forced to rescind various draconian measures he had tried to introduce. Perhaps as a result of this, Gregory recalled that the Council of Trent had given him the power to leave his mark on history through calendar change, and again a papal letter went out inviting submissions on how the times could be fixed.

By the time that Gregory XIII was taken to the Calendar Room to witness for himself the aberrant position of the Sun, the result was a foregone conclusion: the Church had put off essential calendar change for too long already, and action was inevitable. The hour had come, and so had a man determined to put things in some sort of order.

Lilius and Clavius

Although we may remember Julius Caesar as the man who introduced the legal instruments changing the calendar from 45 B.C. onward, we saw that the design of that calendar was due to the Greek-Egyptian Sosigenes. Similarly, although we associate Gregory XIII centrally with the revision of the Julian calendar, introducing the system which bears his name thus producing immortality for this otherwise rather reprehensible pope, in fact it was others—mathematicians and astronomers, of course—to whom the real credit is due. Foremost among these are Aloysius Lilius and Christopher Clavius.

Earlier we saw that the Easter tables of Dionysius Exiguus were taken by Cassiodorus to a monastery which he founded on the Gulf of Squillace in Calabria, and I wrote that if Italy is shaped like a boot, then this gulf is the dip in the bottom of your foot behind the big toe. Moving backward along the sole of this figurative foot one arrives at the ball, upon which you roll as you walk along. At the rear of that portion of Italy's anatomy, just at the wide opening of the Gulf of Taranto (the pediform instep), is a small seaport named Ciro Marina. This is where Aloysius Lilius was born, around 1510.

We are not sure of his birth year. Nor are we sure of his name, in that it is given in different forms in disparate documents from the time: Luigi Lilio or Giglio are common, but I will stick with the oft-used Latin version, Aloysius Lilius. By profession Lilius was a physician, who practiced largely in Naples, but he had a passion for astronomy and it was his suggestion for the reform of the calendar which was eventually implemented. He did not live to see the day, though, as he died in 1576, and it was his brother Antonio who presented his proposal to the pope for review. In 1577 Gregory XIII submitted a description of Lilius's plan, the so-called *Compendium* actually drawn up by an historian named Pedro Chacón about whom history records little, to the various functionaries of the Catholic Church and the courts of Europe for their consideration.

The intricate details of the calendar revision are somewhat confusing, so we must cover them in their proper place, but at this juncture I should make clear that the alteration in the mean year length (using 97 leap years in a cycle of 400 to produce an average of 365.2425, close to the length needed to keep the vernal equinox on the same date) is a relatively simple part of the overall problem which was tackled. That cycle had been suggested prior to Lilius's proposal— for example, in 1560 the astronomer Petrus Pitatus of Verona had advocated this solution—and the essential novel feature of the scheme described by Aloysius Lilius which won over the pope and his advisers was a new system for calculating the date of Easter. But that we cannot describe until later.

The other major figure we should remember here is Christopher Clavius, whom again I cite with a latinized name. Born in Bavaria in 1538, Clavius lived much of his life in Italy as a Jesuit and died in Rome in 1612. While Lilius was a doctor who dabbled in matters mathematical, Clavius was a major talent who has been termed the "Euclid of the Sixteenth Century." From his late teens he had been a member of the Catholic Church and had been called to Rome to teach at the Pontifical College. As appears to be the case in all ages of science and mathematics, Clavius had many rivals with whom he had acrimonious disputes, his great Protestant rival Joseph Scaliger in particular; but all attested to his singular ability.

In terms of calendar reform, Clavius's great contribution was that he expanded and explicated Lilius's suggestion and the resultant Church documents which stipulated how dates should henceforth be calculated, in a work stretching for eight hundred pages, including the dates for all movable feasts until A.D. 5000. His aim was to ensure that the Gregorian calendar reform would be generally accepted beyond all argument—and in that this calendar is now the global standard, one would have to say that Clavius was successful. (As to whether the system which Clavius so vehemently advocated is indeed the best possible solution is another matter, to which we will turn attention in the next chapter.)

Although Gregory XIII has his name attached to our calendar, in fact posterity does accord Aloysius Lilius, its inventor, some recognition. The first day of the Gregorian calendar was October 15, A.D. 1582, and in order to provide a convenient framework for comparing events in countries using different calendrical systems since that time historians sometimes use what is termed the *Lilian Date*. Like the Julian day number, this is simply a count of days, that first day being counted as Lilian day 1.

The *Inter Gravissimas*

The form of missive which put forth the instructions from Pope Gregory XIII is called a papal bull (from the Latin word *bulla,* signifying a seal on a document). It is usually called the *Inter Gravissimas* because those are the first two words, in Latin, in the bull; they mean "amongst the most serious." The full leading sentence reads, in translation, "Amongst the most serious tasks, last perhaps but not least of those which in our pastoral duty we must attend to, is to complete with the help of God what the Council of Trent has reserved to the Apostolic see."

One can see encapsulated there perhaps two distinct things. One is the desire of Gregory XIII to complete a matter with which he had personally been involved for almost four decades, since the Council of Trent convened in 1545. The second is the significance accorded to calendar change: one can see here a resurgence of the Easter Controversy of the previous millennium, with the Catholic Church in its Counter-Reformation mode employing an altered calendar as a weapon against the Protestant churches, in the same way as the Jews, and then the Quartodecimans, and then the Columban Church and so on, had been marginalized in the past. In these contexts scientific reality—the true positions of the Sun and the Moon—is of little consequence, and as we will see the reformed calendar certainly did not put those objects back where they should be in terms of dates and associated liturgical practice.

One may judge for oneself whether it is incongruous or not that the date of the *Inter Gravissimas* was February 24, A.D. 1582, which the percipient reader will recognize to be the bissextile day in a leap year (which 1582 was *not:* but it's a neat little thing to notice). There are a few other interesting points which we might look at before passing on to the nitty-gritty of what the reform really meant.

One of these is that a reading of the *Inter Gravissimas* allows another blow against the notion of papal infallibility to be struck. For example, it states that the rules for calculating Easter had been defined by the Nicene fathers, and we know that this was not the case: they had not done much more than agree with the Council of Arles that all Christians should celebrate Easter at the same time, that the appropriate time should not coincide with the Jewish Passover, and that the Alexandrians were the best people to ask to calculate dates for Easter.

Another point worth noting is that even back in 1582, the pope was worried about whether our computer programs and printed calendars would correctly carry a February 29 in the year 2000. Don't laugh: several commercial spreadsheet programs needed to be recalled a few years back because the authors hadn't read the pope's missive and in consequence erroneously made 2000 a common year, and when the software was reissued it was found that the programmers had added a February 29 to the year 1900, whereupon the responsible companies (which I'd better not name for fear of libel action) simply said "Oops." Getting back to the *Inter Gravissimas,* the end of one longish paragraph stipulating the leap-year rule states, translated of course:

> So in the year 2000, let a leap day be intercalated in the customary manner, so that February might contain 29 days and by the same token the mechanics of inserting and intercalating the leap day every four hundred years be retained in perpetuity.

Now, look again at the end of that quotation. The Christian churches preach that time on Earth is ephemeral, and that Judgment Day will surely come. The "in perpetuity" seems to contradict that notion; or perhaps it is meant to imply that this leap-year rule should also apply in Heaven and Hell.

Within a few years Gregory XIII was dead. For better or for worse he is remembered largely through the calendar reform he pushed through in the few years prior to his expiration in 1585, rather than his earlier fervent anti-Protestant assaults.

The Reforms Introduced

Many accounts make it appear that the only change made by Gregory XIII was to alter the leap-year rule slightly, such that three were to be dropped from every four centuries. The tinkering with the leap-year cycle was only a minor alteration, which was to have no effect until 1700, well beyond the lifetimes of the people of the time (clever, that). We will see at length that the leap-year rule chosen was significant—it happens that quite deliberately it was not the best available—but for religious rather than chronological reasons.

If anything else is mentioned, it is often claimed that the reform altered the date for New Year's Day to January 1. In fact, the papal bull made no such change to the year's beginning: January 1 was already New Year's Day on the civil calendar in many countries, such as France, Spain, and Germany, and ecclesiastically it was well established as a special celebration to start the year—as the Feast of the Circumcision—with implications we shall soon come across. But that feast was actually referenced to the Christmas Day just before. As a matter of fact, the start of the ecclesiastical year of the Catholic Church was at Christmas, as it continued to be until 1911, the formal change to January 1 being instigated through a papal letter of Pius X dated December 8, 1910.

The more important revisions introduced with the *Inter Gravissimas* were twofold. The first was the deletion of ten counting days from the calendar in 1582. Again we need to look at the ecclesiastical significance of this, the reform being based upon religious (rather than astronomical) necessity. The second revision was in various technical changes in the calculation of Easter, and in that regard I hardly need to emphasize that religious questions were central to the reform.

Why was it necessary to delete ten days from the calendar? The question is tied up directly with Easter, and it is not a scientific question. Many have represented the Gregorian reform as being based simply on astronomy, but that is clearly not the case. Due to the Julian year being slightly too long, the date of the equinox was gradually coming earlier in the annual calendar, oscillating by almost eighteen hours around midnight on March 10/11 in the late sixteenth century. The rate of slipping was by one day every 130 years. Some argue that this is a significant amount, with the seasons shifting accordingly, but from that *civil* perspective the slippage is nothing to worry about. So what if summer moves by one day earlier over several generations? This is simply not a notice-able effect for nonreligious concerns. As a contrary example, think about the date of perihelion, which moves later by one day every 57 to 58 years on aver-age, and yet is not a matter addressed in fixing the calendar (and, as you will find in chapter 23, it is indeed of immense importance with regard to the sea-sons).

Those ten days were deleted not for civil reasons, but for ecclesiastical ones. As the astronomical equinox moved earlier in the year, the ecclesiastical equinox—fixed as March 21, remember—was moving seasonally later (toward summer), and so did Easter.

Most of the movable feasts of the Church throughout the year are tied against Easter. (One can broadly think of the movable feasts being those reck-oned by the Moon, the immovable feasts following the Sun and hence on reg-ular dates during the year. An exception is Advent, which moves but is linked to the Sun rather than the Moon: it is the fourth Sunday before Christmas, or the nearest Sunday to Saint Andrew's Day.) These movable feasts may vary in perceived importance from sect to sect, but examples are Ascension Day (forty days after Easter) and Pentecost, or Whitsunday (seven weeks after Easter). As Easter slipped later in the year these were in danger of coinciding with, say, the summer solstice, leading to various problems, not the least of which would have been the pagan festivities linked with midsummer's day.

One solution would have been simply to redefine the ecclesiastical equinox from March 21 to the tenth or eleventh and have it pegged against the astronomical equinox, but then the feasts leading up to Easter would have been in danger of running up against the start of the year as the equinox regressed. January 1 had long since been accepted as the start of the liturgical year—after overcoming initial avoidance because it was viewed as another period of pagan debauchery through its association with the New Year on the Julian calendar—the Roman Church having incorporated the Feast of the Circumcision from the ninth century. Leading up to Easter is Lent, the season of fasting and penitence which starts on Ash Wednesday and lasts for forty days (understood not to include Sundays, so that in all there are actually 46 days in Lent). Even before Lent, in fact on the third Sunday before, is the earliest movable feast in the year, Septuagesima. That is nine weeks before Easter. Redefining the equinox to the first half of March would have led to Septuagesima sometimes occurring in the first two weeks of the year with immediate effect, and as the equinox moved back over the subse-quent centuries matters would have got worse. Something had to give, and the chosen solution was the scrubbing of some days so as to recalibrate the equinox closer to an acceptable time of year.

The Missing Ten Days

Much attention is paid to the deletion of specific dates from the calendar, a matter which seems to intrigue people. For the Gregorian reform of 1582, the ten dates which were jumped in Rome and other parts of Italy (and elsewhere in Catholic states immediately adhering to the papal bull, such as Spain, Portugal, and Poland) were October 4–14 inclusive. It seems that October was chosen for the process because this month contained the least number of Church feasts and other business, leading to the minimum disruption; October is also about six months away from Easter, that pivotal point in the Church year.

Some other nations followed soon thereafter, such as France and parts of Belgium (then known as the Spanish Netherlands) in December, and a few in the following years. They were hurried along by a *motu proprio* (papal decree) from Gregory XIII dated November 7 in 1582, in which he gave instructions as to how all good Catholics should swiftly make their calendars coincide with that of the Holy See, preferably before Easter in 1583. Others were more tardy in falling into line, like Britain and other staunchly Protestant states.

The actual wording of the *Inter Gravissimas* is interesting in this respect. Because the pope held authority over the clergy of the Catholic Church, the bull states that he *orders* or *instructs* all bishops, priests, and so on to observe the new calendrical system, but only that he *asks* all civil leaders to do likewise, bringing their national calendars into line with the new religion-based calendar. It took Britain almost 170 years to accede to this request, and even then (as we will see in chapter 16) it did so without actually admitting the consequences or basis of its actions.

There was no alteration in the weekday sequence, so that in Rome in 1582 Friday the fifteenth followed Thursday the third of October, and similarly when other places adjusted their calendars. This is important in another way: the reform had to allow for an adjustment of the Dominical Letter. In chapter 7 we saw that this is normally the letter assigned to the first Sunday of the year given that January 1 is A, but altering the date without skipping the sequence in the days of the week required a technical change. This is not a simple matter, because Easter was and is calculated, before and after the Gregorian reform, on the basis of the Dominical Letter.

Why 1582?

While asking all these questions, like the number of dates skipped, and why October was chosen, I might additionally ask for you the simple question: why 1582? That is, one might have expected a calendrical reform to have been post-dated and hinged on some later round date, like 1600, or even later. Why give yourself the headache of actually implementing a calendar change? Remember that the change had immediate effect, and it was not, as many accounts suggest, with no ongoing consequences until 1700. I would hope that by now you are in agreement with me that the mere skipping of leap-year days in some century years is a minor thing compared to altering the Easter *computus*.

Why 1582 then? One part of the answer would seem to be that Gregory XIII had a personal interest, and with his life clearly drawing to a close there was some motivation to accomplish at least one positive thing during his papacy. But there was also an important paschal consideration.

Right through to the present there has been a dispute between the Eastern (Orthodox) churches and the Western (Catholic and Protestant) churches regarding when Easter should be celebrated, the Eastern sects having refused to abandon the Julian calendar. This means that most often the Orthodox churches celebrate Easter some time later than is the case in the Western sects. The *Inter Gravissimas* appeared in early 1582, deleting ten days from October. The relevant *motu proprio* was issued in November, urging all states which had yet to adjust their calendars to do so in the following February. This was all happening in a hurry to get the necessary changes made before Easter of 1583. The reason was that in that year a propitious event was to occur: Easter Sunday was on the same day in both calendars (March 31, Julian, and April 10, Gregorian), and it would be several years before such an opportunity again presented itself.

Why Ten Days Deleted, Not Eleven?

But why were *ten* days deleted? This is not so simple a question as it sounds. Numerous present-day reports say that the equinox at the time of the Council of Nicaea was on March 21, whereas in fact it was on the twentieth, and the sixteenth-century calendar reformers knew that this was the case. Given that the astronomical equinox just prior to 1582 was occurring mostly on March 10, why didn't they delete eleven days, thus moving the equinox to an average date of March 21? Let us first look at the actual effect of the reform instigated, and then consider the alternatives.

The leap-year cycle which the Gregorian calendar employs allows the astronomical equinox to drift by a substantial amount. The full amplitude of the drift can be found by considering the earliest and latest times for the equinox within each year over the two centuries which bridge a century year which *is* a leap year, like A.D. 2000. The latest vernal equinox in this period occurred at about 19:00 U.T. on March 21 of 1903, while the earliest will be at around 14:00 U.T. on March 19 in the year 2096. The difference between the times in 1903 and 2096 is almost fifty-three hours, and that is the largest gap which occurs in the four-hundred-year Gregorian cycle.

Why 53? The difference between the mean vernal equinox year of around 365.2424 days and the Julian year of precisely 365.25 days is about 0.18 hours. Over the 193-year time gap a total difference of about 35 hours accumulates. On top of that, before the following leap year pulls the equinox back by 24 hours, in the three-year progression 1901–1903 another 6 hours per year adds on, making 18. The total is therefore 35 + 18, or 53 hours.

By deleting just ten days, the effect was that usually the equinox occurs before March 21, the figurative (but erroneous) Nicene date, but it is never *later* than the twenty-first. That is the essential point, because Easter is based on the new moon *following* the equinox, and by defining the ecclesiastical equinox to be the last possible date when the astronomical equinox could occur it was not possible to get a hiccup with the paschal moon preceding it. More to the point, Passover was avoided, at least most of the time.

Why Not Twelve?

Ten days rather than eleven were deleted, then. But why the fixation with March 21? For example, why not delete twelve days? That would have put the equinox back at the place where it was at the start of the era, the time of Jesus Christ, soon after Julius Caesar had introduced his eponymous calendar with the phase being such that the equinox was oscillating between March 22 and 23, and one might imagine that the start of the *anno Domini* era should be taken to be of greater significance than the Council of Nicaea in A.D. 325.

Although we are not sure of the answer, there are several considerations which might be argued. One is that there was a rival calendar reform scheme which might have been favored by such an adjustment, and we will meet this in the next chapter. A second is that Christopher Clavius is thought to have looked into the question of whether A.D. 4 was a leap year, leading to some confusion about the date of the equinox at the time of Christ's birth, whereas he knew very well that the equinox was on the twentieth in A.D. 325. A pertinent point is the fact that if 1 B.C. and A.D. 4 were observed at the time as common years, between the death of Herod in 4 B.C. and the reform in 1582 there were ten leap days observed in excess of the required number to maintain the equinox around the "correct" date, and Clavius may well have considered that to be significant. That mix-up by the Senate in the decades after Julius Caesar's assassination, erroneously deploying leap years once every three years, has had long-lasting knock-on effects.

Why Not Fifteen?

Deleting ten days from the count shifted the astronomical equinox to an average time on March 20, but one oscillating by fifty-three hours, the ecclesiastical equinox being defined as the twenty-first. Now here's a question. Julius Caesar seems to have thought that he had amended the calendar and phased it such that the equinox would occur on March 25, that hugely significant traditional date which apparently stems from the idea that the early Roman republican calendar of Numa Pompilius, in the seventh century B.C., had the equinox and the start of the year on that date. That may not have been the real date of the equinox at that time on *any* calendar, but that is not important: what is significant is what Julius Caesar believed, and many people since then. So much of our calendrical history hinges on this received wisdom.

The question, then, is this: why didn't the Gregorian reform instead delete fifteen days, shifting the equinox to the traditional date of March 25? Yes, the rules for Easter would have needed to have been changed a little, but there was much religious attachment to that date—for example, in the belief that it was the date of not only the Annunciation but also the Crucifixion—and Clavius and the other papal advisers were needing to alter the Easter *computus,* anyway. One might think that the case for March 25 was almost unanswerable.

The answer is nine months later. If the equinox had been shifted to March 25, then the winter solstice would similarly have been moved to December 25, which was by then (in the sixteenth century) a major Christian feast. By allowing Christmas and the solstice to coincide once more, the Church would have walked into trouble. Christianity had successfully pinched the solstice festival of the pagan religions over twelve hundred years before, through the edicts of Constantine the Great and his papal successors, and it was not about to hand it back.

For all of the above reasons, ten days rather than some other number were deleted. For reasons of astronomy, logic, and history one could argue for a different adjustment, but for the purposes of the Catholic Church it was necessary that ten be the number. The effect of this is that the ecclesiastical equinox (defined as all of March 21) may be one or even two days later than the astronomical equinox (which is at some instant on March 19, 20, or 21).

Return to Easter

We are discussing the actual reforms of the calendar introduced by the papal bull of 1582. I said that one alteration was the 97 leap-years-in-four-centuries rule. Another we have mentioned was the Dominical Letter adjustment. The date of New Year did not need shifting because that was January 1 already, having been so in the Church calendar for some centuries, and on the civil or legal calendars of many nations (but not Britain) for a decade or two. The two most complicated reformatory steps were the deletion of ten days as just discussed, and the technical change in the Easter *computus* we must now describe.

When reading the previous chapter you might have wondered why I introduced the term *golden number* (G) for the value in the sequence 1 to 19 in the lunar/Metonic cycle, as used in Scaliger's Julian Period system. The answer is that the golden number is fundamentally important in defining Easter.

Prior to the Gregorian reform the golden number was used to calculate directly when Easter would be celebrated, under the effective presumption that the Metonic cycle is exact. Easter then followed a 532-year cycle, its date being given by the golden number (value 1–19) and the Dominical Letter coupled with the leap-year cycle, rendering the solar cycle of 28 years: each of those has a single Dominical Letter, except that leap years have two, one for prior to February 29 and one for thereafter. That *computus* was quite simple, and monks like the Venerable Bede and his successors could easily draw up tables showing when Easter should be celebrated throughout Christendom.

But we have seen that the Metonic cycle is *not* exact. By the sixteenth century not only was the Sun ten days ahead of its nominal place (or more, if you prefer to reject the Catholic Church's connivances against all rivals), but the Moon was also four days ahead of itself.

In the same way as three leap days needed to be dropped from each four centuries so as to keep the Sun in line, so a system needed to be devised to maintain the ecclesiastical moon in some specified relation to the astronomical moon. Note that I did not write "correct the four-day aberration" there: just as the Church deleted only ten days from the calendar, and so did not pull the equinox onto any date later than March 21, so the lunar adjustment did not fully compensate for the four-day gap which had built up over the preceding millennium or so.

Shifting the Moon

To correct the Sun, the leap-year cycle produces an average of 365 plus 97/400 days in a year, or 365.2425. This compares with the real vernal equinox year of 365.2424 days (to four decimal places, although this duration gradually changes in the real universe). Not a bad approximation, differing by only about eleven seconds, although we will see in the next chapter that the Church could have done better. To correct the Moon, a similar sort of analysis is needed.

The question is the inaccuracy of the Metonic cycle. Nineteen mean Julian years make 6,839 days and 18 hours, but 235 mean lunations last for 6,839 days, 16 hours, and 31 minutes, a difference of 89 minutes. The Moon therefore gets ahead of the Sun by 89 minutes in 19 years, or a full day in 308 years. Using the time of the Nicene Council as the reference point, in the 1,257 years between 325 and 1582 just over four sets of 308 years had occurred, meaning that the ecclesiastical moon was four days or so behind the astronomical moon. For example, the Church calendar might indicate that the new moon was to be marked on November 9, whereas the astronomical new moon, plain to the eye of any observer, had already manifested itself on the fifth or even the fourth.

To keep things in order, the Church needed to introduce some cycle which would correct this ever-increasing discrepancy, or at least stop it from getting worse. One solution would have been an initial shift in the golden numbers, followed by later amendments as a day's error accumulated. The idea of deleting ten days to correct the solar cycle in effect could be viewed as shifting all the golden numbers by ten downward. To re-zero the lunar cycle, the golden numbers could also be shifted by four upward, giving an overall change of six downward. Then, in every centenary year which was not a leap year (1700, 1800, 1900, 2100 . . .) the golden numbers would need to be dropped by one, thus adjusting the solar cycle, while in every third centenary year (every 300 years, near enough to 308) those numbers would be enhanced by one to facilitate the lunar cycle remaining in step. Complicated, don't you think?

Lilius and Clavius certainly thought so, and so it was decided to drop the golden number system as the direct avenue through which Easter would be derived in combination with the Dominical Letter, and instead introduce a different system, called the *epacts*. The golden numbers were maintained as a perennial nineteen-year cycle—as I wrote in the previous chapter, the golden number can be calculated for any year A.D. by dividing it by 19 and adding one to the remainder, and the fact that such a rule still applies indicates that the Gregorian reform did not change it—but in addition the concept of a table of epacts was brought in, these being altered systematically whenever it was necessary so as to keep the ecclesiastical sun and moon more or less in step (but not coincident) with their celestial counterparts. We will now consider the nature of these adjustments.

The First Table of Epacts

The word *epact* is derived from a Greek term signifying the age of the Moon at the start of the year. In effect, the epact for any year is the phase for the *ecclesiastical* moon on January 1. In reforming the calendar the four-day discrepancy between the astronomical and ecclesiastical moons was largely remedied: the phasing defined through the initial table of epacts is such that the new moon of the ecclesiastical calendar is usually one or two days, but sometimes three days, behind the real astronomical moon. The reasons for this we will come to at length.

The first set of epacts, which was valid until 1699, looked like this:

Golden Number:	1	2	3	4	5	6	7	8	9	10	11	12	13	14	15	16	17	18	19
Epact:	1	12	23	4	15	26	7	18	29	10	21	2	13	24	5	16	27	8	19

If you run through the epacts, you will see that what is happening is that each one is 11 larger than that preceding, except that when the value exceeds 30, that number is subtracted. Thus 1 becomes 12; 12 becomes 23; 23 becomes 34, which is curtailed to become just 4; and so on. The progression is a series of additions of 11 with the subtraction of 30 when necessary. We need to consider the source of those particular numbers.

Working to the nearest day, the solar year is 365 days long, whereas the lunar year of twelve lunations lasts for 354 days. The difference is 11 days. For golden number 1, the first year in a 19-year cycle, the lunar phase or epact is 1 on January 1 from the table above. At the start of the next year, golden number 2, the twelfth lunation of the preceding year finished 11 days earlier, and so the lunar phase is 11 higher, and the epact is 12. At the start of the fourth year, golden number 4, the calculated phase would be 34, but that has no meaning in itself: it means one complete lunation plus another 4 days (with a 30-day lunar periodicity assumed), rendering an epact of 4. What has happened, then, is that the third year has effectively had an intercalary month—a thirteenth month, of 30 days—added to it.

This happens with every third year in the sequence, such that for the years corresponding to golden numbers 1 to 18, twelve years may be thought of as having 354 days and the other six 384 days each. (I am ignoring the extra days in leap years for simplicity.) Looking at golden number 19, the epact is 19. If one adds 11 to that, one gets 30 (or, equivalently, 0), whereas this is supposed to be a cycle such that the following year is golden number 1 again, with a repeated epact of 1 being due. The rule therefore is that the final year has twelve added to its epact—19 plus 12 equals 31, hence epact 1 for the next year—and it also has an intercalary month added with only 29 rather than 30 days.

With the above rules in mind we can now tally up what is happening over the complete cycle of nineteen years. Twelve of those years have 12 lunar months each, whereas seven of them have an additional intercalary month. Thus in nineteen years there are (12 × 12) plus (7 × 13) = 235 months, as required in the Metonic cycle.

The twelve short years have 354 days each; of the years containing intercalary months, six have 384 days each while the final year has only 383, because its extra month is only 29 days long. Add that lot up and one gets 6,935 days. But we have so far been ignoring the leap-year days. In the nineteen-year cycle there are either four or five leap years, resulting in the final total being either 6,939 or 6,940 days, again as expected for the Metonic cycle.

The reason for the ecclesiastical moon most often being 1, or 2, or sometimes 3 days later than the astronomical moon may now be apparent. Six of the intercalary months have 30 days, longer than the average lunation (29.53 days), while one has only 29. On the other hand, twelve mean lunations last for 354.37 days, longer than the integer 354 days used to derive the 11-day shift in the epacts from year to year, although the average year is also longer than the integer 365 days employed, compensating in part. On top of that, note that the *mean* moon gives a lunation lasting just over 29.53 days, but that in itself is a theoretical construct, with the real astronomical moon sometimes being eight hours behind or ahead of the mean moon.

All this causes some skipping around, and there is an additional amount resulting from the leap years: three-quarters of the golden number/Metonic cycles will have 5 leap-year days and thus 6,940 days in total, while one-quarter will have only 6,939 days.

How do the epacts listed above compare with the effective values prior to the reform? In essence the question here is how Lilius and Clavius arrived at the values in the first table. By deleting ten dates from the calendar, the following January 1 was arrived at ten days earlier, effectively decreasing the figurative epacts by ten. However, the error in the lunar phase—the four days between real and ecclesiastical moons which had accumulated over the previous twelve or thirteen centuries—also had to be accommodated. The overall shift in the epacts was therefore by six. For example, take 1583 itself, the year of the first Easter under the Gregorian rules. The golden number for that year is seven: divide by nineteen, take the remainder and add one. For golden number seven, the epact is also seven, but prior to the reform the value would have been thirteen.

Coinciding with Passover

This first epact table had its values defined specifically to try to make the ecclesiastical moon never agree with the astronomical moon, in the same way as the astronomical equinox can never occur later than the ecclesiastical equinox of March 21. Why? It all comes back to trying to avoid Passover. The Jewish year is based upon the real cycles of the Moon and starts with the month of Tishri (around September), and so does not hinge upon the vernal equinox. To try to ensure that Easter was not coincident with Passover, the ecclesiastical phenomena (equinox and new moon) are defined to be later than the real astronomical phenomena.

Unfortunately (from the perspective of the Christian churches) it does not always work, because the real moon can trail the mean moon by up to eight hours, and that, combined with particular phasing in the leap-year and golden number cycles, can result in Easter occurring on the Sunday of an astronomical full moon, the ecclesiastical full moon being calculated according to the epact rules as having occurred on the preceding day. When this happens Easter and Passover coincide: this was the case in 1805, 1825, 1903, 1923, 1927, 1954, and 1981, and it is due again in 2123.

Updating the Table of Epacts

In our discussion above only the first table of epacts was considered, valid until the end of 1699. In the year 1700 the Gregorian reform stipulated that there should be no leap-year day, thus breaking the once-per-four-years cycle which had persisted for seventeen centuries. This had the effect of knocking one day out of the above calculations for the epact versus golden number cycle, meaning that all the epacts needed to be decremented by one. From 1700 onward an updated table of epacts was therefore necessary:

Golden
Number: 1 2 3 4 5 6 7 8 9 10 11 12 13 14 15 16 17 18 19

Epact: * 11 22 3 14 25 6 17 28 9 20 1 12 23 4 15 26 7 18

The asterisk symbol * is usually employed to denote the numerical values of 0 or 30 when they result in epact calculations: saying the "zeroth day of the moon" when *one* is defined as being the new moon makes little sense, and saying that the Moon is thirty days old when a lunation lasts for 29 days plus a fraction is equally nonsense.

When the year 1800 came around, one might imagine that the epacts were again dropped in value, but this was not the case. In each non-leap-century year the epacts are indeed figuratively reduced by one to accommodate the Gregorian leap-year rule, keeping the dates and thus the epacts in step with the Sun. But the lunar cycle, requiring one day to be dropped about every 308 years, must also be considered. In the event the suggestion of Lilius was that eight days should be dropped from every 2,500 years, producing an average correction rate of one day every 312.5 years, a pretty good approximation. The way in which this was enacted was that once every 300 years the epacts were to be incremented by one, starting with the year 1800, but after 2,100 years a jump of 400 years instead is deigned sufficient. One may therefore think of a solar equation and a lunar equation, the effects of which counteract in some events:

Solar: Drop epacts by one in 1700, 1800, 1900, 2100, 2200, 2300, 2500, 2600, 2700, . . .

Lunar: Increase epacts by one in 1800, 2100, 2400, 2700, 3000, 3300, 3600, 3900, 4300, 4600, 4900, 5200, . . .

Overall: Drop epacts by one in 1700, 1900, 2200, 2300, 2500, 2600, . . . *but* increase epacts by one in 2400, 3600, 5200, . . .

Note the peculiarity produced here. One is familiar with the leap-year rule being such that three out of four century years are lost from the simple Julian four-year leap cycle. But the rules for adjusting the epacts are such that the ecclesiastical moon gets shifted both backward *and* forward: in 2100, 2200, and 2300 the lack of an intercalary day means that the epacts are dropped by one day, but in 2400 the epacts are enhanced again by one, before dropping again in 2500, 2600, and 2700, whereas there is no change in 2800.

From the perspective of the past, and the immediate future centuries, the effect is that the above table of epacts applied from 1700 to 1899, whereupon the epacts were again decreased by one, the new table as follows applying from 1900 through to 2199:

Golden Number:	1	2	3	4	5	6	7	8	9	10	11	12	13	14	15	16	17	18	19
Epact:	29	10	21	2	13	24	5	16	27	8	19	*	11	22	3	14	25†	6	17

The perceptive reader will notice that there is a dagger symbol placed next to the epact of 25 under golden number 17. This is a technical point which I will not trouble you with too much here, but merely mention why it is necessary. Obviously the epacts make use of integer arithmetic, whereas the actual duration of a lunation involves a fraction of a day. The epacts can therefore never be more than a rough guide to the lunar phases. In some cases it happens that an epact of 25 will result in a new moon being indicated on the same date twice within a nineteen-year cycle, and in reality this cannot happen. To obviate this drawback, in some cases an epact of 26 is used instead of 25, and the dagger on the 25 indicates that this may be the case.

In total I have given above three tables of epacts—one for 1583–1699, one for 1700–1899, and the last for 1900–2199. Since there are thirty possible values for the epacts, there are thirty possible tables. Taking into account the frequency of increments in the epacts due to the solar equation, and the decrements due to the lunar equation, one finds that the full set of thirty tables is circulated through in 7,000 years. Thus the first epact table would come into use again in the year A.D. 8500, presupposing that no calendar revision has occurred before then. If you look in, say, the Church of England's *Book of Common Prayer,* you will find that the tables necessary to calculate the dates of Easter are given through to that year.

Recapitulation on the Change of Century and Millennium

Now let us mention a point here which relates to people's misconceptions about when the new millennium begins. That is, strictly, with the start of the year 2001, because until then the full two thousand years have not elapsed since the zero point of the dating system (the instant of midnight between December 31, 1 B.C., and January 1, A.D. 1), arbitrary though that may be. But the fact that the epacts are shifted in what we tend to call the century years—the years ending 00—and that those are also the years which are common years in three out of four cases, rather than leap years, all adds to most people viewing those as being the pivotal points in the calendar. Indeed they are, in that they are the years in which the epacts may be shifted.

On the other hand, even our contemporary pope and his advisers seem to be confused on this matter of when the new millennium begins. In a 1994 apostolic letter entitled *Tertio Millennio Adveniente,* through which a Great Jubilee of the Roman Church is proclaimed, Pope John Paul II initially assigns the year 2000 correctly to the second millennium, but later allots it to the third, in error. In fact, the Great Jublilee is proclaimed to last from Christmas Eve in 1999 until the same date in 2000, which means that it will terminate on the day before the two thousandth anniversary of the (traditional) birth date of Jesus Christ, and seven days prior to the beginning of the third millennium of the era.

How Long Is the Easter Cycle?

The idea that the dates of Easter have been gazetted for over sixty-five hundred years into the future leads one to ask: how long will it be before a complete cycle has been circuited, if the rules are indeed obeyed "in perpetuity" as Gregory XIII instructed?

Under the Julian calendar, the Easter cycle was merely 532 years long. This is just the solar cycle times the lunar cycle: $532 = 19 \times 28 = 19 \times 4 \times 7$. (The nineteen is the Metonic cycle, the four is the leap-year cycle, and the seven is the number of days in the week.) Thus, after 532 years the almanac for the year would repeat itself: the day-of-the-week combinations with all dates would repeat, and Easter would reappear on the same date, as would all the movable feasts. The only problem is that the Sun and the Moon gradually shift out of kilter.

One may therefore ask how long the Easter cycle lasts under the Gregorian calendar. Let's work it out. There are five distinct cycles which must all return to their initial conditions for the Easter cycle to be completed: (1) the solar cycle; (2) the lunar cycle; (3) the Dominical Letters; (4) the epacts; and (5) the golden numbers.

The leap-year cycle (the solar equation) lasts for 400 years, and the moon phase cycle (the lunar equation) for 2,500. But since there is a common factor of 100 in both of those, only one of the two factors need be kept, giving an overall cycle for this part of our investigation of 10,000 years ($4 \times 2,500$ or 400×25, whichever way you prefer to look at it).

The Dominical Letter equation would introduce a factor of seven, but we know that 400 years contains an exact number of weeks, so that the seven is again a common factor which can be left out.

So far we have accumulated 10,000 years, which is the overall cycle time for features (1), (2), and (3) above. Next consider the epacts.

During these 100 centuries, the epact associated with any particular golden number will have been increased 32 times due to the lunar equation (4×8: eight changes in every 2,500 years) and will have been reduced 75 times due to the solar equation (25×3: three changes in every 400 years). The overall effect is a decrease of 43 (75 − 32) for that epact, but we know that all calcu-lated epacts in excess of 30 are truncated by that amount. So the answer is 13.

This means that after 10,000 years all the epacts have dropped by 13, and that happens to be an awkward number. In order for the epacts to return to their original values, the epact tables will have to be cycled through 30 times, and that takes 300,000 years.

After that length of time the same epacts are associated with the golden numbers, but 19 is another awkward number, and it does not divide precisely into 300,000. Thus the time required for the golden numbers to come back to their original state is 19 × 300,000 = 5,700,000 years! This has much im-pressed some people; for example, Philip Whitwell Wilson, in *The Romance of the Calendar* (London, 1937), wrote:

> *It is found that 5,700,000 Gregorian years or 2,081,882,250 days are almost exactly 70,499,183 lunar months. The error in such a month is only a millionth of a day, and in 300 years it will cor-rect itself. This period of 5,700,000 years is thus the ultimate cy-cle that embraces day, month and week with true astronomical exactitude.*

Wilson forgot to mention that it is also exactly 297,411,750 weeks. But the no-tion expressed in that quoted extract from his book is nonsense, the reductio ad absurdum of calendrical prognostications. As we will see in chapter 22, it makes no sense to try to plan any calendar to last for more than a few thousand years, because there is no such thing as "true astronomical exactitude" for periods longer than that. Over longer time spans the Moon recedes from Earth signif-icantly, making the month longer, while the terrestrial spin rate decreases, mak-ing the day longer. The agreement to within a millionth of a day only lasts for a short time, before inexorable astronomical shifts change things forever. Wil-son's book seems aptly titled: the idea he expresses is romantic, but unrealistic.

On Which Day Were You Born?

The Dominical Letter has been mentioned several times, and it is required for calculating the date of Easter and the other movable feasts, as below. One needs some method of deriving it, then. The way to do this can be illustrated by ask-ing the question: on which day of the week were you born? The following is a simple algorithm which allows you to answer that question, through a few triv-ial sums.

Let d be your birthday, in month m and year y. Then go through these steps:

$$a = (14 - m)/12 \text{ (drop any remainder; thus } a = 1 \text{ only if } m = 1$$
$$\text{or } 2, \text{ else } a = 0)$$

$$b = y - a$$
$$c = m + (12 \times a) - 2$$
$$z = d + b + b/4 - b/100 + b/400 + (31 \times c)/12$$
$$\text{(and again throw away any remainders from each of the divisions)}$$

Now divide z by 7 and take the remainder; call it x. That result gives the day of the week, $x = 0$ being Sunday, $x = 1$ for Monday, and so on. For example, I was born on June 11, 1955; thus:

$a = (14 - 6)/12$ (and the value to carry is $a = 0$)
$b = 1955 - 0 = 1955$
$c = 6 + (12 \times 0) - 2 = 4$
$z = 11 + 1955 + 1955/4 - 1955/100 + 1955/400 + 124/12$
$\quad = 11 + 1955 + 488 - 19 + 4 + 10$
$\quad = 2449$

Dividing 2,449 by 7, one gets 349 with a remainder $x = 6$. Thus I was born on a Saturday. Note that the above algorithm works only for the Gregorian calendar, so you would get a wrong answer if, say, you were born in Moscow in 1915 (when Russia was on the Julian calendar).

The Great Easter Table

Having rather tortuously thought through the preceding description of how the calculation of Easter is based, one might finally like to be able to see how to derive the date for that holiday in some particular year. There are just two things you need.

The first is the Dominical Letter for the year in question. As aforesaid, this is the letter for Sunday given that the first day of the year is A, and it can be calculated as we did above for your birthday, except that now the date in question is January 1: the answer x you derive is subtracted from a sequence running backwards from A. If $x = 0$, then the letter is A; if $x = 1$, then the letter is G; if $x = 2$, then the letter is F; and so on. For example, 2001 starts with a Monday, so the Dominical Letter is G, and 2002 starts with a Tuesday, so the letter drops back one to F. On the other hand, a leap year is called that because it leaps a letter: 2000 starts with a Saturday, so initially the Dominical Letter is B, but after February 29 the letter becomes A. Easter, of course, always comes after the intercalary day, so the second of the Dominical Letters is used in working out when that Christian feast arrives in a leap year.

The second thing required is the epact, which is found from the golden number, a quantity we have already calculated several times. For 2000 it is 6, for 2001 it is 7, and for 2002 it is 8. From the last table given for such parameters (see page 175), the epacts for these years are 24, 5, and 16 respectively.

Given the epact and the Dominical Letter, the following *great Easter table* shows the date of Easter under the Gregorian rules (M26 means March 26; A14 means April 14; etc.):

Dominical Letter

Epact	A	B	C	D	E	F	G
*	A16	A17	A18	A19	A20	A14	A15
1	A16	A17	A18	A19	A13	A14	A15
2	A16	A17	A18	A12	A13	A14	A15
3	A16	A17	A11	A12	A13	A14	A15
4	A16	A10	A11	A12	A13	A14	A15
5	A9	A10	A11	A12	A13	A14	A15
6	A9	A10	A11	A12	A13	A14	A8
7	A9	A10	A11	A12	A13	A7	A8
8	A9	A10	A11	A12	A6	A7	A8
9	A9	A10	A11	A5	A6	A7	A8
10	A9	A10	A4	A5	A6	A7	A8
11	A9	A3	A4	A5	A6	A7	A8
12	A2	A3	A4	A5	A6	A7	A8
13	A2	A3	A4	A5	A6	A7	A1
14	A2	A3	A4	A5	A6	M31	A1
15	A2	A3	A4	A5	M30	M31	A1
16	A2	A3	A4	M29	M30	M31	A1
17	A2	A3	M28	M29	M30	M31	A1
18	A2	M27	M28	M29	M30	M31	A1
19	M26	M27	M28	M29	M30	M31	A1
20	M26	M27	M28	M29	M30	M31	M25
21	M26	M27	M28	M29	M30	M24	M25
22	M26	M27	M28	M29	M23	M24	M25
23	M26	M27	M28	M22	M23	M24	M25
24	A23	A24	A25	A19	A20	A21	A22
25	A23	A24	A25	A19	A20	A21	A22
26	A23	A24	A18	A19	A20	A21	A22
27	A23	A17	A18	A19	A20	A21	A22
28	A16	A17	A18	A19	A20	A21	A22
29	A16	A17	A18	A19	A20	A21	A15

Easter Algorithms

The calculation of the date of Easter, done by hand using the system and tables above, is clearly quite complicated in practice, but in principle it is simple. In fact, a competent computer programmer should be able to write a little routine to derive the date of Easter with a modest amount of effort. It is generally reckoned that good programmers produce about twenty lines of debugged code in a day's work, and that's about all that would be needed in this instance.

But sixteenth-century Jesuits did not have mechanical or electronic computers. The way forward was to draw up the various tables required such as those presented above. For those who could not even work their way through the few simple sums, the vast compendium of dates of Easter and all movable feasts through to A.D. 5000 was laboriously compiled by Christopher Clavius.

Nevertheless, as time has passed people have become interested in inventing algorithms enabling the quick computation of the date of Easter using either the Gregorian or the Julian rules, the latter still being employed by the Eastern Orthodox churches. Basically these algorithms are just alternative ways of making manipulations of integer quantities, although some may be regarded as being better than others. For example, one system might only work for a few millennia into the future (but then, so what?).

The most famous method is called the *Gauss algorithm*. If you have ever studied any physics, you will have heard of Gauss. In fact, even if you have no acquaintance at all with that science, you will have heard his name many times, although you probably have not recognized it. On most computer monitors is a little button which you press to neutralize any residual magnetic field around the picture tube which may distort the image: this is called de*gauss*ing. When ships were still navigated using a compass, it was necessary to degauss the bridge, the metal hull producing a magnetic field which could swamp that of Earth. Carl Friedrich Gauss was a big man in magnetism and other areas of physics, mathematics, and astronomy.

If you pick up one of the several biographies which have described his life and achievements, you will find that he was born in Germany on April 30, 1777. But that is more than Gauss himself knew, at least during his early life. Quite naturally, he asked his mother to tell him his birth date, but she did not know. All she could tell him was the year and that he had been born on a Wednesday, eight days prior to Ascension Day. This may seem astonishing to us now, but in the past people did tend to structure their lives more around religious holidays rather than specific dates.

Gauss, though, wanted to know his birthday, and in his late teens he set about deriving the date given the information he had in hand. Knowing the year was 1777, and that Ascension Day, or Holy Thursday, comes forty days after Easter Sunday, he was able to find his birthday once he had found out when Easter had fallen in that year, and that got him interested in the general problem of the *computus*. In several publications written from 1800 onward he showed how to derive Easter on both the Gregorian and the Julian calendars, and also the date of Passover, for several centuries into the future. Gauss did make a few mistakes which he later corrected, and by 1816 he had worked out how to compute the dates beyond A.D. 4200.

But will the Western churches still be using the Gregorian rules for Easter more than two millennia in the future?

How Long Will It Last, How Well Does It Do?

Many works will tell you that the great Easter table given above is perpetual, which it is, so long as the rules for Easter are not amended; recall Gregory XIII's instruction that the rules "be retained in perpetuity," and our analysis which shows that the master cycle does not even start to repeat itself for 5,700,000 years.

On the other hand, there are reform proposals which have been discussed in recent years among various Christian organizations, such as the World Council of Churches (WCC). The idea is to calculate Easter by some mutually agreeable system which might finally achieve the wish of the early Church fathers (that Easter should be celebrated throughout Christendom on the same day). The proposals have been largely based around either giving up on any connection with the biblical account of the Crucifixion (for example, by making Easter always the second Sunday in April regardless of the lunar phase), or else using a more veracious astronomical basis.

One such proposal was announced by the WCC in a press release dated March 24, 1997. It is suggested that its adherent churches all move to a new *computus* based upon astronomical reality, with effect from 2001. As in the timing of the Gregorian reform, the year chosen is significant: not only is it the start of a new millennium, but also Easter occurs on the same day whether one uses the Gregorian or Julian scheme, or the new astronomical basis. On the Gregorian calendar, Easter is April 15 that year; but in 2000 it is a week later in the Eastern Orthodox churches than in the West, and in 2002 Easter is five weeks later in the East.

This astronomically based suggestion for reforming the *computus* would lead to the possibility of an Easter Sunday on March 21, earlier than has been hitherto allowed (the range of dates being March 22 to April 25). Note that instants of time would be used, rather than whole days, so that it is not feasible for the equinox and the full moon to coincide (occur at the same *instant* as opposed to on the same *day*, which event happens in about one year in thirty when whole days are used). For Easter to be on March 21 under such revised rules one would need (1) March 21 to be a Sunday, and (2) a full moon between the instant of the vernal equinox and the start of March 21, Jerusalem time.

So, should you hold your breath waiting for that special day to arrive? If one takes Jerusalem time to be two hours ahead of Universal time (one has to define a meridian or time zone in order to stipulate the day of the week, and Jerusalem is the suggestion of the WCC), and assumes that Earth's spin continues to slow down at the present rate (affecting the number of leap seconds to be inserted), then those conditions can be checked against the movements of the Sun and the Moon. On such a basis it is found that Easter may be expected to occur on March 21 in A.D. 2877, but not before. This calculation is predicated upon this new astronomically based Easter *computus* being introduced and accepted. Given previous history, this seems unlikely.

Gregorian Hiccups

Even with the present Easter *computus*, comparing the results obtained with what the Church might desire throws up some discrepancies. Earlier we mentioned years when the Gregorian system has led to Easter and Passover coinciding, despite the intention having been to get the Christian feast to always postdate the Jewish festival.

One can also play the game of seeing when the Gregorian rules lead to Easter being a month later than it would be if we adhered to astronomical reality. For example, consider the year 2019. The golden number is 6, so that the epact is 24. January 1 is a Tuesday, so that the Dominical Letter is F. From the Easter table, one finds that Easter Sunday will be on April 21, whereas if the epact had been one less, then Easter would have been a month earlier, on March 24; but the epact is almost always made too large, tending to postpone Easter rather than let it agree with Passover.

In fact, in 2019 the astronomical equinox is at 21:59 Universal time on March 20, and the astronomical full moon will occur at 01:43 U.T. on March 21, so that in reality the full moon comes some hours after the equinox. Because Jerusalem (the appropriate longitude for time definitions associated with Easter) is more than thirty degrees (or two hours) east of Greenwich, these two events will both occur on March 21, Jerusalem time. If the real sun and moon were followed, then Easter would be on March 24; nevertheless, the Gregorian algorithm insists that the date of the new moon is March 20, and Easter is delayed by a month over what the heavens themselves would indicate. A similar hiccup occurred in 1962.

If nothing else, this tells you that the Gregorian reform of the calendar—an ecclesiastical reform, remember—has led to less than perfect results. One could merely say that the cycles of the Sun and, in particular, the Moon are far from simple, making the perfect representation of their movements through simple constructs like the epacts and the Easter table an impossibility. But the question arises: was the reform stipulated in the *Inter Gravissimas* the best possible solution under the circumstances? In the next chapter I show that it was not.

I should not leave you believing, however, that the calendar used by the Roman Catholic Church has been unchanged since 1582. Although the leap-year cycle and the Easter *computus* remain, alterations have nevertheless been made to the internal structure of the liturgical year. In considering Easter one focuses upon the paschal moon, but the Church supplies a lunar calendar for the whole year. Various reforms to the annual round were made by the Second Vatican Council (1962–1965), resulting in the calendar contained in the present Roman Catholic missal.

Chapter 14

❧

THE PERFECT CHRISTIAN CALENDAR AND GOD'S LONGITUDE

Let me tell you about coffee. Coffee is a wonderful thing. At school I read H. G. Wells's novel *Kipps*, on which the movie *Half a Sixpence* was based, and I recall (although likely imperfectly) his line about "an obnoxious liquid which only an Englishman would admit to being coffee." Perhaps this shame over the local standards of coffee preparation is one of the reasons why I felt the urge to leave the land of my birth, although it must be said that I made the mistake of moving to somewhere that the coffee is even worse: the United States. Someone once told me that the coffee is good in any country if the emphasis in the local lingo is on the second syllable rather than the first, and I've found it a useful rule of thumb.

Coffee is an immensely important thing in the lives of many people, and few aficionados would admit to anything less than a high standard of preparation for their favorite brew. But . . . some time ago I read in the newspaper about a survey of coffee drinkers, which led to the surprising result that the majority of such people surveyed in a blind tasting test preferred instant coffee granules over various types of "proper" coffee derived from fresh-ground beans and prepared by the "best" possible techniques.

Why? How could this extraordinary result come about? The answer was manifest to the pollsters when they cross-checked their tasting trials against the responses the interviewees gave when asked what sort of coffee they drank at home. Overwhelmingly, those people who drank instant coffee at home chose the instant coffee in the blind test, even though they would always demand a proper espresso or cappuccino when in a coffee bar, abhorring any suggestion that a spoonful of granules taken from a jar and then plunged into hot water might represent a palatable beverage. The point is that in the blind test their responses correlated excellently with that to which they were habituated, rather than what they would logically argue to be the best given full knowledge of what they were imbibing.

What has this to do with the calendar? What I am going to argue here is that the calendar to which we are habituated is far from the "best" solution. I am not arguing for reform, I am not arguing for more logic, nor for astronomical veracity. My argument in this chapter is simply that in addition to the calendrical reform introduced by Pope Gregory XIII being not the best which could be done astronomically speaking (and I hope you are convinced of that already), it also was not the optimal solution on the basis of potential agreement with what is written in the Bible about the Crucifixion and hence the fixing of Easter, and the apparent desires of the Nicene fathers in this regard. That is, I argue that the Gregorian reform was deliberately designed so as to lead to a second-rate calendar.

The Target Leap-Year Cycle

The Gregorian calendar uses a leap-year cycle such that there are ninety-seven leap-year days added to each set of four hundred years, producing an average length of 365.2425 days. I doubt whether many of my readers alive as I write this book will themselves have experience of a year A.D. divisible by four which is not a leap year (i.e., 1900 or 2100). We are used to having a leap year every fourth year, and the fact that A.D. 2000 is also a leap year means that the familiar quadrennial cycle will not be interrupted in our lifetimes (barring radical calendar reform).

But this does not mean that this system is the best possible, let alone perfect. Actually it's the equivalent of instant coffee: a not entirely satisfactory solution (I hope you appreciate the pun). One can quibble over whether the implemented approximation to the desired mean year length based on astronomical reality is as good as might be achieved—and this is the major focus of the present discussion—but also we have seen that having so long a cycle (four hundred years) leads to the real equinox occurring at times spread over fifty-three hours and thus on three possible dates (March 19, 20, or 21), whereas the ecclesiastical equinox is defined as a solitary date (March 21). The point is that by using a leap-year cycle which is shorter, it would have been possible to reduce the wandering in the time of the equinox, and this has important ramifications as we will see below.

But first let us limit our investigation to the leap-year cycle from the perspective of astronomical truth. The Gregorian calendar reform was fundamentally a religious imposition. One of its intentions was to keep the astronomical equinox on more or less the same date, coming before, or coinciding with, the ecclesiastical equinox. The mean year length which was therefore the target length for the Gregorian calendar was the vernal equinox year (365.2424 days) rather than the tropical year (365.2422 days). Here I will state those lengths only to four decimal places, having described in detail their values in appendix B, and discussed how they are mean durations averaged over many years, but also changing with time.

I hope that the preceding paragraph made sense to you. Because that simple point seems to have been beyond almost every person who has ever written a book or magazine article about the calendar. They all seem to assume that the tropical year is the measure against which the mean Gregorian year of 365.2425 days should be judged. Balderdash. Untrue. Let me give some examples of how that misconception leads people astray.

First, let me mention an example of such misguided thinking. Back in 1923 the Eastern Orthodox churches, having been poorly advised by various astronomers, almost adopted what is called the *Revised Julian calendar*. A consequence would have been that those churches would have A.D. 2800 as a common year, but 2900 a leap year. The mistaken thinking was that by using a nine-hundred-year cycle with 7 leap-year days missed out, their calendrical performance would be closer to astronomical reality than the Gregorian system. Seven days missing in nine hundred years produces a fractional part of $218/900 = 0.242222$ . . . days, which is indeed close to the length of the tropical year. The idea is misguided, though, because the real target is the vernal equinox year with a fraction of near 0.2424 days, and with reference to that the Gregorian $97/400 = 0.2425$ is actually superior.

Second, a question often asked is: isn't there a four-thousand-year rule for the leap-year days being missed? (Short answer: No. The "in perpetuity" quote from the *Inter Gravissimas* in the previous chapter tells you that). The fallacy leading to this erroneous belief is as follows. Take the difference between the tropical year and the mean Gregorian year, and see how much of a discrepancy this accumulates to produce over the Gregorian four-hundred-year cycle: the answer is a shade under three hours. On that basis it would take almost thirty-three hundred years for a total difference of a whole day to add up. This has given rise to the incorrect notion that a leap-year day should be skipped either after 3,200 or after 4,000 years (i.e., in one of the year numbers divisible by four hundred which is scheduled to be a leap year, like A.D. 2000, 2400, 2800, 3200, 3600, 4000, . . .). Let me point a finger of accusation at one of my favorite astronomers as being the inventor of this absurdity: John Herschel, whom we met in chapter 12 as the nineteenth-century originator of the Julian Date scheme. It was in the same book (*Outlines of Astronomy*, 1849) that he made this specious suggestion.

The fallacy here stems directly from missing the point that it is the vernal equinox year which is significant, not the tropical year. If you compare the vernal equinox year with the mean Gregorian year, the discrepancy grows to a little over a single hour in four hundred years, and it would be almost eight thousand years before a full day accumulated, if (I repeat *if*) the vernal equinox year stayed the same length. In reality the vernal equinox year is changing (but at the slowest rate of the other pertinent years: those measured between each of the solstices, between autumnal equinoxes, and the average of the four which happens to be equal to the tropical year), and the Moon is receding with a concomitant slowing of the terrestrial spin, so that the concept of planning for a leap-year day being missed in perhaps the year A.D. 8000 or 10,000 is absurd. *Trying to design calendars for use on timescales of more than a millennium makes no sense astronomically, logically, or sociologically.*

Now another example of people's misconceptions. By mistakenly making use of the tropical year as the target year length, one would arrive at an "error" in the Gregorian calendar of three days over ten thousand years. I have already pointed out that extrapolating such calculations over such periods makes no sense, but this is the sort of thing you will find authors writing, even in quite erudite tomes penned by respected scholars. They may then go on to point out that over a millennium ago the ancient Mayans of Central America were avid watchers of the skies who produced a calendar with an average year length which leads to an "error" of only *two* days over ten thousand years. On that basis it is claimed that the Mayans knew better than the great Renaissance astronomers such as Nicolaus Copernicus, Tycho Brahe, and Johannes Kepler, resulting in a calendar more precise than our own. Balderdash again. With regard to its desired result, keeping the equinox steady, the Gregorian calendar is better than that of the Mayans, although this is a case of comparing apples with oranges.

Now let's get back to reality. The target year duration of the Gregorian reform needed to be about 365.2424 years, but it does not necessarily follow that the designers of the reform knew that. The figure I quote here is a modern value based upon high-precision measurements. Most of the astronomical learning in sixteenth-century Europe was in the Protestant nations, which is hardly surprising if the Catholic Church persecuted as heretics anyone who dared to contradict the Ptolemaic system, which had Earth unmoving and the Sun, the Moon, the planets, and the stars supposedly circuiting us. The prime source which Aloysius Lilius seems to have had available in order to deduce the length of the year was the Alfonsine tables, a set of planetary positions which had been compiled for King Alfonso X of Castile in the middle of the thirteenth century. Castile, part of Spain, was of course a good Catholic kingdom.

Because Lilius's own notes have not survived, or at least have not yet been found, or maybe not been made available, one has to go to Christopher Clavius for the year length deduced. Clavius gives a value corresponding to a shade over 365.2425 days, but there are reasons to believe that he was being less than open in his published account. Certainly he was attacked in writing by many other knowledgeable men, on the grounds that various celestial calculations he presented were known to be wrong. The twentieth-century U.S. author Henry L. Mencken once opined that "there is nothing more incomprehensible than a wrangle among astronomers." This is a fairly good example from some centuries back, with vehement arguments on each side powered by both scientific and religious considerations, greatly complicating matters. But there is more than a hint that Clavius was trying to keep some matters secret, and was not being wholly forthcoming in his reasons for promoting the Lilian system.

The balance of evidence (if one has a suspicious nature) says that Clavius actually knew that the desired mean year length should be slightly less than that given above, but he stated a value such that one day every 134 years would be the accumulated error of the Julian calendar. On that basis he could argue that three days needed to be dropped from each 402 years, and that a round cycle of three skipped leap-year days in 400 years would be near enough, and that was what the Catholic Church determined to use.

The thing to note is that there seems to have been a subterfuge being acted out, a covering up of superior knowledge. In reality Clavius and others knew that the desired year length to use was closer to 365.2424 days. The difference is critical. Let us see why.

Rational Number Approximations

A *rational number* is one which may be expressed as the ratio of two integers. Since year numbers are integers, any leap-year cycle must be a rational number. Thus the Gregorian cycle produces a mean year length of 365 plus 97/400 days, that fraction being a rational number. Expressing the length in decimals, as 365.2425 days, one has an *irrational* number, which may nevertheless be written in a rational form.

Real astronomical year lengths in units of days are irrational numbers, and the challenge in designing a leap-year cycle is to find a good rational number approximation to employ. The closeness of the approximation is one consideration, but the brevity of the cycle is also a criterion. In the rational number above, 97 is the *numerator* while 400 is the *denominator*. A short cycle is produced if the value of the denominator is low.

The first thing to decide upon is the target. Here our target is the vernal equinox year of 365.2424 days, which we know with some precision and which the Catholic Church (and others) also likely knew in the sixteenth century. The 97/400 cycle employed in the Gregorian reform is good, with an accumulated error of only a little over an hour in four hundred years, but the effect of such a large denominator is that the equinox time oscillates, so that a shorter cycle might be preferable. Is there one?

Let us do some simple sums. The Julian cycle results in an additional fraction of ¼, or 0.25, days, and we need a value slightly less than that, and as close to 0.2424 as possible. You can quickly break out your pocket calculators and verify the following:

$$2/9 \quad = 0.222222$$
$$3/13 \quad = 0.230769$$
$$4/17 \quad = 0.235294$$
$$5/21 \quad = 0.238095$$
$$6/25 \quad = 0.240000$$
$$7/29 \quad = 0.241379$$
$$8/33 \quad = 0.242424$$
$$9/37 \quad = 0.243243$$
$$10/41 = 0.243902$$

As both numerator and denominator increase further, the result converges upon 0.25, which is too high. The value obtained from 9/37 (nine leap years spread over a thirty-seven-year cycle) is also too high. There is no doubt which cycle would give the best match whether the target were 0.2424 (the vernal equinox year) or 0.2422 (the false target of the tropical year): it's a cycle of eight leap years spread over a thirty-three-year cycle. It performs *much* better than the 97/400 cycle we actually use.

Note that one could make a case for other cycles. One which has been much advocated by those enthusiastic about reforming the calendar is 31/128. This has some obvious appealing qualities, and was suggested in 1864 by Professor Medler (an appropriate name) of the Dorpat Observatory in Tartu, Estonia, as an alternative to Herschel's recent idea of dropping one extra leap day from every 4,000 years. We have previously noted a slippage rate of around one day per 130 years, and the denominator here is close to that. In fact, it is a nice binary number, being thirty-two times the quadrennial cycle, thirty-two itself being divisible by four, and so leading to regularity if one is habituated to counting in twos, fours, and eights. All that is needed is a deletion of one leap-year day every 128 years rather than three every four centuries. Once in every couple of lifetimes. No big deal.

There are three points which weigh against it. The first, which should always be paramount in one's mind, is that Easter will be affected: imagine having to alter the epacts in eight century years spread over 2,500 years for the lunar equation, as we do now, but additionally every 128 years for the solar equation rather than in three century years out of four. The second is that $31/128 = 0.2421875$, and its supporters are most often those who believe that the tropical year of 365.2422 days is the target, although the changing lengths of the variously defined years over millennia make the distinction largely insignificant. The third objection is that the denominator is large, and we see below that a small denominator is much favored if one wants to keep the equinox on close to a single date. On that basis, the thirty-three-year cycle wins out again.

When you first taste it, this "Brand 33" type of coffee may not impress you, but don't spit it out yet. Let me convince you that the thirty-three-year cycle has much to recommend it.

The Wonderful Thirty-three-Year Cycle

Even in the sixteenth century, the idea of a thirty-three-year cycle containing eight leap years was not new. In the previous century an English chaplain named Richard Monk, about whom little is known, suggested a thirty-three-year calendar. His manuscript is preserved in the Bodleian Library in Oxford.

But other traditions outside of Christianity have known about the possibilities of a thirty-three-year cycle for much longer. The great eleventh-century Persian writer Omar Khayyam is best remembered for his poetry, but he was certainly multitalented. He had great attainments in mathematics and astronomy, although he largely made his living from tent making. In about A.D. 1074 he was one of eight astronomers asked by Sultan Jalaal-ol-Din Malik-Shah Saljuqi to consider a suitable reform for the Persian calendar. The system they presented to the sultan five years later, using the 8/33-year system, is called the Jalali calendar. Although this was employed for some time thereafter, it fell into disuse; but it was reintroduced as the official calendar of Iran by Reza Shah Pahlavi in 1925, and it is also used nowadays in several surrounding regions, including the central Asian republics and some Kurdish areas. On this calendar the year begins right after the vernal equinox, and in eight years out of thirty-three an intercalary day is introduced to coincide with the equinox (March 20, Gregorian).

I cannot let mention of the calendar of Iran pass without pointing out the following. Many Persian scholars of the present day have been misled by putting too much faith in the ability of the astronomers of the United States and the United Kingdom. In appendix B it is shown how the definition of the tropical year given in the monumental *Explanatory Supplement to the Astronomical Almanac* (1992), a joint publication of the governments of the United States and the United Kingdom, is incorrect: it confuses the tropical and vernal equinox years.

Iranian scholars have assumed, using that source, that the actual mean year length to be employed in order to get the equinox to persist on the same date is 365.2422 days (the tropical year), rather than the slightly longer 365.2424 days (the vernal equinox year). The latter (correct) version would allow the Iranian thirty-three-year calendar to continue without alteration for centuries, even millennia.

But by erroneously believing that the *Explanatory Supplement* is correct, the Iranians have been misled into thinking that a 128-year cycle is required: three complete 33-year cycles followed by a 29-year cycle with 7 leap years, thus producing the $31/128 = 0.2421875$ fraction which is close to the current value of the tropical year. Wrong. Their target should be 0.2424, and their current 8/33 cycle hits that admirably.

Unless they realize soon that the U.S. and British astronomers gave a mistaken definition, the Iranians will adjust their calendar by skipping a leap year, in the belief that things will thus be improved in terms of obeying the cycles of the sky, whereas in fact they will be making things worse. Iran should trust Omar Khayyam, and not Her Majesty's Astronomer Royal, or the U.S. Naval Observatory.

Elsewhere, other cultures have also used the 8/33-year cycle. Rhetorically I ask, isn't it difficult to work out when leap years are due with such a system? There are two arguments against such a notion. One is that most people use printed tables or almanacs anyway, not worrying too much about leap years until the extra day dawns, and in any case computers are taking over almost all time calculations. That latter argument did not apply four centuries ago, when the calendar was being reformed, but the first part certainly did: people were especially dependent upon reproduced tables back then, hence the epacts and the Easter table.

The second argument comes back to the coffee consideration, and the role of familiarity. We are habituated to the fact that the year A.D. is a leap year if it is divisible by four, and I've previously noted that this is purely by chance (unless Dionysius Exiguus fiddled with things to make it that way). If there were a thirty-three-year cycle with eight leap years to be interspersed over it, how would you know which were common years and which leap? Dividing by thirty-three in your head and looking for the remainder seems pretty complicated to most people, and that is used as an argument against such a scheme.

Actually, the reality would be that there would simply be four-year gaps between leap years as at present, and once or twice in a lifetime (or maybe three times for the long-lived) there would be a gap of five years. No big problem.

On the other hand, one might like to be able to calculate which years are leap years for the distant past or future. Given that the thirty-three-year calendar was *not* adopted by the Christian churches, one could make up any phasing against the *anno Domini* framework that one liked, as I'll do below. But one should note that calendars such as the Iranian which *do* employ the thirty-three-year scheme *have* set their phasing. From A.D. 1964 to 1992 the leap years on the Gregorian calendar also contained the intercalary days for the Iranian leap years, but then a hiatus from 1992 until 1997 was due for the Iranian calendar (which uses a different numbering system and the aforementioned equinox-to-equinox year), such that it will be A.D. 2096 before the A.D.-year-divisible-by-four rule will again hold for that calendar, whereupon the Gregorian calendar will immediately upset the agreement because 2100 is a common year.

Now, what phasing system shall I use in my thirty-three-year-rule calendar? The obvious answer actually stems from the traditional life of Jesus, with his death in A.D. 33 when thirty-three years old. In fact, this was a major consideration in the sixteenth century, a vote in favor of the thirty-three-year rule, with the layman of that time more likely to be impressed by such a divinely inspired cycle and undaunted by arithmetical objections: to such a person, the calendar was a holy instrument, for which the Church held responsibility.

This concept of using the traditional lifetime of Jesus Christ has been much discussed in religious literature, and it even has a separate name: the *anni Domini* system. Is that not the same, you ask? Not if you have even the rudiments of Latin. I was always bottom of the class, but I do recognize *annus* or *anno* to be singular forms, while *anni* is the plural. Thus *anni Domini* means "years of the Lord"; the thirty-three years, to be exact. Stepping back to the 1570s, it seems that this argument was contemplated within the upper echelons of the Church, and suppressed on the grounds that although it would have great appeal, the Catholic Church hierarchy was determined to use another scheme. In an election campaign, you keep secret any information which might give your opponents an advantage.

The traditional life of Christ, then, sets the phasing to use. The years A.D. 1 to 33 inclusive is the first cycle, A.D. 34 to 66 the second, and so on. The rule is: take the year A.D., subtract blocks of 33 until the answer is equal to or below that limit, and if the answer is divisible by four, it's a leap year. Thus A.D. 4, 8, 12, 16, 20, 24, 28, 32, 37, 41, 45, 49, . . . are leap years. The first extended gap is taken up by A.D. 33 to 36, the latter being a common year in this scheme.

Too complicated for large numbers like 2010? Here is an alternative system. Take the last two digits and add them to the number of centuries. Thus, 2010 produces $20 + 10 = 30$. Often a three-digit number results, so repeat the process: for example, 1999 renders $19 + 99 = 118$, from which $1 + 18 = 19$. If your result is larger than 33, subtract that number and repeat until the answer is 33 or less. For example, 2015 creates $20 + 15 = 35$, so subtract 33 to get 2 as the final answer; 2052 creates 72, so subtract 33 twice to get 6 as the value you need. All such results are between 1 and 33 inclusive, and if that result is divisible by four then it's a leap year. Piece of cake.

The Epacts for a Thirty-three-Year Cycle

Let me assume that, as a reasonable person, you've been persuaded that the thirty-three-year cycle has a lot going for it. However, after the complicated description of the calculation of Easter in the previous chapter you may well now be wondering what alternative system to the golden number–epact combination could be used when a thirty-three-year cycle is employed. To look at it another way, so far I've only made the case for the thirty-three-year system on the basis of the Sun. What about the Moon?

A quick refresher. The Gregorian calendar uses three basic cycles: 4 years, 100 years, and 400 years. All the epact changes occur in century years: three times in every 400 for the solar equation, and eight times in every 2,500 for the lunar equation. This, rather confusingly, can lead to the epacts oscillating. Looking at the Metonic cycle, the difference between 19 mean Julian years and 235 mean lunations leads to a full day's difference accumulating over 308 years, which the Gregorian calendar approximates with a $2,500/8 = 312.5$ year mean correction time.

Now get out your calculator again. The Metonic cycle one would use in the thirty-three-year system would contain on average 19 years, each having a mean length 365 days plus a fraction ($8/33 = 0.242424 \ldots$). That makes 6,939.6060 days. The 235 mean lunations last for 6,939.6884 days. The difference is 0.0824 days (over 19 years), which accumulates to a full day in 230.6 years. Call it 231.

That may seem less accurate than the 308 years for the Gregorian calendar, but wait. In the 33-year system the solar equation is self-correcting, so that the epacts do not need to be shifted on *that* account. Only the lunar equation is left, requiring a shift of one in the epacts every 231 years. Does that number look special? It should, because $33 \times 7 = 231$. Seven is the number of days in the week, and 33 is the cycle of years. Rather than a hugely complicated master table of epacts with 30 lines corresponding to the years A.D. 1600–8600, mostly dropping down a line in some century years but not others, and sometimes retreating upward, the wonderful 33-year cycle leads to a simple rule with a jump of one in the epacts every 231 years. The complete rule can be expressed in one sentence. For example: whenever a Sunday, February 29, follows five years after a Monday, February 29, increase the epacts. That's it.

Could the Thirty-three-Year Cycle Be Implemented Now?

If you have been convinced by the above that the thirty-three-year cycle for leap years is superior to the Gregorian system which we actually employ, and in addition is a better solution for shifting the epacts, you might ask whether it would be possible to implement it now. The answer is *yes*.

Step back to the year 1980 (coincidentally, the start year for the calendar algorithms on many personal computers). Divide that year number by 33; the answer is precisely 60. Because that result is divisible by 4, it follows that a complete number of (4 × 33 = 132-year) grand cycles finished in that year, the implication being that every fourth year since on the Gregorian calendar—the leap years—would also be a leap year in the *anni Domini* (8/33-year cycle) system. This situation continues until February 28, A.D. 2016. In the Gregorian calendar the next day would be February 29. In the reformed 33-year system, A.D. 2016 would be a common year, the intercalary day (the leap year) being postponed until 2017.

Is this a possibility? One can list several counterarguments. One is that the Christian fundamentalism represented by having a calendar based on the traditional life of Jesus Christ would be resisted in many quarters. Another, of course, is the expense involved in altering all our computer software to accommodate *any* calendar change. On the other hand, with the possibility in the air of some change in the Easter *computus* so as to bring all the Christian churches together at last (a matter about which I am frankly cynical), and the present World Council of Churches proposal being seemingly too soon (a change from 2001), scheduling a change in Easter from 2017 seems a better bet. In 2016, Easter is five weeks later in the Eastern churches than in the Western; but in 2017 their Easters occur on the same day, it happens, as is the case in 2001.

Holding the Equinox on One Day

I wrote above that the 33-year leap cycle is wonderful, and we've met some of the ways in which that is true, including in particular the fact that it renders the closest approximation to the vernal equinox year. But the most significant feature in its favor has yet to be described: it has the capacity to hold the equinox on one calendar date. Ecclesiastically speaking, one would have imagined that this was a major feature, given the complications of the Easter *computus* which result from having the equinox wander by 53 hours under the Gregorian 97/400 system, and the concomitant failure of the *computus* to avoid Passover, and so on. For comparison, the 31/128 system mooted previously leads to a wander of 39 hours. The desideratum is a wander of below 24 hours, and the 33-year system can provide this, as we'll see.

Why the Catholic Church decided not to employ this system is therefore a bit of a mystery, although I make some suggestions below concerning that puzzle. First we need to understand how the 33-year rule would make an equinox perennially on one date a possibility.

The vernal equinox year is about 365.2424 days long. The excess over 365 is equivalent to about 5 hours and 49 minutes (plus a few seconds). In a quadrennial leap-year cycle, imagine that the equinox occurs at 03:00 on a certain day. The next year it is at 08:49, and the following year at 14:38. The 4th year it *would* be at 20:27, but the insertion of an intercalary day a few weeks earlier means that in this leap year the equinox occurs at that time, but 1 day earlier on the calendar. In the 5th year in the sequence the equinox occurs at 02:16, which is 44 minutes earlier than 4 years before. Over 400 years these sets of 11 minutes accumulate to just over 73 hours, hence the 3 days dropped and a residual error of a little more than an hour.

Now imagine instead a 33-year leap cycle. As shorthand, denote the time difference of 5 hours and 49 minutes as X. Within the sets of four years the equinox time moves later by $3X$ and then in the 4th year (1) drops back by 1 day/24 hours, and (2) moves later by X. After one quadrennial cycle the equinox is coming 44 minutes earlier, as above. There are seven such cycles, so one expects a shift of around 308 minutes in all; it is actually closer to 305 minutes if one accommodates all the seconds in the sums.

But next is a 5-year sequence. The equinox time moves later by $4X$ and then in the 5th year—the leap year—drops back by 24 hours less X. After that quinquennial cycle the equinox would be coming 55 minutes earlier than it was five years previously, except that there has been an extra period of near six hours (or 360 minutes) inserted, and overall the 305-minute gap is compensated.

The essential point is this. In the whole sequence the greatest distance the equinox moves from some reference time is $4X$. Because X is less than six hours, that movement is less than 24 hours. For example, at the start of a five-year segment let the equinox occur at 22 minutes past midnight; two years later the equinox is at midday; two years later again it is at 22 minutes before midnight on the same day; in the following year it has dropped back to 05:27. From there it slips back by a bit less than 11 minutes a year over the next 28 years until again it is at 22 minutes after midnight when the next quinquennial sequence is due. *The equinox is always on the same day.*

The problem reduces to this question: How does one get the equinox to occur, on the first year of the five-year sequence, soon after midnight? The only answer to that is to pick your longitude. Previously we have talked about local time according to the Sun, and often we have worked in Universal time, which may be simply thought of as the time according to the mean sun on the Greenwich meridian, but in other contexts alternative time systems are appropriate. In the previous chapter we mentioned the fact that it might be thought appropriate to use the meridian of Jerusalem in making calculations of the date of Easter, that being the location of the Crucifixion. In chapter 18 we will see that there have been many claimants for the prime meridian for world mapmaking and timekeeping, and until the last century the reference most often used was the far western part of the Canary Islands. If some cleric in England in the year A.D. 700 were using one of the Easter tables of Dionysius or Felix, he would perhaps have needed to take into account that those would have been calculated for the longitude of Italy, or maybe for Judaea. Nowadays it is necessary for there to be separate Islamic calendars for the Middle East and North America (say), because that religious calendar requires an actual *sighting* of the new moon, and that depends on where you are, although most Muslims tend to accept the situation as seen from Mecca as being the ultimate authority.

What, then, is the longitude which would maintain the equinox on one day? Due to such things as the equation of time causing some wander (see appendix A), and the fact that although the vernal equinox year when figuratively averaged over many years is reasonably constant, in actuality it varies by some minutes from one equinox passage to the next, there is only a quite narrow band of longitude within which the local time is such that the equinox would remain within one calendar day on a thirty-three-year cycle. At most it's a couple of degrees wide. And it lies about seventy-seven degrees west of the Greenwich meridian.

If you get out your atlas you will find that this line slices along the Atlantic coast of North America, cutting the far east of Virginia and North Carolina, and is then in open ocean before reaching the Bahamas and Cuba, Jamaica, then Colombia, Ecuador, and Peru, before passing into the Pacific. In the 1570s the northern parts of that meridian were unsettled, but the southern were in the Spanish domain and thus under Catholic control. With a thirty-three-year cycle the equinox could have been kept on a single day if the time on that line of longitude were adopted as the standard. It may be thought of as being *God's longitude*.

Why Was the Thirty-three-Year Rule Not Adopted?

When the committee appointed by Gregory XIII to consider the relevant questions was deliberating, they did so with fairly good knowledge of what had gone before, both in East and West, within Christendom and without. Thus they were not in ignorance of the potential of the thirty-three-year rule. Indeed, they received a presentation from N'ammat allah, who was a delegate from the Syrian Christians, concerning the excellent astronomical observations and analysis from Persia and other Arabic states, these clearly demonstrating that the vernal equinox year was quite consistent when compared to the other marker points. Omar Khayyam had not been forgotten.

In fact, N'ammat allah strongly advocated the thirty-three-year system, but was overruled. The problem, at least in part, seems to have been that if one accepted that the vernal equinox year was of a near-constant length, then it would follow that the Council of Nicaea, over thirteen hundred years before, had erred in its allocation of the equinox to March 21. To the Church, the open admission of fallibility on the part of the Nicene fathers would have been as bad as calling the pope a liar. What Clavius seems to have done is to have fudged things, claiming that the vernal equinox year had attained a value slightly longer than in reality, thus destroying the appeal of the otherwise clearly superior thirty-three-year system, but without explicitly saying so.

But the reasons go deeper than that. We must keep in mind that this was a time of great dispute between the churches. In England and other Protestant nations, the pope was viewed as being the Antichrist, and no communication with the Roman Church was allowed. The pope held sway over most Catholics, but in some areas that control was not firm, with further defaults to the Protestant side being distinctly possible. All religious matters were potential grounds for point scoring and the swaying of others to "the one true Church" (whichever that might be), the matter of the calendar being a major weapon. When the Council of Trent met, it was for Counter-Reformation purposes, and that is what made calendar reform a prominent topic for discussion and immediate action, at least on the timescale with which the Catholic Church moves. And when the actual reform was to be considered, an important point to contemplate was: what alternative could the Protestants come up with?

In some ways one can think of this as being a chess game. You might see that your next move could involve taking your opponent's queen, and then a rook, and then you have check; but thinking through the moves with which your foe can respond you recognize that your short-term gains will be countered by the loss of your queen, a rook, a knight, and a bishop: checkmate. And so you must content yourself instead with merely grabbing a pawn. A gain is a gain, even though a bigger prize is lost: the aim is to win the war, not just one battle.

The Catholics had to convince themselves and their fellow travelers that the second-best solution was the one to adopt, rather than the 8/33 proposal from the Syrian N'ammat allah, which was superior on grounds of both astronomy and religion: it was the closest approximation to the cycles of the Sun, and gave the opportunity to keep the equinox on one day as the early Church fathers had wanted. The 97/400 solution which was adopted made it impossible for the equinox to remain within a twenty-four-hour period, so that the question could be ignored, and any problem associated with setting up a specific geographical longitude to provide a standard time zone was obviated, so long as your rival did not know of this possibility.

For this reason the information from the East seems to have been suppressed, no mention of the thirty-three-year concept appearing in Clavius's very extensive published notes. In the many arguments Clavius had with other mathematicians and astronomers who accused him (with just cause) of recommending an inexact calendar, he needed to gloss over—even lie about—alternative solutions. Clavius knew that the 33-year cycle was better on various grounds, but if the Catholics had used that, the Protestants would undoubtedly have resisted it, and likely brought in their own immediate reform using the 97/400 system, which was well known (recall that it had been invented by Petrus Pitatus a couple of decades before) and would have had great popular appeal due to it being such a slight alteration from the Julian scheme. So the Catholics went with the popular appeal rather than the superior system. What they wanted to avoid was any counterreformation of the calendar by the Protestants.

Clavius was presenting a fait accompli, the Protestants being left, in the Catholic leaders' minds, with two alternatives: either stick with the Julian calendar, a pagan invention which was obviously wrong, or fall in with the Catholic calendar revision. Both would clearly be unpalatable to the Protestants (although that has not stopped the Eastern Orthodox churches from championing the Julian system ever since). But had the Catholic party underestimated their opponents?

What Did the English Know?

As I wrote above, the introduction of the reformed calendar allowed the Catholic Church to claim superiority over the Protestants, especially since the choice made was for the second-best scheme: by keeping the potential thirty-three-year system up its sleeve it had a fallback position, and the advantage was maintained so long as its rival was in ignorance.

The situation is similar to modern-day conflict and warfare, whereby new advanced-technology weapons and intelligence about your opponent's capabilities give you a great boost. I am a physicist, fundamentally. Most physicists in the world are employed one way or another in developing either new weapons technologies or surveillance systems, and spies are just intelligence gatherers feeding back information about what your opponents can and cannot do. But it all falls apart if your enemy knows something which you assume he doesn't, and that seems to have been the situation here.

If you think that things were so much different four centuries ago, let me give you an example which involves one of the men who is centrally involved in the present story; as yet we will leave him nameless. If you pick up any elementary astronomy book, or even a specialist volume describing the history of the telescope, you will find the claim that the first use of such an instrument was by Galileo Galilei around 1610, although a Dutch spectacle maker named Hans Lippershey may have arranged lenses so as to construct a telescope just a few years earlier. But the following words were written some decades before, by an Englishman in 1570. This man wrote about an optical system which makes

> . . . *thynges, farre of, to seeme nere: and nere, to seeme farre of. Small thynges, to seeme great: and great, to seeme small. One man, to seeme an Army. Or a man to be curstly affrayed of his owne shadow.*

That sounds much like a telescope to me. The problem with regard to the free dissemination of this information was that it had obvious military usefulness. The same writer had the following to say (I will not trouble you further here with the archaic spelling he actually employed): "The captain may wonderfully help himself (in estimating the forces of the enemy) by the use of perspective glasses." If the British Admiralty had this facility to survey its opponents available, then obviously it would not make that information (and the secret of how to duplicate the ability) accessible to the many enemies it faced. For that reason, even our present-day histories of optics seem to be far from the truth. Galileo may have been the first to have turned a telescope towards the heavens and looked in wonder at the craters and mountains of the Moon, and the four large satellites of Jupiter which collectively carry his name, and then been able to speak about his experience, but others may have gone before him and just kept their mouths shut for reasons of security.

The Englishman quoted above was John Dee (1527–1608), onetime of Cambridge University, and a remarkable man he was. Although he was one of the most celebrated and controversial personages of the Elizabethan Age, with the private ear of the queen, to a large extent he has been forgotten by history. He was a scientist and mathematician, and yet he cast horoscopes; he was an antiquarian and a philosopher, and yet he dabbled in magic, sorcery, and conjuring; he was an early devotee of the Copernican system of the heavens, with Earth orbiting the Sun, and yet he believed he could converse with angels.

Most of all, in the present context, he was an expert on the movements of the Sun and the Moon, and their calendrical consequences.

In England, at the same time as the Catholic Church was considering its calendrical reform, there was a group informally known as the School of the Night. This included such characters as Dee, Sir Walter Raleigh, Sir Francis Drake, Thomas Harriot, William Cecil (the First Baron Burghley), and others. This was the party which was involved with considerations of the development of a peculiarly English calendar reform, with Dee in particular taking a leading role.

Dee knew, for example, that the length of the year depended upon the point it was measured from, so that the tropical year (in our parlance) was not good enough; rather, the fundamental turning point of the vernal equinox was necessary to provide religious one-upmanship. This was vital: if the English could design a calendar which was demonstrably superior to that of the Roman Catholics, it could then be used to enhance not only their relationship with the other European Protestants, but perhaps even to convert some wavering Catholics to their cause.

But Dee needed to work in secrecy. The commissioners in Rome—Clavius and his colleagues—will have known that action was afoot in London, but the details of Dee's plan could not be revealed. Certain aspects were common knowledge, but the Romanists did not know how much the English knew. This was one reason that the Gregorian reform was rushed through, in addition to those matters mentioned in the previous chapter: the gauntlet was being thrown down by the Catholics in the belief that the Protestants were unable to pick it up.

The English Alternative

John Dee had great attainments in many fields, and in that he was greatly assisted by his vast personal library, which was recognized to be the best in Britain. He had many books pertaining to the calendar question, and in particular he had access to the work of the aforementioned Richard Monk, who had shown in the previous century that a thirty-three-year leap cycle had much to recommend it. At precisely the time that the pope was chivying the remaining Catholic nations into making the necessary date adjustments, Dee was finishing a treatise on how to correct the calendar, which he delivered to the Privy Council (a special committee of advisers to the sovereign in Britain) in February of 1583, a year after the papal bull appeared. Back in 1582 he had already had many discussions with the relevant people, and had written much of his report; note that for clarity the year numbers I am using here are based on a January 1 New Year rather than the one actually in use in England.

As we shall see, there were still some things Dee could not put in writing, but his alternative calendar is clear enough. His work was never printed, although it is preserved in the Bodleian Library; but at the time Dee's suggested reform created quite a sensation among those who knew of it. The title is quite a mouthful: *A playne Discourse and humble Advise for our Gratious Queen*

Elizabeth, her most Excellent Majestie to peruse and consider, as concerning the needful Reformation of the Vulgar Kalendar for the civile yeres and daies accompting, or verifyeng, according to the tyme trewly spent.

Dee's treatise starts with a verse:

> *I shew the thynge and reason why;*
> *At large, in breif, in middle wise,*
> *I humbly give a playne advise;*
> *For want of tyme, and tyme untrew*
> *Yf I have myst, commaund anew*
> *Your honour may. So shall you see*
> *That love of truth doth govern me.*

Why the controversy over Dee's report? The answer seems to be because at first sight his suggestion appears to be a conformation with the Gregorian calendar, which in fiercely Protestant England just would not do. Indeed, some historians today mistakenly have stated on this basis that England very nearly capitulated and decided to implement the Gregorian calendar immediately, but this is simply not the case. In reality, the English were rather fervently working on an alternative, but they needed to shroud it in secrecy. In that regard, Dee's plan was a masterstroke.

Here's the story. Dee's real calendar plan, which he could not reveal in toto, was for a shift of eleven days which would put the equinox onto March 21, the date the Nicene fathers had mistakenly assumed to be correct astronomically. But there's more to the eleven-day shift than that. Dee looked at the length of the year as measured from winter solstice to winter solstice (the traditional date of Christ's birth), showed that this is different from the year measured from equinox to equinox, and that the eleven-day shift would also put the solstice back to the "correct" time—the point in the year at which it was occurring at the start of the era. Thus his proposed shift would not only correct for the time of the equinox in A.D. 325, linked to the Nicene fathers' decision about Easter, but also shift the time of the solstice to that when Jesus was born. Any Catholic reasoning that the movement of the solstice would be the same as that in the equinox was wrong. That makes the score 2-0 in favor of Dee over Clavius and colleagues, enough perhaps to make other nations decide to back the English system.

John Dee's Secret Cycle

But the prospective clinching advantage Dee could not reveal. *He wanted to bring in the thirty-three-year cycle for the leap year.* We have already seen that this is an excellent cycle to use if one is concerned primarily with maintaining the equinox within a period of twenty-four hours, and Dee knew this very well. But while the Spanish controlled the longitude for which the local time allowed the equinox to remain on a single date, Dee could not reveal his intention. How, then, could he possibly bring in a reformed calendar without letting on which leap-year cycle was to be used? Read on.

When Dee was drafting his treatise in 1582 he wrote on the basis of an eleven-day shift, but by the time that he presented it to the Privy Council it was 1583, and he added a note saying that a ten-day shift was now appropriate. It is this which has fooled people into thinking that he was suggesting capitulation to the Catholic foe. Earlier we saw that the year 1980 was the last in one of the cycles of thirty-three years which would apply in an *anni Domini* system, making 1981 to 2013 the next set. Not only that, but 1980/33 = 60, which is divisible by four, so that the leap years from 1984 to 2012 inclusive are the same on both the Gregorian calendar and the thirty-three-year system. Now get out your calculator again and consider the late sixteenth century. Dividing 1584 by 33 one gets 48, so that 1584 was the last year in a 33-year cycle. The thirty-second year is a leap year, so that in a thirty-three-year system 1583 would nominally be a leap year, and 1584 would not. Under that circumstance, a ten-day correction in 1583 with 1584 being made a common year would produce a shift of eleven days against the Julian calendar. Dee suggested adding a few days to each of the middle months of the year in 1583, rather than a single ten-day block as was the Gregorian correction.

But if the English had done this—that is, if they had declared 1584 a common year—then they would have effectively revealed their hand to the Catholics. Instead they bided their time, Queen Elizabeth telling Dee that "what is deferred shall not be aborted" (it sounds better in Latin: *quod defertur non aufertur*).

The deferment was essential, because a great opportunity presented itself. Due to the fact that 1584/33 = 48, and 48 is divisible by 4, as in our present epoch (1984–2012), it followed that the years 1588, 1592, 1596, 1600, 1604, 1608, 1612, and 1616 would be leap years on both the Gregorian and the 33-year systems. The English could in principle introduce a calendar in 1585 or soon thereafter without defining their leap-year rule, thus giving themselves three decades for furtive attempts to found an English colony (New Albion) on the seventy-seventh meridian, God's longitude.

This seems to have been the idea. But were they successful? Certainly Dee's calendar reform, an effective eleven-day shift with an unrevealed thirty-three-year cycle, was deferred pending progress being made on the other parts of the plan. John Dee left England in September 1583 for the Continent, and many history books make it sound like he did so in shame, with criticism from his countrymen ringing in his ears for daring to suggest that England should take up the papal reform of the calendar. In reality it seems that he was engaged in a covert operation, trying to persuade the Protestant states and others that the mooted English or Elizabethan calendar was superior to the Gregorian not only in terms of mathematics and astronomy, but most importantly religion. In his absence Raleigh and his colleagues in Parliament continued to lobby for calendar reform, but without specifying anything other than the shift proposed against the Julian calendar. The Catholics could be kept in the dark concerning the leap-year cycle which they intended to use.

The European Campaign

John Dee spent the following six years on the Continent, but things did not go smoothly. He had hoped that his alternative calendar plan would be recognized as being superior by the Protestant princes, striking a solid blow against the papists, and in particular he needed to convince the (formally Catholic) Emperor Rudolph II of this. This Habsburg, the leader of the Holy Roman Empire, was certainly not pleased that the news of the reformed calendar had been issued from Rome rather than the proper imperial headquarters in Prague, and thus Dee was hopeful that Rudolph would be amenable to being persuaded of the superiority of the English proposal.

Just how significant calendrical matters were in these days can be seen from a quick perusal of a history of astronomy in the period. Such histories written today often make it appear that the establishment was heavily against the heretical Copernican view of the universe, as exemplified by Galileo's forced recantation. But there's more to it than that: to the Catholic Church, Galileo had dangerous knowledge. It was *because* that knowledge was correct that it had to be suppressed. Histories make it appear that, for example, the great observer Tycho Brahe was heavily funded on the island of Hven (in the strait between Sweden and Denmark) by various parties such as the Danish king simply on a whim, or for astrological purposes, but in fact the information Tycho was collecting was essential if the cycles of the heavens were to be precisely defined and thus the best possible calendar designed. It was no accident that Tycho later found a haven in Prague, and that Johannes Kepler was then supported in using the former's observations to deduce the laws of planetary motion and draw up the Rudolphine tables, named for his patron.

But back to the 1580s and Dee's European campaign. In the event it seems that agents of the Vatican in Prague intervened, making it impossible for Dee to reveal his plan to Rudolph at that stage: England had yet to break Spain's monopoly on the seventy-seventh meridian, a matter which Dee had left in Raleigh's hands, and until New Albion was founded on the crucial longitude the English had to keep their devious conspiracy a secret.

Roanoke Island, the Reference Point

While Dee was absent on his mission in Europe, his collaborators in England were moving swiftly on their part of the subterfuge: the establishment of an English colony in the unclaimed North American sector of the seventy-seventh meridian.

The great advocate of the British transatlantic colonization was Richard Hakluyt (1552–1616). Nowadays there is a Hakluyt Society based in London, recognizing this great geographer who made major contributions by editing and publishing the accounts of various explorers. As the saying goes, the job is not done until the paperwork is finished, and Hakluyt ensured that the knowledge gained by England's great exploratory expeditions in that era were not lost just because great sailors were not necessarily great writers. He also campaigned vociferously for more voyages of exploration and colonization, writing many pamphlets which he addressed to Elizabeth, Raleigh, Sir Francis Walsingham (the secretary of state), and other influential men, who then made the policy decisions which eventuated in the first English settlements across the Atlantic. It has been said that if it were not for Hakluyt, perhaps the United States would not exist.

Hakluyt's true motivation, however, may have been misunderstood until now, if the bare-bones story I relate here is correct: that the English desire to colonize Old Virginia was fueled largely by a need to seize the calendrical meridian which would allow the equinox to be repeated on one date, and thus facilitate a calendar demonstrably superior to that of the Roman Catholic Church. Let us start with the history of European incursions into the area. A map indicating the various locations mentioned is presented as figure 3, on page 202.

The first attempt to found a European colony on the Atlantic seaboard appears to have been by a group of about 150 French Huguenots in 1562. Their meager settlement, called Charlesfort, was occupied for less than a year, and its ruins have only recently been identified beneath a golf course near Beaufort, South Carolina. In 1564 a second group of Huguenots arrived and set up a stockade near the mouth of the Saint John's River in Florida. Spanish missionaries had been present along this coast since 1513, and around the Chesapeake Bay area for some decades, but that region was not regarded as being of great importance by the Spanish crown, which had found more of interest in Central and South America, and the Caribbean. Around the northern coastline of the Gulf of Mexico, the Spanish had been exploring since 1519. A major expedition ventured inland in 1540, covering parts of what is now Texas, New Mexico, Colorado, Arizona, and possibly Oklahoma and Kansas, and many missions and settlements were established. A century thereafter the French were exploring northward from the eastern end of the gulf (hence the name of Louisiana), and claiming all the land drained by the Mississippi. But this is all quite distinct from the activity on the Atlantic coast, which is our present interest.

The Spanish were alarmed by the earlier arrival of the French in Florida, with the result that in 1565 they established their first permanent presence just down the coast at San Augustin (now Saint Augustine, the anglicized name), and then further north a small colony known as Santa Elena, right over the abandoned (or annihilated) Charlesfort. It seems that the Spanish were less than gentle in dealing with the French Protestants.

The English were not disinterested in the region, and from this time various expeditions from that quarter arrived in the Caribbean and points north, led by Sir John Hawkins, who battled with the Spanish in 1568, and then by Sir Martin Frobisher, in 1576–1578. The Spaniards were continually harassed in this atmosphere of religious rather than merely national rivalry by such as Sir Francis Drake, who as an admiral in the Royal Navy had a free license to take all he could of their bounty and sink their ships; he was the archetype of the buccaneers. Nowadays Drake is remembered most commonly for his great circumnavigation of the world, completed in 1580—which was inspired by Dee's book *The Perfect Art of Navigation*—but most of his time at sea was spent in murderous rather than peaceful pursuits.

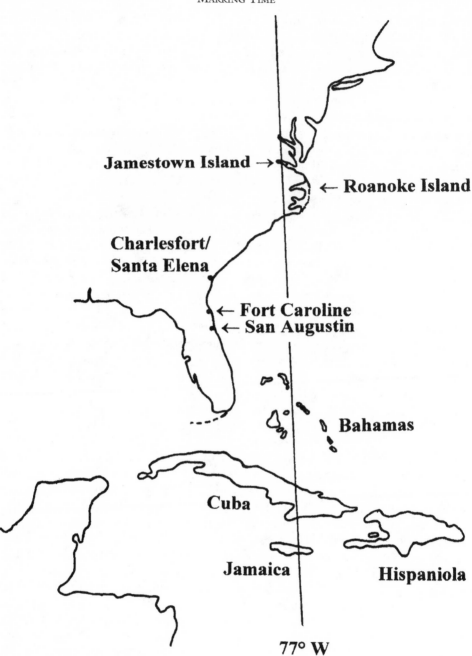

Fig. 3. The locations in the eastern Caribbean and southern regions of North America where the Spanish, French, and British established presences during the sixteenth and early seventeenth centuries. Also shown is *God's longitude*, the meridian seventy-seven degrees west of Greenwich.

The British resolve to squash Spain was emboldened about this time when they signed a treaty with the other great seapower, the Netherlands, with specifications to that purpose. In the following few years there were several more forays to the Americas led by Sir Humphrey Gilbert, ending in disaster when he was returning from Newfoundland in 1583, his fleet being sunk in a storm.

The major English ambitions were further south than that, however. Note how I phrased that statement: one would tend to think of going *south* as one travels down the Atlantic Coast of North America, but from the present calendrical perspective the significant point may have been that this also entails movement towards the *west* and thus nearer the seventy-seventh meridian. In 1584 the English expeditioners Philip Amadas and Arthur Barlowe reached Roanoke Island, on the Atlantic seaboard of North Carolina, an area better known nowadays for the aeronautical endeavors of Orville and Wilbur Wright in 1903. Amadas and Barlowe remained there for several months before sailing home. Why? Why spend such a length of time on a comparatively small and barren island before hurrying home prior to the onset of winter?

The answer seems to jump out: *to determine its longitude.* The necessary astronomical observations, before the availability of marine chronometers, were protracted and could not be accomplished in short order. In particular, noting the local time (according to the Sun, that is) of a celestial event which could also be witnessed in England was an essential step forward, and a lunar eclipse late in 1584 provided such a facility.

Once a reference point had been established on the western side of the Atlantic, its meridian ascertained with decent precision, it would swiftly be possible to choose where to site New Albion on the basis of how much further west the founders would need to go: Fifty miles? A hundred? More? What better place to use as your reference point than an undesirable and uninhabited, but easily defensible, island on the eastern seaboard near the desired longitude? Surrounded by sea and with the advantages of cannon and musket, the early Englishmen who were sent to this outpost were much less likely to be hounded by the natives than if they had attempted to carry out their secret mission on the mainland.

Yes, a New Albion was to be founded, but it does not follow that this was to be on Roanoke Island, where the Fort Raleigh National Historic Site stands today. In essence it was just a convenient piece of dry land on which the necessary instruments could be set up, at the eastern edge of the continent. Once they had determined that location's meridian, they would know how far inland they needed to go to reach God's longitude.

Raleigh's American Adventure

It is a matter of record that Sir Walter Raleigh was, in 1584, given seven years' rights for exploiting the coasts of what we now call Virginia and North Carolina. He was given the power to impress both men and ships for his voyages (that is, seize them in British ports and force the sailors into his service), in return for which Queen Elizabeth was to receive 20 percent of all gold and silver found (or stolen from the Spanish). Most often this episode is portrayed in history books simply as an attempt to set up a base from which to attack Spanish shipping in the Caribbean, although various other outcomes are also popularly ascribed to Raleigh's American adventures, such as the importation to England of tobacco, coffee, tomatoes, and turkeys. In reality, it seems that the fundamental purpose of the expedition was not to be revealed lest the Catholics realize the English calendrical plan.

In mid-1585, Raleigh's expedition arrived at Roanoke Island: seven vessels, under the command of Ralph Lane, containing over a hundred men, but ill-equipped for actual colonization purposes. Their mission has remained a secret, but I would wager that they carried more astronomical than farming implements. They stayed only a year before Sir Francis Drake arrived from Florida and returned most of them to England. That was in June 1586; by August, Sir Richard Grenville and three ships had returned to the island, resupplying those who had remained, assumedly to continue their survey. The makeup of the personnel on these first expeditions has long been a puzzle to historians, with a radical lack of people able to scratch a living from the land, like farmers and artisans, but the inclusion of men like Thomas Harriot who were skilled in surveying the heavens.

The Lost Colony

In 1587 came a new wave to Roanoke Island: an expedition led by John White reportedly consisting of 150 men, women, and children sent by Sir Walter Raleigh with the aim of founding a city to be named for him. (These names should not be confused with the present-day towns of Raleigh, North Carolina, and Roanoke, Virginia, both of which are far inland.) The story is that problems with food and the local Croatoan Indians led to White soon returning home for supplies and reinforcements, but such an idea is predicated upon the belief that this was meant to be a self-sufficient colony, whereas the planning and composition of the group gives the lie to this notion.

Other matters now intervened, resulting in what is now referred to as the *Lost Colony*. It was anticipated that White or some other captain would return with relief ships in 1588, but that succor never came. The reason is obvious to any English schoolboy who knows more than the date of the battle of Hastings, 1066. The other dates said youngster is likely to know are 1815, for the battle of Waterloo, and 1588, for the Spanish Armada. This was the fleet sent against England by King Philip II, which was conclusively beaten and scattered by the Royal Navy with its Dutch allies, and then wrecked by following storms, producing one of the most crushing maritime debacles of all time.

An unrecognized consequence of the dismemberment of the Spanish navy, unless one understands the present perspective, was that there was no great pressure any more to seize quickly that part of the landmass on the seventy-seventh meridian which was yet unclaimed. The Spanish, it seems, had not realized what was by chance within the grasp of themselves, and hence the

Catholic Church. With the destruction of the Spanish Armada, they could no longer vie with England on this front.

White eventually returned to Roanoke Island in 1590, having been delayed not only by the war with Spain but also by shocking weather, to find the site abandoned. The word *Croatoan* was carved on a tree, perhaps indicating trouble with the indigents, but that does not seem to be the whole story, and we may never know the truth: remember that this whole episode was conducted as a deception. White had earlier written that their objective meridian would be about "50 miles into the main"—that is, fifty miles west, into the mainland—and this is where he followed. Obviously the astronomical observations had enabled them to work out just how much further they needed to strike west to arrive on the important meridian, the fiducial line, and rather than simply wait the members of the Lost Colony had started toward the ultimate objective of these expeditions: Roanoke Island was just an intermediate step.

The mainland directly west of Roanoke Island, and to the south of there, is covered largely with rather dismal swamps. Thus the Lost Colony members would have headed further north, to the area where the city of Norfolk now stands, at the mouth of Chesapeake Bay, where there are excellent natural harbors. Recent climatological research has shown that the years 1587–1589 produced the worst drought in the region in eight hundred years, and that coupled with the unsuitability of the expedition members to eke out a living from the land and the hostile natives *may* have led to their expiration. But their fate remains a mystery.

James I Unifies the Crowns

For the time, at least, the quest for God's longitude was a lessened priority, and just what did happen in this respect over the next two decades has yet to be teased out by historians, although there are plenty of hints of activity of one sort or another in Virginia—named, of course, for the Virgin Queen, Elizabeth I, who was sponsoring the various sorties into the area. England was not necessarily in any great hurry. At any time until the era of potential agreement between the Gregorian leap years and the thirty-three-year cycle of leap years terminated in 1620, the English could suddenly reveal Dee's superior calendrical plan, delete eleven days from their dates, and claim divine approval through having taken possession of the important meridian, producing what might be called the *perfect Christian calendar.*

Perhaps they hoped to grab not only Old Virginia, but also ultimately to chase the Spanish out of the other lands along that meridian line, down through the Caribbean region. During the 1590s this was certainly the zone which received the most attention, with flotillas led by Christopher Newport, Amias Preston, and George Somers patrolling the West Indies.

It was not until 1602 that we hear again of interest in the North American portion of the meridian. In that year Samuel Mace voyaged to Virginia under instructions from Raleigh, to search for survivors of the Lost Colony—without success, we are told. The following year Bartholomew Gilbert was also sent on a voyage to Chesapeake Bay, but he and four others were killed by Native Americans.

The year 1603 marked a turning point back in Britain, when Queen Elizabeth I died. Elizabeth being without heir, the throne of England was taken by James I. He had previously been king of Scotland as James VI since 1567, when he was just thirteen months old, when his mother (Mary, Queen of Scots) had been taken captive by the English. After his hurried ride south following Elizabeth's death (a person for whom he had no love; she had his mother executed in 1587 on a trumped-up charge), he seized the English crown as James I, and was thus both James VI and James I. Elizabeth, then, was the last of the Tudors; James was the first of the Stuarts, who were to rule a unified Britain until 1714 (except for the eleven years of the Commonwealth, during which Britain was a republic under the Roundheads).

This change in dynasty at the start of the seventeenth century was certain to cause a hiccup in any British activity in North America, the main man (Raleigh) having been Elizabeth's favorite. But the outcome was a renewed effort, with the only real difference being the substitution of the name of James for that of Raleigh.

Jamestown Island

In June of 1606 King James granted a charter to the Virginia Company, the published aims of which included a British settlement in that region, a search for gold, and also a route by sea to the Pacific and Asia. On December 20 a fleet carrying 104 men and boys set sail from London under Christopher Newport's command. After some perambulations, the expedition arrived at what is now Hampton, Virginia, on April 30, 1607. Over the next two weeks they surveyed what they called the James River, and by May 13 had decided upon an island upstream in that waterway—further west and close to the desired meridian—as the site for the fledgling colony to be named Jamestown. Again historians have long puzzled over why they chose that island, sixty miles from the mouth of Chesapeake Bay and thus very inconvenient for shipping, as the place to stop, but the quest for the vital longitude provides a viable explanation.

The trials of the settlers were just beginning. Not only were they largely unsuited to the task of chiseling out a new colony, but the Algonquian natives were hostile, and the Europeans quickly had to build a wooden fort and palisades to defend themselves against numerous attacks. On June 22 Newport departed with his ships to return to England, leaving the colonists to fend for themselves; when he returned the following January with a hundred new recruits he found only thirty-eight survivors, with disease, hunger, internecine strife, and the continued hostility of the Algonquians having taken their toll. One of the leaders of the colony, John Smith, had been captured by the natives, but the chieftain, Powhatan, released him soon after Newport reappeared in early 1608.

This was not the end of the troubles. Soon the fort burned down, and in April Newport again set sail for England to bring back yet another wave of settlers, who surely could not have known what they were facing. That summer the Jamestown fort was rebuilt and extended, numerous cannons being mounted to rebuff the indigents. With Smith now appointed commander, some mettle was instilled into the men which had previously been lacking, and he instigated regular drilling and a constant lookout for marauding natives.

In October Christopher Newport was back with more supplies, and the first women to be sent. Newport soon started back east toward Mother England, his ships laden with various valuable items produced by the colonists (pitch, tar, soap, silk, sassafras, and frankincense among them), encouraging those in Britain who would try their hand in a new land. In August 1609 seven ships carrying several hundred men, women, and children arrived at Jamestown, one vessel having been wrecked in a storm on the voyage over, but with all hands surviving and struggling onto the beach at Bermuda. In September, the Virginia Company having decided to give near-dictatorial control to one man, George Percy replaced John Smith, and the latter returned to England, as it happens leaving Jamestown just when a leader of his caliber was needed most.

Those on board the *Sea Venture* when it foundered on Bermuda in 1609 turned out to be the lucky ones. By the time that they eventually reached Jamestown in May 1610, having built two smaller boats from local cedar and salvaged items, they found that all but sixty of the original five hundred colonists were dead. Disease and native attacks were significant contributors, but it was starvation which had taken the major toll. As with the Lost Colony of Roanoke Island twenty years before, the Jamestown colonists had arrived at the time of a tremendous drought lasting from 1606 to 1612, the severity of which has only recently been identified through climatological detective work. The colonists were harried by the Algonquians such that they could not cultivate wide tracts of land, being in mortal danger should they venture even a stone's throw from their meager fortifications, and their efforts at intensive farming were stymied by the lack of rain. In June the decision was made to abandon Jamestown, but the arrival of Thomas West, Lord De La Warr, the freshly appointed governor (of what would eventually become the state of Virginia), gave fresh heart to all.

Deciding that attack was the best form of defense, De La Warr launched an offensive against the natives, culminating in the massacre of the queen and her children. In March 1611 he felt able to return to England, leaving the colony relatively stable, although many decided to leave with him.

The Virginia Colony Expands

Over the next few years two main factors led to the eventual expansion of the colony. One was that the British realized that a greater investment (in men, ships, and supplies) was needed so as to overcome not only native resistance, but also to be able to repel any attack by another European power.

The other was that peace and prosperity were aided by the marriage between Pocahontas, the favorite daughter of Powhatan, and John Rolfe, a tobacco entrepreneur who had been a survivor of the Bermuda shipwreck. Pocahontas, a name known to most young children today, did not live long—she, Rolfe, and their baby son departed on a trip to England in 1616, where she became ill and died at the age of twenty-two—but the truce between Algonquians and Englishmen occasioned by this union enabled the latter to gain a better foothold (for example, in Henrico, present-day Richmond, Virginia).

This truce did not last forever, and in 1622 the disenchanted Algonquians killed over three hundred of the settlers in the now widely spread plantations. This provoked King James into revoking the charter of the Virginia Company, turning the region into a Crown Colony rather than a privately owned venture. Britain really did now own part of God's longitude in the Americas, with Jamaica (in particular the capital, Kingston, and Port Royal) and various parts of the Bahamas being additional possessions along the seventy-seventh meridian.

Aim Achieved, Britain Lets the Matter Rest as a Latent Threat

In this interpretation of history the English settlement of Old Virginia stemmed not from a simple desire for more land, for gold, for trade, or any such obvious thing. Their persistence and repeated attempts around 1584–1590 and again in 1602, and then 1606 onward, were driven by an imperative to grab land on the seventy-seventh meridian in order to be able to introduce a novel calendar which would be demonstrably superior in terms of mathematics, astronomy, and religious veracity to the hated product of the Catholic Church.

This imperative led to the English being forced to fight for land which otherwise would not have been deemed desirable. The flat, swampy region near the seventy-seventh meridian down the North Carolina coast was not attractive, and was too close to the Spanish settlements and passing ships. The areas around Chesapeake Bay itself had much to recommend them, but the site for New Albion needed to be further west, making passage up the James River into fanatically defended Algonquian territory a necessity.

The later success of the Virginia Colony undoubtedly owes itself to many factors, not the least of which was the arrival of the first Africans (as indentured servants rather than slaves) at Jamestown in 1619, but one needs to look into the context of the times to discover England's prime initial motivation for its ventures in that part of the world. In this chapter I've argued that it was the pursuit of God's longitude.

Why has this not been recognized before? Why did the English not proclaim the perfect Christian calendar once Jamestown was successfully founded? One possibility is that they did not feel themselves sufficiently secure before the deadline of 1620 had passed (although thereafter its introduction was still possible), and in any case by then the School of the Night comprising Raleigh, Harriot, Dee, Burghley, and colleagues had passed on, as had the Tudor dynasty. Perhaps the Stuarts had different calendrical ideas, as exemplified on a trivial level by the fact that Scotland started using January 1 as New Year's Day from 1600, and even today Hogmanay (New Year's Eve) is a much bigger celebration in Scotland than it is in England.

Stepping back to the period around 1600, though, the reality may have been that the British felt that the *threat* of a rival calendar reform, the Spanish occupants of the rest of the crucial meridian being in eclipse, may have been a more effective weapon against the Catholic Church in its attempts to destabilize Britain and Ireland than its actual implementation. We know now that the English had great tribulations at Roanoke Island and Jamestown, but the Catholics had no such intelligence, only spies in London reporting regular well-equipped convoys heading to the Americas carrying everything needed for a major presence to be established there. In Rome the pontiff may have been bluffed into believing that Britain had all necessary weapons up its sleeve, and neither side wanted to provoke a confrontation.

Chapter 15

❧

ARCHBISHOP USSHER
AND THE AGE OF THE EARTH (A.D. 1650)

After the reform of the calendar promulgated by the Roman Church in 1582—the introduction of the Gregorian calendar—most Catholic nations, and some Protestant ones besides, fell into line with the new dating system either immediately or over the subsequent few decades. As we will discuss in chapter 16, Britain did not change until 1752 (and some others, such as Russia, held out until the twentieth century).

It does not, however, follow that Britain was necessarily backward in its learning when it came to matters calendrical. Nor does it follow that it refused to adopt the Gregorian calendar simply out of antipopish animosity and bloody-mindedness. In the previous chapter we saw that a strong case can be made for the English having devised a superior calendrical scheme which, if all the necessary conditions had slotted into place, might have led to a counter-reform, a perfect Christian calendar rather than the Gregorian mishmash, which would surely have unified many states against the papal position.

For example, if England could have established New Albion on the seventy-seventh meridian in 1585–1587, before its energy and attentions were diverted by the imperative of defending its own shoreline against the Spanish Armada, then the news might have been relayed to John Dee, who would then have been able to reveal a fait accompli to Emperor Rudolph in Prague, perhaps with the effect that the Holy Roman Empire would have been lured to the English side and the Roman Church marginalized. Many other states would then surely have joined that consortium, rather than just Britain and the Netherlands allied against Spain and the hard-line papal Catholics. Subsequent history would have been quite different.

Despite the fact that Dee's calendar conspiracy never saw the light of day, one should realize that what happened in Britain between 1582 and 1752 was not a simple matter of the Julian calendar being adhered to in the face of the rest of Europe changing to the Gregorian system. Even within the framework of the Julian calendar, the British had their own annual cycle, which was peculiar to them and them alone. In some respects it represented their own long-term, deep-seated beliefs about the turning points in the year, while in others it gave rein to significant recent events which needed to be commemorated for the sake of national (and religious) unity. Let me give one example of the former, and then the latter.

March 25 as New Year

A long-term aspect of the calendar was its perceived turning point at the vernal/spring equinox. We have seen before that this was a continuing tradition dating back to at least the seventh century B.C. and the early Roman republican calendar.

In fact, the Greek poet Hesiod (whose dates are uncertain, although he seems to have lived in the late eighth or early seventh century B.C.) emphasized this matter. Best known for his oral composition entitled *Works and Days,* which was not written down until long after his death, Hesiod gave a framework for the year in terms of what is seen in the sky, which phenomena of nature occur on sea and land, and what forms of work humans should turn to as the year itself turns. His pivotal juncture is clearly the equinox, when (he says) the Pleiades star cluster disappears from view for forty days before its eventual return in its heliacal rising. In chapter 4 I argued the case that *any* annually repeated phenomenon could be used to measure the length of the year, giving the example of the recording of when a cuckoo is first heard each spring, and it is interesting to note that over twenty-six hundred years ago Hesiod was telling his listeners that spring was here when the cuckoo started to call. Regarding the significance of the equinox, Hesiod was quite specific: "Keep all these warnings I give you, as the year is completed and the days become equal with the nights again, when once more the earth, mother of us all, bears yield in all variety."

In view of this persistent tradition of the equinox marking the beginning of the year, it is not surprising that there was a disinclination to abandon it. The Church also provided an example to follow with regard to when the year began. Both the Catholic and the Protestant churches had calendars which made use of both January 1 and March 25 as year starts in different aspects. Although Scotland had gone over to using January 1 as New Year's Day for civil purposes from 1600, England continued to use March 25 until 1752, but again this was not simply a reflection of anti-Catholic stubbornness: the uptake of January 1 as New Year was *not* a part of the Gregorian reform. It makes no more sense to criticize the English of that time for persisting with March 25 than it does to criticize present-day Canadians for having their Thanksgiving on a different date than the United States. The English position on the equinox (and hence March 25, even if the date were off by a few days) being the appropriate juncture about which to reckon the year was given quite nicely by Lancelot Andrewes, the bishop of Winchester, when he preached to King James in 1619:

> For once a year, all things turn. . . . Now at this time is the turning of the year. In heaven, the sun in his equinoctial line, the zodiac, and all the constellations in it, do now turn to the first point [of Aries]. The earth and all her plants, after a dead winter, return to the first and best season of the year. The creatures,

> *the fowls of the air, the swallow and the turtle and the crane and*
> *the stork, know their seasons and make their just return at this*
> *time every year.*

Why Andrewes omitted to mention the cuckoo I do not know.

Why should the English change? Clearly their year start around the equinox made a great deal of sense to them, and the fact that others had shifted over time to an alternative system was not necessarily a strong argument for falling into line. The first time I went to the United States was in 1976, to work as a counselor (and general factotum) on a summer camp in upstate New York. Of the several hundred kids, mostly from the city, all but two seemed to be Yankees fans. That pair of Mets supporters had a miserable summer, especially since the Yankees were running away with the American League East pennant, but I admired them for their loyalty to their team. Similarly, the whole of the rest of the globe could have decided to use January 1 as New Year, but if the English felt good about March 25—and everyone will know that traditions die hard in England—why should they change?

Remember, Remember

Now let me give an example of the introduction into the British calendar of significant anniversaries commemorating relatively recent events.

As a boy living in England I looked forward to the month of October not because it ended with Halloween, an apparently American festivity of which I was aware only through movies and television programs, but because it was the lead-up to Bonfire Night. For our bonfires we would collect over the preceding several weeks huge piles of wastepaper, wood, and other inflammable junk, looking forward to the time when it would be set ablaze to the accompaniment of fireworks and other pyrotechnic displays. We would get old clothes and stuff them with screwed-up newspaper to produce a human effigy, walking the streets with it mounted on an old perambulator, pleading with passers-by for a "penny for the Guy," and singing:

> Remember, remember, the Fifth of November,
>
> Gunpowder treason and plot,
>
> I see no reason why gunpowder treason
>
> Should ever be forgot!

I was a pupil at a Church of England primary school, and I always thought it odd that when most of my class marched down the road for a service at the big church, one girl was collected by her mother and taken to some other place of worship, which was much smaller and less impressive. I felt sorry for her, and could not fathom why she didn't join us. Since then I have generally found that the contrary seems to be the usual case—that Catholic churches tend to be larger and more ornate—but my point is that I never put two and two together. I never associated Bonfire Night and burning the tormented Guy with an anti-Catholic stance; it was just something we did, and great fun, too.

Actually the Gunpowder Plot commemorated on November 5 each year stems from 1605, when a group of Catholic gentlemen, Guy Fawkes being the best remembered of them (hence the Guy on the bonfire), plotted to blow up the Houses of Parliament in London, and kill King James and his government. This conspiracy, which was thwarted just in time, resulted from despair that the king would keep good his promise to emancipate the Catholics. Almost four centuries later it is not clear that James's assurances have yet been fully enacted in all parts of the United Kingdom, and I bear some responsibility for that: my unwitting childhood glee in incinerating the Guy, the representation of a Catholic malcontent, in one small way contributed to the continued subjugation of all Catholics in my country of birth. I am contrite: it may not have been a big thing compared to the Inquisition, but no one is innocent.

My point here is that the great Fifth of November anniversary was a feature of a peculiarly British calendar which quickly became established as a marker point, and it continues through to today. Another similar anniversary which has not fared so well is November 17. This was the date of the accession of Elizabeth I in 1558, and within about a decade it had been accepted as being a day on which celebrations were warranted, the English determination in this respect being redoubled when Pope Pius V excommunicated their queen.

In the American colonies, these dates were picked up to differing extents. November 5 became the time for a major carnival in New England by about 1700, with fires being lit and the ridiculed and despised effigy being supposed to be that of the pope rather than Guy Fawkes, the day becoming known as Pope Day. Eventually the new Americans found, through the lapse of time and the rise of local concerns, their own events which needed to be commemorated, and after the Revolution the separate festivities of November 5 and 17 became amalgamated to produce Thanksgiving. While this is now regarded as a time for families to reunite, in the past the season had a quite different familial connection; Ned Ward, an English writer who visited Boston in 1682, wrote that "it is observed, there are more bastards got in that season than in all the year beside, which occasions some of the looser saints to call it rutting time."

A Reduction in Holy Days

Obviously one cannot introduce new holidays without limit, else no work would get done. In seventeenth-century England and the fledgling American colonies, the introduction of festivities in November was not too disruptive because the harvest time was over and there was no urgency with the routine jobs of late autumn and winter, but over the year the number of holidays needed to be kept more-or-less constant.

The celebrations of November 5 and 17 may have stemmed originally from anti-Catholic sentiment, but fundamentally they were secular occasions rather than sacrosanct days of the Church. This suited the Puritans, who looked askance at the activities which were conducted on the putative holy days—back in chapter 7 it was mentioned how Christmas Day was supposed to be a time of fasting and repentance rather than festivity—and gradually Parliament removed the holy days and replaced them with secular anniversaries.

This meant that the religious holidays, previously gazetted as red-letter days in the printed almanacs due to the long-gone Roman custom, systematically disappeared, and the term took on a new meaning. In some Puritan regions any woman found guilty of adultery was forced to wear a large scarlet A on her clothing, visible to all; hence the term *scarlet woman*. In the mid–seventeenth century English schoolboys would chant:

> Since the mis-called holy days, profanely spent,
>
> Are justly now cashiered by parliament;
>
> For that the scarlet garment that they wore,
>
> Was but a rubricked badge of the Roman whore.

The British calendar, then, was changing as time went by, adopting a distinction from the Gregorian calendar which went far beyond the trivial maintenance of the Julian system. Historian David Cressy began his 1989 book *Bonfires and Bells* with the following thoughts:

> *Under Elizabeth and the early Stuarts the English developed a relationship to time—current time within the cycle of the year, and historical time with reference to the past—that set them apart from the rest of early modern Europe. In many ways the calendar of seventeenth-century England had less in common with Renaissance France or Spain, and more with twentieth-century America or Australia. It was based on, and gave expression to, a mythic and patriotic sense of national identity. Though founded on Christianity, purged of the excesses of late-medieval Catholicism, the guiding landmarks were taken from recent incidents in English history. The calendar became an important instrument for declaring and disseminating a distinctively Protestant national culture. It served, sometimes, as a unifying force, binding the nation to the ruling dynasty and securing it through an inspiring providential interpretation of English history.*

The Roundhead Reform

The Puritan party, which came into power after the English Civil War and the execution of Charles I (son of James I) in 1649, was derisively termed the *Roundheads* because of their style of cutting their hair short. Their opponents were the Royalist party, the *Cavaliers*. Best remembered of the Roundheads is Oliver Cromwell (1599–1658).

Cromwell and his cronies tried to instigate major reforms in Britain, one of them being a radical alteration of the calendar. For example, the *anno Domini* dating system was to be abandoned, with 1649 being referred to instead as the "first year of freedom." The idea, however, soon fizzled out, and this Puritan calendar reform was rapidly forgotten by all but the Quakers, who maintained a calendar purged of all un-Christian associations; for example, the months in the Quaker calendar are numbered rather than named.

The rule of the Roundheads—during which England was a republic—did not last much longer than their transient calendar. The Restoration (the reestablishment of the monarchy) came about in 1660, Charles II ascending to the throne to which he had aspired since his father's downfall eleven years before. In the Interregnum he had been hidden in various Royalist safe houses, one of them in the town of my birth. I wrote above of trundling along with my primary school class to church services. That church is named for Saint John the Baptist, the feast day of whom is Midsummer's Day, hence the name of the town: Midsomer Norton, a Norman town associated with midsummer. The church tower proudly displays a statue of King Charles II, who presented two bells for the peal in gratitude to the congregation which had protected him in his time of need. One can hear those bells whenever there is a special event such as a wedding, like that of my parents in 1952.

The 4004 B.C. of Archbishop Ussher

It is against the above overall perspective, as a start, that we should place the ideas of James Ussher (1581–1656), archbishop of Armagh and Primate of All Ireland. Most readers will have heard of him, his notoriety resting on his much-misinterpreted statement that the world began in 4004 B.C.

In fact, many scholars of his era had made similar calculations and arrived at similar ages for Earth, as we will see, but his own work gained prominence to our present day through a connection with what I have described above. The relatively young Anglican Church needed to differentiate itself from its Roman rival, and of course there were many ways to accomplish this quite apart from calendrical questions. One was the preparation of an edition of the Bible set up for distinctly British purposes, and such a volume was published in 1611. Formally called the *Authorized Version* (thus demonstrating the control of the state over religious matters compared to the secular system in the United States), this Bible is most often referred to as the *King James Version,* because James I was on the throne at the time. Ussher's book containing his age of the world appeared at the middle of the century, and thereafter the new printings of the King James Bible had his chronology given in the margins, the book of Genesis starting with 4004 B.C.

There is another aspect to the background which we should also consider, however, a point which has been missed by most writers of modern-day science textbooks who delight in ridiculing Ussher, and that is the scientific knowledge of his era regarding the age of Earth. To a modern, scientifically educated person, it is ludicrous that anyone could believe that our planet is only a handful of millennia old, but such was the unchallenged consensus in Ussher's time, when the Bible was presumed to be literally true. Many misrepresent Ussher as being a buffoon railing against superior scientific knowledge, but that is a gross misunderstanding of the context of his period, in which he was a respected scholar simply applying the contemporaneously accepted methods of historiography.

This is not to say that there was no earlier suggestion that the timescale of Earth history was rather longer than a mere handful of millennia—for example, the Greek philosopher Xenophanes (ca. 560–ca. 478 B.C.) had recognized that changes in the fauna he recognized in fossils implied that the inhabitants of the planet had changed over time, and within a century his compatriot Herodotus (ca. 484–ca. 425 B.C.) was writing that the thick accumulations of Nile sediments indicated the passage of a great deal of time—but in the climate of the late Middle Ages, dominated by religious concerns, any thought that the world had existed for longer than the Bible suggested could not be entertained. Nowadays we may refer to a certain definitive book as being the *bible* of a certain discipline (as in "the *Audubon Society Field Guide* is the bible of North American bird-watchers"), but the point is that back in Ussher's day the Bible was indeed taken to be the literal truth and not to be doubted. Sorry to say, much of modern science also "progresses" in this way, with certain "facts" not to be doubted; consider how the geological fraternity resisted the idea of continental drift for so many decades, and how it still vehemently opposes the notion that asteroid and comet impacts might significantly affect the face of our planet.

The reason that Ussher's name and his date of 4004 B.C. have been remembered and subject to misplaced ridicule is simply that it was his system rather than some other which was chosen for deployment in the King James Bible. Ussher's fame derives from the fact that his was one of the best analyses of the question as posed at the time, and not that people recalled it with scorn. Such disdainful and supercilious remarks made by modern-day writers speak more to the knowledge of those writers than they do to the merits of Ussher's work.

Back in chapter 10 we discussed the interpretation of Psalm 90, which speaks of a thousand years being as a day in the sight of the Lord, and how that led to the idea that the end of the world would come after six thousand years had elapsed, with recurrent millenarian movements being provoked as a result. Such an interpretation gained in popularity through the writings of men such as Julius Africanus, who in the early third century A.D. wrote a book called *Chronographia* in which he argued that the Second Coming was due within a few hundred years, given that Earth had been formed about fifty-five hundred years before Christ's first incarnation. That age was derived through the Septuagint, the first translation of the Old Testament from Hebrew into Greek, which was carried out in the third century B.C. by a group of Alexandrian Jewish scholars at the request of the Ptolemaic Egyptian rulers, leading to the *annus mundi* ideas previously described. The Septuagint, meaning "seventy" in Latin, gets its name from the seventy (or perhaps seventy-two) translators who were involved.

There are various reasons why later clerics decided that a dawn of the world in around 5500 or 5200 B.C. was a mistaken belief, leading to a rescheduling of the Day of Judgment. Let me again refer you back to chapter 10. The crux here is that the Christian churches brought about a generally agreed notion that Christ appeared at about the end of the fourth millennium of the world, with the sabbatical millennium therefore not being due until about A.D. 2000, the dating being inexact because the various estimates for the Creation around 4000 B.C. varied by some centuries. For example, the Jewish/Hebrew era starting from 3761 B.C. is well known. On the other hand, on the basis of the

Egyptian wandering year, with its cycle of 1,461 years over which the seasons shift through all dates, a start in 4241 B.C. could be argued. Martin Luther (1483–1546) came up with an estimate of about 4000 B.C. for the Creation. The Venerable Bede calculated 3952 B.C., while Joseph Justus Scaliger thought Earth was two years younger than that.

Ussher therefore lived in an era when the idea that the world was formed in about 4000 B.C. had been almost universally accepted for well over a millennium. The paradigm shift whereby we gradually realized that Earth is not only millions of years old, but actually billions, had to await various scientific advances.

In the late Middle Ages the concept that Judgment Day was due within a few centuries was a useful construct, because it put the Apocalypse far enough into the future that the religious leaders did not need to explain its nonappearance during their lifetimes, but close enough to put common people in awe. Something was expected. For example, the great Leonid meteor storm witnessed over the east of North America in 1833 was taken by the Seventh-Day Adventist Church as being a sign that the Second Coming was imminent, a doctrine which it maintained for a century (the maximal lifetime for a human), until it was clear that the signs had been misinterpreted.

Such cynical comments on my part aside, in Ussher's era the common belief was that the universe had formed (when God said "Let there be light") about five and a half millennia previously. His estimate, in that climate, was therefore entirely reasonable. Indeed, if it had not been thought to be so—if his estimate had been wildly discrepant compared to other evaluations and contemporary scientific knowledge—then it would have been quickly forgotten, relegated to the marginalia of history, and we would likely never have heard much about him.

Neither should one believe that the way in which Ussher arrived at his 4004 B.C. figure was as portrayed in many recent accounts, which erroneously give the impression that he simply sat down and counted his way through all the "begats" in the Old Testament. In fact, Ussher brought to his task the very best techniques of historical research which were available in his time. On a trivial level one can see that the false picture painted by many twentieth-century authors is nonsense, from the mere fact that the Old Testament brings the date up only to the fifth century B.C. Thereafter, Ussher needed to bridge a gap in his chronology, using such things as Egyptian king lists; and earlier gaps also occurred. To understand what Archbishop Ussher actually achieved, I would recommend the excellent essay on this topic by Stephen Jay Gould in his collection entitled *Eight Little Piggies* (1993), in which Gould severely, but properly, deals with those who have offended against the memory of Ussher.

Other Aspects of Ussher

There are three other points I would like to make about Ussher, before moving on. The first is calendrical and takes a while to describe, while the concluding two are briefer and will follow thereafter.

The calendrical point first, then. Ussher wrote in 1650 that the world had come into being at God's behest at about midday on October 23 of 4004 B.C., a precision of statement which again has been made the target of persiflage and scorn by some ignorant authors, as if Ussher had managed to deduce this simply on the basis of the lifetimes of biblical characters. We might gain a clue as to Ussher's intention if we ask ourselves the simple question: to which calendar is that date referred?

James Ussher was archbishop of Armagh and Primate of All Ireland. In this context "primate" is not a zoological classification, the order of beasts containing monkeys, gorillas, humans, and chimpanzees, among others, but the leading churchman within a specific organizational zone, in this case All Ireland. Note that this is an old title with a specific meaning, not a modern revision meant to imply the whole of Ireland despite the political division into Northern Ireland and the Republic of Ireland; for example, the archbishop of Canterbury is the Primate of All England, whereas the archbishop of York has to make do with being the Primate of England (only). This usage of the term *primate* is another matter with which Gould has humorously dealt.

Ussher, then, was the leader of the Protestant Church in Ireland, and he was a rabid anti-Catholic at that. With that in mind, one would obviously tend to assume that Ussher would reject the heretical Gregorian reform and insist upon the Julian calendar. And one would be correct to do so.

In very recent years (the late 1990s), various people have stated that Ussher said that Earth was formed on October 23, 4004 B.C., and thus the sabbatical millennium is due soon because six thousand years have elapsed. They make a number of mistakes quite apart from believing that the Bible is literal truth. The first simple error made by many is in not recognizing that there is no year zero, so that the six thousand years had not passed until some time late in 1997: 1996 was a year too soon. Even if they know that 1997 was the year in question, they err also in allotting the anniversary to October 23.

Here is why. Let us ask the question: What is an anniversary? It is when an exact number of years has passed, so that to calculate an anniversary one needs to specify how long a year might be.

Obviously, over short periods like human lifetimes, we tend not to worry about leap years adding a day between birthdays, say, but there are various questions which arise. If your birthday happens to be on the day when the clocks in your time zone are turned forward in the spring, does this mean that you are only allowed to celebrate for twenty-three hours? Was George Washington right to keep marking his birthday as February 11 after 1752, making him always eleven days early? I was born in England, but I live in a time zone nine and a half hours ahead of U.K. time, so when should I celebrate my birthday? The obvious answer to such a question is that a complete number of years has passed—and thus it is my birthday—when Earth has reached the same point in its orbit as at the time when I was born. That can lead to some complexity, but the overall truth of my statement seems clear.

That is for short timescales. For longer periods of time, like the thousands of years back to 4004 B.C., the definition used for the year length obviously will make a great difference. We have already seen an example of this: at the start of chapter 13 I asked you when the winter solstice occurred relative to January

1, 4713 B.C., the reference point for the Julian Date scheme of Scaliger, and we saw that the fact that the date there is given on the Julian calendar is a major consideration. We could do exactly the same sorts of sums for October 23, 4004 B.C., and work out the corresponding date on the proleptic Gregorian calendar. Actually, though, we can sidestep the tedious calculations.

Ussher gave his date as being the Sunday immediately following the autumnal equinox. With that in mind one could calculate, using astronomical formulas, the precise instant of that equinox in the year −4003 (astronomers' parlance), and from that calculate when the six thousandth anniversary would be in A.D. 1997. There is a drawback again, though: Which year length should one use? The mean Gregorian year (365.2425 days)? The vernal equinox year (with varying values through those six millennia)? What I would maintain to be the correct answer obviates any such confusion and complicated computations: the anniversary was at the autumnal equinox in 1997, or on the Sunday thereafter if you prefer. About a week before the end of September will do.

I implied above that there are two other matters connected with Ussher which I would like to mention, and these will bring us full circle. The next is to make the point in another way that Ussher's work was an honorable effort in the context of his times. Rather than dogmatically trying to hold back an avalanche of scientific advance, James Ussher made various contributions to learning which aided our eventual enlightenment regarding the age of our planet and the rest of the cosmos. He promoted an atmosphere in which his successors as leaders of the Protestant church in Ireland could promote scientific examinations of the natural world, and the universe above. In particular, in 1790 Archbishop Richard Robinson founded an astronomical observatory in Armagh, its motto being "The Heavens Declare the Glory of God." Over the two centuries since then the Armagh Observatory has proven to be a frontline research institution, of international repute. James Ussher deserves some of the credit for that fact.

And that brings me to my third point. At the time of the bicentennial of the founding of the observatory, it was decided that a biennial lecture series should be instigated, commemorating Robinson. The first lecturer in 1992 was the Astronomer Royal, Sir Arnold Wolfendale. He was followed in 1994 by Professor Rodney Davies, a radio astronomer at the Jodrell Bank facility of the University of Manchester. Subsequently, the new director of the Armagh Observatory, Professor Mark Bailey, invited me to give the third Robinson Lecture, in 1996. I was pleased to accept not the least because one of my scientific heroes, Ernst Öpik, a pioneer in my research area of asteroids, comets, and meteors who died in 1985, was an anstronomer at that institution. I have the magnificent Robinson Medal, one of my most-prized possessions, before me as I write.

As in the past, the present archbishop of Armagh and Primate of All Ireland, Dr. Robin Eames (now Lord Eames), is the chairman of the board of governors of the Armagh Observatory. When I went there to give the lecture in November 1996 I was delighted to meet him, and found that he had a keen interest in matters astronomical. Not only that, but he had a detailed knowledge of James Ussher, his predecessor from more than three centuries before.

The date on which we met is significant: it was a few weeks after October

23, 1996, when those with poor arithmetical ability had been expecting Armageddon, and Eames's office had received plenty of inquiries, from the media in particular. How should one answer such queries? If you merely say that it is not due for another year, then you might be understood to be implying that Judgment Day is indeed at hand; such questions are hard to answer, like "Have you stopped beating your wife?" My answer would have been to suggest to all callers that they wait until October 23, 1997, which naively is six thousand years after Ussher's date. Then in the week or two previous to that day dawning, one could simply point out that the autumnal equinox—the *real* six thousandth anniversary of Ussher's beginning of the world—had passed some weeks earlier, without incident. So far as I can tell from the newspapers, the world did not end in September 1997.

The Age of the World from the Salt in the Sea

As I hope has been made clear, in the seventeenth century of James Ussher there was no available evidence to make even the most learned believe that the world was more than several millennia old. The Bible was regarded as being a complete history since Earth was molded by God, and Adam and Eve placed in the Garden of Eden.

If this were not the case, how would you know? How could you date the planet, given the technology of the time? One of the very first suggestions for a possible technique for reckoning the age of Earth was put forward by Edmond Halley (1656–1742), the English scientist best known for the comet which bears his name (although he did *not* discover it). Halley suggested that the salinity of the oceans might be used to provide a limit upon how long Earth had existed. His reasoning was that the salt content of the sea must be ever increasing because rivers bring salt down to the ocean, rainwater having leached it from the soil and rock, but when seawater is evaporated by the heat of the Sun the salt stays behind. By comparing the salinity of the sea with river water, Halley suggested that the time over which the saltiness had accumulated in the open ocean could be derived, and that must give an indication of how old Earth could be. If you are interested to know more, Stephen Jay Gould also describes this contribution by Halley in *Eight Little Piggies.*

When he wrote, Halley still had in mind an age for the world measured in but thousands of years. Much later, in the last decade of the nineteenth century, Halley's basic method was employed by John Joly, an Irish geologist, to arrive at a value of 100 million years for the age of the Earth. That is much less than the real value we now know, but there are problems with the basic assumptions inherent in the salt-in-the-ocean method. For example, it is assumed that the evaporation of water leaves the salt behind (correct) and thus the oceanic salinity is ever increasing, but in fact there are various ways in which the salt is removed from solution in the sea.

I could also nominate another problem with the basic premise of Halley's suggestion, this one humankind induced. In South Australia, the River Murray is used to irrigate orange orchards in the east of the state, and as the water is repetitively pumped out onto the land, some draining back into the river, the salinity level is elevated far above that which was the case before Europeans arrived here in the nineteenth century. This means that Adelaide's water supply, especially in the summer when the flow is low, tastes distinctly salty. The increased salinity also has meant that orange growers further down the river receive water which is too salty for their trees to withstand, and in recent years many orange groves have been uprooted and replaced with grape vines, which thrive better with the available water quality. That's one of the reasons that excellent South Australian wines are so cheap.

Cyclic Time, Finite Time

There is another way in which Halley contributed to Earth chronology, albeit in a backhanded way. While in the present chapter we have been working on an assumption of common belief a few centuries back that the universe was only a handful of millennia old, an alternative but growing concept was that time and space were essentially infinite, this movement culminating in the work of the great German philosopher Immanuel Kant (1724–1804). An aspect of this belief in the minds of many was that time is cyclic, with humans being placed on Earth by God only during the present cycle, the end of which may or may not be at hand. An example is Sir Isaac Newton, friend and colleague of Halley. Newton viewed time in a cyclic sense and was sure that the end of the world was nigh. He computed that the great comet of 1680 had passed uncomfortably close by our planet, and viewed it as being a harbinger of awful things to come. But he could not be open in his opinions. He interpreted the biblical books of Daniel and Revelations as being evidence of past catastrophes from the sky, but was not able to have his writings in that regard published during his lifetime.

Let me give a later example of this thinking, from the lips of the poet Lord Byron. Byron seems to have been the first to have imagined that humankind could save itself from a devastating comet impact by diverting an impending impactor. Halley and many others had earlier recognized that comets on Earth-crossing orbits (like his eponymous object) would cause catastrophes should they strike our planet, as they surely must from time to time. In conversation with a friend in 1822, and thus in the era of the steam-powered Industrial Revolution, Byron said the following:

> *Who knows whether, when a comet shall approach this globe to destroy it, as it often has been and will be destroyed, men will not tear rocks from their foundations by means of steam, and hurl mountains, as the giants are said to have done, against the flaming mass?—And then we shall have traditions of Titans again, and of wars with Heaven.*

In our context of cycles of time, Byron's "as it often has been and will be destroyed" is the phrase to note, although the whole of his commentary is interesting, given the fact that humankind is only just now recognizing that asteroids and comets pose a significant hazard to our long-term survival, and yet we have the power to intervene; indeed, this was the subject of my Robinson Lecture at Armagh in 1996.

The connection between all this and Edmond Halley is through his appointment to the University of Oxford, a university which was organized on a religious basis (recall my comments about the founding of University College, London in chapter 8). There are two professorships within that university which are called the Savilian chairs. Halley had applied for the one in astronomy when it fell vacant, but enemies within the institution blocked his candidature. In the following years Halley worked on an analysis of the timing of ancient solar eclipses, the location of observations of which leads to information about the position of the Moon in times gone by and hence its orbit about our planet. Halley was able to show on this basis that the speed of the Moon was changing, which he interpreted as implying that the lifetime of the universe is finite. This allowed him to regain the religious bias initiative, with the result that he was successful in gaining appointment as the Savilian professor of geometry.

The Cooling Sphere Approach

We should return to estimates of the age of Earth based upon alternative scientific techniques, apart from the ocean salinity already discussed. One of the foremost French scientists of the eighteenth century was Georges-Louis Leclerc, Comte de Buffon (1707–1788). Buffon was mainly a naturalist, and a favorite of the royal family—he was largely responsible for the development of the regal forests and gardens at Versailles—but he also made major contributions in mathematics and the physical sciences. He developed a theory for the formation of the planets through the impact of a huge comet with the Sun (in those days many thought comets to be much more massive than they really are), a filament of solar material being ejected like an elongated blob of liquid bouncing up when you drop a sugar cube into a cup of coffee, the vast body of sun stuff then agglomerating into separate globules which are now the familiar planets.

Each of those spheres would start off with an extremely high temperature, and Buffon knew about how hot the Sun's surface happens to be (actually near six thousand degrees Celsius). If Earth started out in this way, it must have gradually cooled to its present ambient temperature, and Buffon reasoned that if he could work out the present cooling rate, he would then be able to estimate the age of Earth. There are problems with this technique, as we will see, but nevertheless one must admit that there is a great deal of sense in it, given the knowledge base available to Buffon. He knew from the evidence of volcanoes that the deep interior of the planet is very hot. His experimental technique was to heat up spheres of different sizes and compositions (rock and metal), and then see how long it took them to cool. By scaling up from his measurements he made various estimates for the age of Earth, and published a favored value of seventy-five thousand years.

That may seem ludicrously short to anyone with a little knowledge of geological and astronomical timescales, but in Buffon's time his result was revolutionary: he was saying that the world was much older than biblical history would attest. So strong was the sentiment against this that he felt unable to make public other evaluations he had derived in quite different ways. For example, he and others in the fledgling science of geology had correctly deduced that certain types of rock were the result of the gradual laying down of sediments and subsequent compaction. When they measured the rates of sedimentation in present-day aqueous environments they found that some millions of years would be required to explain even the simple strata they had to hand. In private conversation Buffon expressed the opinion that Earth must be at least 3 million years old, but he could not say so in print.

In fact, the cooling globe method of Buffon was predicated upon a false assumption. It was presumed that the outward flow of heat from the center of the Earth was the tail end of the cooling from an initially white-hot planet, but that is not the case. Buffon and his contemporaries could not know the truth because they did not know about radioactivity, the secrets of which only started to be revealed in the late nineteenth century, with several decades of work by various physicists being required before the significance of nuclear reactions over the entire cosmos was recognized. Let me not give you the impression that this was a case of physicists to the rescue, however: in fact, the work of some in that discipline actually obfuscated the reality, as you will see below.

The Age of the Sun

Stepping back just over a century, there was a major problem not only with the age of Earth, but also that of the Sun. At our orbit one astronomical unit (A.U.) from the sun, the flux of sunlight is about 1.3 kilowatts per square meter, and this is termed the *solar constant*. Knowing that value and the distance to the Sun, one can easily calculate the rate at which our local star is emitting radiant energy, and given its mass one can then calculate how long it would take to cool down, given an assumption that it is not generating any energy itself. The answer is only a matter of some tens of thousands of years.

One must therefore look to possible energy sources for the Sun. Physicists, the great William Thomson (later Lord Kelvin; 1824–1907) prominent among them, turned attention to the possibility that meteoroids from far out in space could supply the energy, knowing that they would heat and deposit their considerable energy into the solar atmosphere as they collided with it, in the same way as we see a much smaller flux of meteors in our own atmosphere. Kelvin's sums predicted a short solar lifetime, of the order of thirty-two thousand years.

Hermann von Helmholtz (1821–1894), a notable German physicist, made an alternative suggestion. If the Sun were gradually shrinking under its own gravitational attraction, then that would lead to an energy source—the Sun's own gravitational potential energy. Think of a waterfall. As the water cascades down, it is losing potential energy and speeding up so as to gain kinetic energy at a rate exactly compensating for that loss. When the water hits the ground below, it loses that kinetic energy, but since energy cannot be destroyed, it is converted into various other forms: sound energy (what you hear, representing air vibrations which carry energy) is one type, but the major sink of energy is in heating the water. If one does a controlled experiment, one finds that the water at the bottom of a waterfall is slightly hotter than that at the top.

In the case of the Sun, the gradual shrinkage could heat its constituents, that heat energy then being radiated away. In fact, just this process is occurring in the gas giant planets of the outer solar system. Since it was formed 4.5 billion years ago, Jupiter has shrunk a little, and in the present epoch that planet is radiating about 2.5 times as much energy as it receives from the Sun. When you look by eye through a telescope you see Jupiter through the sunlight it reflects; but if you use a suitable radio telescope, that planet is especially bright in the far infrared region of the spectrum, where its emission rate is high, because of its gradual contraction and consequent cooling through radiation.

Applying von Helmholtz's idea to the Sun, Kelvin showed that although one could twiddle with the sums and justify an age as great as 500 million years, more likely the Sun was less than 100 million years old. This brought von Helmholtz into conflict with people like Charles Darwin (1809–1882), who had found that the chalk beds of Kent, where he lived, required 300 million years to be gradually laid down. While I am not entirely sympathetic with Darwin, who was influenced to adopt a gradualist line (that is, the assumption was that the chalk deposition continued at a sensibly constant rate throughout, whereas I would maintain that major floods and so on would be huge contributors) by his geologist friend Charles Lyell (1797–1875), this is neither the first nor the last instance where physicists have displayed profound arrogance and later been demonstrated to be wildly incorrect through advances in their own domain of science.

The great advance was the aforementioned discovery of radioactivity. In the case of Jupiter, the internal energy source certainly seems to be gravitational settling: come back in 10 billion years and that planet may well be considerably smaller. Note that exogenous sources such as meteoroids and other debris are fairly small contributors. Stepping back to July 1994, when numerous fragments of the broken-up comet Shoemaker-Levy 9 slammed into Jupiter and gave it a bloody nose, the planet absorbed more energy from the Sun over the week of the impacts than it received from the kinetic bombs the comet lumps represented. One was spread out, the other concentrated; one represents a gradual influence (like Darwin and Lyell's ideas), the other a catastrophic change (which type in general I believe is far more important than most natural scientists realize).

Turning to the Sun, the center of that orb is at an extreme of temperature (over 10 million degrees Celsius), and a phenomenal pressure, forcing the nuclei of hydrogen atoms so close to each other that fusion occurs, and this is the predominant energy supply for our mother star. Planets the size of Jupiter never achieve the interior conditions necessary for the onset of fusion, so they can never become stars in their own right.

Radioactive Heating of the Earth

While nuclear *fusion* is the solar energy source, it is nuclear *fission* which heats the interior of Earth. I mentioned above that the solar radiation field at Earth has a flux of near 1.3 kilowatts per square meter. Such sunlight as reaches the ground heats the surface, but the timescale over which a wave of summer heat penetrates the ground, due to the low thermal conductivity of the surface materials and the nightly cutoff in sunlight, is such that once you go down more than a few meters the temperature remains constant, year in, year out. Let me give a personal anecdote to illustrate this.

As a teenager in Somerset I would often go spelunking on Sundays under the Mendip Hills, which are built up of limestone and so honeycombed with caves. We would need to dress warmly because the temperature below ground was always chilly, going down a couple of hundred feet to where the pulse of solar heating never infiltrates. Years later in Australia, I was once passing through Coober Pedy, an opal-mining town in the northern part of South Australia, a largely barren area which is now pockmarked with mine shafts. In the summer there the temperature may top fifty degrees Celsius, while in the winter it can be surprisingly cold. The solution for the miners is clear: live below ground where the temperature is cool and constant. Indeed there is a rather splendid subterranean hotel there, simply chiseled out of the soft rock with the interior surfaces then sprayed with a rubber sealant so as to stop further crumbling and dust. My experiences, then, said "below ground is cool."

A short while later I was visiting a colleague in Broken Hill, the famous town in western New South Wales where lead, zinc, and silver are mined. It's a great place, if you ever get a chance to visit, with road or street names in the town center like Argent, Wolfram, Chloride, Iodide, and Sulphide. This colleague was working for the University of Sydney on a research project aimed at detecting the neutrinos emitted in the fusion reactions occurring in the middle of the Sun. Why Broken Hill? The problem is with background radioactivity, such as cosmic rays from deep in space. By going half a mile down a mine shaft, the rock above you shields out these particles, but not the neutrinos, which mostly pass right through the entire planet without being stopped. This makes them difficult to detect, but a few do get intercepted, the instrument consisting of a huge, dark vat of purified water, minute flashes of light being detectable when a neutrino is absorbed. Going down the mine, however, produced other problems: the metal ore contains some radioactive isotopes, which can cause their own flashes in the water, and so the entire vat (the size of a small swimming pool) needed to be surrounded by a wall of bricks made of lead. To say the least it caused some bemusement to the miners that these academics from the University of Sydney wanted to take processed lead blocks back down a lead mine.

Now to the point of my story. I fronted up with warm clothes, expecting it to be cold down there. I used to live near Denver, the Mile High City, where winters are frigid. This mine was a mile *deep*. My friend quickly disabused me of the notion that I needed thick clothes, and insisted that a jockstrap under a one-piece boiler suit was more than adequate. He was right. At the bottom of the shaft, the rock temperature is sixty degrees Celsius. To make it workable for the miners, cooled air is blasted down the shafts, and heaven forbid that their air-conditioning ever fails. The neutrino detector was higher up, in a disused side passage, where the temperature was a mere thirty-eight degrees Celsius.

Back in nineteenth-century Britain, Lord Kelvin also went down various mines and measured the heat flow rate upward, from that arriving at an estimated age for Earth of 20 to 40 million years. Unfortunately, his analysis was predicated upon the assumption that this heat flow is just the gradual cooling of the terrestrial interior since it came into being as a molten orb in space, and that is not the case. The inside of our planet is heated as a by-product of nuclear decay, with various isotopes having half-lives measured in billions of years. Since our planet agglomerated about 4.5 billion years ago, only a fraction of the long-lived isotopic content has decayed. As each atomic nucleus comes to the end of its life, it emits not only a radioactive particle but also some heat. Quite a lot of heat, actually. It is such decays which keep Earth's interior hot, and produce an outward heat flow of 0.06 watts per square meter. That is not a large flux, but due to the low thermal conductivity of rock (that's why we build houses out of the stuff) and the great distances involved, it is distinctly toasty if you drop a mile down into the bowels of the planet. Let me tell you, it's true.

From the early 1900s physicists have developed a good understanding of radioactivity, and have determined the half-lives of numerous isotopes. Knowing how a particular isotope (say uranium 238) decays and what daughter products it produces, and then how those daughters gradually decay, and so on, it is possible to determine the ages of different rocks based upon the ratios of the different isotopic species they contain. That's a pretty dismissive explanation, but a full description would require a whole book. The point is that we can work out how old a specific rock is, within some bounds, using such radioactive dating.

Does that mean that we can measure the age of the planet? Not quite. Some rocks are obviously young, like the topmost layers in a limestone formation. You show me a stratum containing a dinosaur fossil, and I'll show you rocks which are less than 215 million years old: a long period on the human scale, but less than 5 percent of the age of Earth. The oldest macroscopic fossils—due to the first polycellular life—are about 570 million years old: the Ediacara formation, first identified in an outspur of the Flinders Ranges of South Australia. There is some dispute over the first monocellular fossils. There are stromatolites (fossilized beds of blue-green algae) from near Shark Bay in Western Australia which have been dated to about 3.5 billion years, but there is evidence of primordial life in some Greenland rocks which appear to be about 3.8 billion years old. Terrestrial rocks of *any* type do not seem to be much older than that. So is this the age of our planetary home?

The Ages of Meteorites

The answer is *no*. In the beginning our planet was very hot—pretty much molten throughout—and so one does not expect to find rocks which have survived intact from its formation. Even after Earth solidified, it was subject to cataclysmic impacts by planetesimals a hundred miles or more in size, with enough energy to dessicate the planet and even evaporate some of the crust. We have no terrestrial rocks as old as Earth itself.

But we do have extraterrestrial rocks as old as our planet. Along with the rest of the planets, our home was formed from the debris left flying around after the presolar nebula collapsed to form the Sun. Some of that debris never accumulated into major planets, such as the asteroids in the main belt between Mars and Jupiter (the huge gravity of the latter kept their orbits so stirred up that no large planet formation was possible), and the comets at the periphery of the planetary region (where the orbital speeds are so slow that there are fragments left behind which have yet to add their mass to a planet). Largely derived as bits and pieces from the main asteroid belt, meteorites occasionally meet Earth and manage to reach the ground intact. When that happens, various age-measurement techniques may be used to investigate them.

First, the tracks produced when cosmic rays penetrate their surfaces allow the age over which the meteorite was exposed in space to be estimated. The basic idea here is that the lump of rock was, for most of its lifetime, just a part of a larger body in the asteroid belt, away from the surface of its parent asteroid and so shielded from cosmic rays. A chance interasteroid collision, however, breaks up this parent and leaves the daughter meteoroids on orbits spread around the inner solar system. This makes arrival at Earth possible. The space exposure ages for such meteorites typically range from 1 to 100 million years, in accord with the sorts of values expected on theoretical grounds (on the basis of dynamical simulations which seek to model how long such a fragment might freely orbit before colliding with a planet or falling into the Sun).

A second point is more of concern to us here, though. Irrespective of cosmic ray impacts, the time elapsed since a meteorite formed from the detritus of the presolar nebula (at the dawn of the solar system) can be determined from its isotopic makeup, just as is the case with Earth rocks. Such determinations for meteorites all show more or less the same age, around 4.52 or 4.53 billion years. We cannot be sure precisely, because the half-lives of long-lived isotopes cannot be measured with absolute precision, and the best values have recently been reduced from about 4.6 to nearer 4.5 billion years. But meteorites are about that age. And assumedly Earth and the other planets shared their genesis, making all these objects about that old.

This evaluation is not contradicted by any other scientific data source. For example, astrophysicists believe that the universe is about 15 billion years old, with argued ages ranging from 10 to 20 billion years. Our knowledge of stellar evolution concurs with the Sun being between 4 and 5 billion years old: it must be much younger than the universe because some fast-evolving stars must have gone through their life cycles and exploded as supernovas, that being the way in which heavy elements are produced, and you don't get planets without heavy elements. The solar system could not have formed early in the history of the universe because the building blocks were not yet available. As the Bible says, in the beginning there was light, and that means energy; then some of that energy was transformed into mass, and in particular hydrogen nuclei; for that hydrogen to be transmuted into elements like carbon and oxygen required nuclear fusion in the core of a star; for elements in the periodic table heavier than iron to be produced, supernova explosions were required so as to attain the phenomenal transient temperatures and pressures at which the lighter elements are pushed together with such force that the nucleus-nucleus repulsion can be overcome and those heavy elements form. We are indeed stardust.

We know, then, the age of Earth. It is around 4.5 billion years old. James Ussher got a value almost a million times less than that, but that is because, like all his contemporaries, he based his analysis on an assumption which was invalid. He could do no better.

The Common Calendrical Start Era

To complete our discussion of Ussher's life and times, we should pay a little more attention to the fact that many disparate cultures have traditions which say that time began a few thousand years B.C.

I have previously mentioned several values around 4000 B.C., derived from Ussher, Luther, Scaliger, Bede, and the Hebrew *anno mundi*, but these are clearly all part of the same basic tradition, handed down through the generations. I mentioned also the start of Egyptian Sothic-year counting, which some argue is from 4241 B.C., but I think on shaky ground; personally I would favor an origin after the next 1,461-year cycle, around 2780 B.C. Whichever one chooses, again this might be connected to the same Judeo-Christian tradition.

What does one make, though, of more distant cultures which also place the start of time around that basic era? Consider the Mayans. Well before the arrival of the Spanish conquistadores in the sixteenth century, the Mayans had an elaborate calendar based upon what is termed the *long count* of 2,880,000 days, about 7,885 years. This they viewed as a repeated cycle, with the present phase having started on September 6, 3114 B.C. (Julian).

Similarly, there is a Sanskrit calendar based on a *great year* of 2,850 solar years, made up of 150 Metonic cycles. The origin point for this is said to be a grand conjunction of all the visible planets on February 17–18, 3102 B.C. There are also other old Hindu calendars with zero points in the same epoch (although the modern Hindu calendar has an origin at 2,163,102 B.C.!).

I could give other examples, but the thing I am trying to emphasize is that many distinct cultures, without communication and without written records, have supposed origins of time in the fourth millennium B.C. One would like an explanation for this. There are various celestial possibilities one could point toward for inspiration.

The first is that it is certainly true that repeated cycles may be invented by considering the times between the bright planets coming into mutual conjunction. Since these all have orbits close to the ecliptic, the only question concerns their relative orbital periods. For example, Jupiter and Saturn come into conjunction about every 20 years. By requiring a multiple conjunction involving Mars as well, the cycle is greatly extended, to 805 years.

Second, diligent watchers of the skies will discern the precession of the equinoxes and thus the approximate time in which the equinoctial sun moves from one constellation into another, from that deriving eras lasting for some millennia. For example, many New Age folk nowadays base their lives around the concept that this is the dawning of the Age of Aquarius. In fact, the solstitial sun moves into that constellation in A.D. 2137. Of course, other civilizations may have had constellations defined quite differently from those which have been inherited by our rather homogeneous Western culture.

The third explanation, and my own, is that something truly spectacular happened in the heavens in the fourth millennium B.C. My opinion is that this was not one discrete event like a multiple planetary conjunction, but rather a continuing phenomenon which persisted for some centuries and that, coupled with the vagaries of myths and legends and the timings associated with them handed down over the generations, led to a spread of start epochs being assigned across the many cultures which witnessed much the same thing.

This concept I foreshadowed in chapter 4. There I opined that we are hampered in our interpretation of what influenced long-gone civilizations by a mistaken belief that what those people saw is similar to the relatively quiescent sky which we observe now.

This is a similar view to the contention that Darwin and Lyell were incorrect in their interpretations of geological and biological evolution in that they presumed that the processes they could see acting on a day-to-day basis were the same processes, in the same magnitude, which added up to produce the eventual results. I maintain that this philosophical underpinning is invalid. When I drive from Adelaide to the Siding Spring Observatory, a distance of nine hundred miles, much of the trip is across a flat, arid plain. It is a continual surprise to see flood-level warning markers beside the road because it looks like the parched land rarely experiences a drop of rain. But the reason that it *is* a plain is that every so often it floods, severely. It may only be once in a thousand years that a real humdinger occurs, but it is such events which have shaped the terrain one sees. Whenever I stop overnight in the Outback agricultural town of Nyngan, I always look at the many photographs in motels and restaurants of the great flood which almost sunk the place in 1990. That was a *catastrophic* event, in the scientific sense—by which I mean that such things occur infrequently, but when they do, they dominate all other processes.

Similarly, the dinosaurs seem to have been finally seen off by a catastrophe. Whether this was an asteroid impact alone, or a combination of whammies involving volcanic eruptions and other sudden changes of the environment, does not matter from the perspective of the present context. The point is that it was a sudden change, not a gradual one.

By the same token, one expects major changes in the environment in the inner solar system to occur suddenly, and have the greatest effects. The total sum of all the meteoroids and dust in the region of the terrestrial planets is only equivalent to a single cometary or asteroidal mass perhaps five kilometers in size. If one adds in all the Earth-crossing asteroids, then the size of the aggregate object is only ten to twenty kilometers across. Now imagine a singular massive comet entering the inner solar system, perhaps an object fifty kilometers or more in size. We know that such things exist (comet Hale-Bopp was one, and we see many more in the outer solar system on unstable orbits), and we know that comets often break asunder. Along with various colleagues (several of whom work at the observatory in Armagh, Ussher's hometown for a while those centuries ago), I have shown that the smoking gun exists diagnostic of such a large comet having arrived some time within the past ten thousand years or so, which as it disintegrated produced a trail of debris which would have intersected Earth at different times in the fourth millennium B.C.

I believe that it is the very active sky in that epoch which got different peoples terribly interested in the heavens—indeed their calendar developments would have been stimulated by the need to predict when next the associated meteor storms would be due—and *that* is the reason that so many civilizations have histories which date back to the same basic era. All one has to do is to read the various mythological accounts of the god Apollo (or Typhon, or angels, or however they named what they saw) returning to the sky at certain times and doing certain things, and think in terms of a cometary explanation, and you will see that I may have a point. The problem is that whether they realize it or not, most ancient historians—like biologists and geologists—still implicitly follow Darwin and Lyell, rejecting the concept of exceptional events being responsible, even though it is such phenomena which are most likely to produce exceptional outcomes.

Chapter 16

∽

LORD CHESTERFIELD'S ACT (A.D. 1751)

The Catholic states in Europe generally went over to the Gregorian calendar as quickly as arrangements could be made, given the speed of communications toward the end of the sixteenth century. Of course the national borders back then were not the same as now, but one could say that Spain, Portugal, and Poland acceded to the pope's request immediately in October 1582, as did most of Italy; France followed in December, along with the Catholic parts of Belgium and the Netherlands such as Flanders, Brabant, Holland and Artois, and Luxembourg; Austria followed in 1583, and Bohemia and Moravia converted in January 1584 (before John Dee had a chance to influence the emperor in Prague); Hungary in 1587. Prussia was a rather belated adherent, in 1610.

Thereafter came a hiatus. Throughout the seventeenth century Britain persisted on the Julian calendar, except for a brief period in the middle when the puritanical Roundheads tried to force their own month- and year-numbering system upon the general populace, a matter which was resisted and then forgotten with the Restoration of the monarchy in 1660.

As the year 1700 approached, however, there was a newfound impetus to questions of calendar conversion in the Protestant states which had declined the invitation of Gregory XIII to adopt the Catholic system. The salient point was that the coming century year was deemed to contain a leap day on the Julian calendar, but not on the Gregorian. During the 1600s people traveling from country to country would have got used to there being a ten-day difference between the dating systems, and that would have caused some headaches with regard to writing contracts of sale and so on, but it was not an insurmountable problem. Neither was the fact that the difference was soon to grow to eleven days, but it did provide something to focus the mind upon the calendrical question, and much of the religious animosity provoked by the schism in the Christian churches in the sixteenth century had been allowed a chance to die down. There is another point worth remembering, though: the growing difference in dates, although only by a one-day step, made it more likely that the Julian Easter would be much later (by weeks) than the Gregorian Easter.

The upshot of this was that various Protestant states decided to convert to the Gregorian calendar from the year 1700; for example, Denmark (which included Norway at that time), the rest of the Netherlands (as recognized today), those cantons of Switzerland which were not Catholic (although Saint Gallen resisted until 1724), and the Protestant sections of Germany.

Why Did the British Hold Out Post-1700?

There was a proposal that Britain should reform its calendar at this stage, a matter which was much discussed in the late 1690s, although it is not clear that the full story has ever been revealed. The simplistic belief is that the proposal was founded upon a conversion to the Gregorian calendar simply for the sake of convenience, but it would seem likely that things were a little more complicated than that.

Within Britain the superiority of the calendar of John Dee had not been forgotten, at least among the numerate members of the intelligentsia, and it was the favored calendar if no other considerations (such as matters of trade and communications with foreign powers) were of import. To the members of the ruling set, the safeguarders of a Britain which really did then rule the waves, not only was the thirty-three-year calendar of Dee viewed as being superior on mathematical, astronomical, and religious bases, but also the fact that the English-speaking colonies in the Americas which had now transcended the Latin America of the Spanish and the Portuguese as the dominant presence on the calendrical meridian (God's longitude) provided another reason not to fall into line with an inferior system.

Although nowadays scientists are not much listened to by governments, economists and accountants being considered of much greater significance, in those days astronomers, mathematicians, and the like had the ear of those directly in power. Foremost among these was Sir Isaac Newton, who would have been able to discern the fact that John Dee's century-old proposal was clearly superior on the grounds previously described. On the other hand, adopting that system would have thrown Britain even more out of kilter with the Continent, because the difference in the number of days between the dates employed would shift around according to the phase of the thirty-three-year leap cycle.

Those are reasons of logic and convenience, but in anything connected with Newton's life one should always be aware that there may be underlying matters connected with personal animosity. With foreign philosophers Newton's most famous tangle was with a German mathematician, Gottfried Wilhelm Leibniz (1646–1716); the argument focused upon their respective claims to have invented the integral and differential calculus. In fact, this dispute not only caused a severe ruckus between the pair in question, but also many of their countrymen took sides. In 1699 it was Leibniz who was largely responsible for prompting the Protestant states of Germany to convert to the Gregorian calendar. Under the circumstances, Newton, who was considered the ultimate authority in Britain, could hardly be expected to do likewise; he would have wanted either to bring in a superior system, or not to play ball at all.

Strange Times in Scandinavia

Before leaving the events around 1700, the curious case of Sweden should be mentioned, for interest if nothing else. Let me start with a brainteaser which you may have met before. You get up one morning, and you walk five miles due south, and then you walk five miles due east. Then you take your rifle and you shoot a bear. Then you walk five miles due north, and arrive back at the point from which you started. Here's the question: what color was the bear? And the answer is: white. Why? Because one can only walk fifteen miles along the directions as stipulated and as a consequence come back to your point of origin if that point is the North Pole, which means that it was a polar bear.

Now to Sweden. Actually I was put in mind of the polar bear poser by the fact that the only place I've ever crossed the Arctic Circle (apart, perhaps, from some transpolar flights) is in the north of that country. Forget that you've guessed already that the answer is Sweden, and consider the following puzzle. You pick up an old calendar, and the almanac for February shows a thirtieth day. Where was it issued? (And a supplementary question: to which year does it apply?) The answers, then: Sweden; 1712.

What happened was that the Swedes decided in the 1690s that they would adopt the Gregorian calendar, but did not want to delete ten days from their calendar. So they thought that a solution was to skip all leap year days (all February 29s) between 1700 and 1740, such that from then onward they would at length be in step with the Gregorian system. Having counted no leap day in 1700, they decided that their plan was not so great, and made 1704 and 1708 leap years. Overall this meant that the Swedes were out of step (by either ten days or one day) with everyone else, so the decision was made to shift back to the Julian calendar. To do this they made 1712 a super leap year with 367 days, containing both a twenty-ninth and a thirtieth of February. In the end, Sweden made the conversion in 1753, when its March 1 followed February 17, assumedly because Britain had capitulated the year before, leaving Sweden on its own in western Europe.

Britain's Eventual Reform

There were several considerations which led to Britain eventually adopting the Gregorian calendar from 1752 onward. One was the Act of Union of 1707. In essence, this brought together Scotland with England (and Wales) as a single entity. There was a dating problem because Scotland had begun to count January 1 as the legal start of the year from 1600. Some writers have assumed this meant that Scotland adopted the Gregorian calendar from that time, but that is not the case. The only change was when the year number was incremented, on January 1 in Scotland, but not until March 25 in England and Wales. There was still an eleven day (post-1700) offset from the Gregorian calendar. Thus a particular twenty-four-hour period might be labeled February 1, 1720, in Scotland, February 1, 1719, in England, and February 12, 1720, in most of the rest of Europe.

By this time all almanac makers and diarists in London (such as Samuel

Pepys) regarded the year as starting with January 1, and that also marked the start of the liturgical year for the Anglican Church. The year starting with March 25 was termed "the lawyer's computation," as disparaging a connection then as it is now. Obviously the time was ripe for change.

In another aspect, think of how leap years were reckoned. Throughout the period when Britain had used March 25 as being the start of the year, it was the *anno Domini* number for the major part of the year (March 25 through December 31) rather than the minor part (January 1 to March 24) which decided whether that was to be a leap year, even though the extra day was injected into that smaller part. Recall George Washington's birthday: using double-dating, he was born on the eleventh day of February 1731/32 (Julian). In Scotland (and much of Europe), that day was part of 1732, but in England it was still 1731. Despite that, even in England and its colonies the month in which Washington was born contained a twenty-ninth day.

We can see, then, that there were some matters which made a calendar conversion in Britain a desirable thing. That does not mean that action was swift. In times of war it is often a matter of "cometh the hour, cometh the man": when great deeds are needed, the person able to accomplish them steps forward. On the other hand, in political matters it is often the converse which is true: the hour does not come until the man (or woman) appears.

In this case the hour was in 1750, and the man was Philip Dormer Stanhope, the fourth earl of Chesterfield (1694–1773), a great statesman and author who came from a long-established political family. Chesterfield (British lords usually sign themselves merely thus: just try reading the letters to the editor of the *Times*) was renowned as a cynic and a wit, and published a satirical magazine called the *World*. He railed at the insularity of his countrymen, and was notorious for importing different facets of fashion and culture from the Continent.

Chesterfield had firsthand experience of the desirability of the more worldly attitudes over the English Channel, as he had served in Paris as the ambassador of the Court of Saint James's (named for the palace in London where royal receptions were held). That experience covered not only the agreeable matters of the salon, but also the difficulties inherent in Britain persisting with a dating system different from almost everywhere else. Although communications were still antiquated compared to our present-day Internet links and satellite television transmission—how times fly: when I was a boy I recall that it took four days for the film of the Cassius Clay–Sonny Liston fight to make its way across the Atlantic, while pop groups wrote instrumental pieces lauding satellites like Telstar—Chesterfield saw that the ever-increasing cross-Channel flow of people and commerce necessitated Britain falling into line. And he determined to do something about it.

Chesterfield enlisted the help of various people in drafting and promoting his act. One of these was James Bradley, the Astronomer Royal, but it seems unlikely that Bradley realized how poorly the Gregorian calendar performs in astronomical terms. Another was Martin Folkes, the president of the Royal Society of London, the foremost scientific group in the land; the same complaint could likely be levied at him. Another member of the House of Lords, the earl of Macclesfield, also laid claim to some astronomical attainments, and it was he who smooth-talked the act through Parliament in this respect.

The full descriptor of the act is: An Act for Regulating the Commencement of the Year; and for Correcting the Calendar Now in Use. A.D. 1751. *Anno vicesimo quarto GEORGII II. CAP. XXIII.* Note that this is quite explicit in differentiating between the two alterations required. One was to change the date on which the year was to commence; I have already emphasized that the *Inter Gravissimas* did not address this question, as there was no need. The second was the action of "Correcting the Calendar"—deleting eleven days and altering the leap-year rule on the civil front, but also adjusting the Easter *computus* from the ecclesiastical perspective.

The bill was debated in the presence of 73 bishops and peers of the realm in the House of Lords on March 18, 1750; that is, a week before the end of that year in the English count, but when it was already 1751 elsewhere. It was passed shortly thereafter in the House of Commons (the elected representatives) with few amendments.

The effect of the passing of the act was that the English year of 1751 was to be an *annus confusionis,* with only 282 days. The thirty-first of December in that year was to be followed by January 1, 1752, the old March 25 system being abandoned:

> That in and throughout all his Majesty's Dominions and Countries in Europe, Asia, Africa, and America, belonging or subject to the Crown of Great Britain, the said Supputation, according to which the Year of our Lord beginneth on the 25th Day of March, shall not be made use of from and after the last Day of December 1751; and that the first Day of January next following the said last Day of December shall be reckoned, taken, deemed and accounted to be the first Day of the Year of our Lord 1752.

There are various points of interest in that paragraph from the act. That word *supputation* we'll come to later. But "Europe?" you ask. Remember that the British royal family is partly German, the House of Hanover—in the person of George I, who couldn't even speak English—succeeding the last of the Stuarts in 1714 because of laws passed banning Roman Catholics from the throne after James II converted to Catholicism. In terms of blood there were over fifty superior claimants to George, their religion disqualifying them, which tells you something of the animosity which the British ruling class still felt towards the Roman Catholic Church.

The Hanoverians ruled until 1901, although the throne of Hanover had been lost from 1837 because Queen Victoria, as a female, was barred from inheriting it, and so that separate European nation was lost to the British Crown. The family name of the British royalty became *Saxe-Coburg,* from Prince Albert's teutonic line, but that was abandoned in favor of *Windsor* (the castle which is the main royal residence, west of London) from 1917 due to anti-German sentiment during the First World War.

This, then, shifted New Year's Day to January 1. Still, though, Britain would be eleven days out of step with most other states. Again for reasons of minimal interference with other business, it was decided that in 1752 Wednesday, September 2, would be followed by Thursday, September 14. This deleted the necessary eleven days (September 3–13), but because 1752 was a leap year it contained 355 days in total, and was perhaps an even more confused year.

Florence, Not Pisa

You may have found it tedious back in chapter 11 when I insisted upon differentiating between the *stilo Pisanus* and the *stilo Florentinus:* recall that these are dating systems with the start of the year being March 25, the former counting from 1 B. C. and the latter from A.D. 1. If you had to say that either was correct, then you'd have to choose the Pisan system, because Dionysius Exiguus had founded his chronological framework upon an assumed date for the Annunciation of March 25 in 1 B.C. The Florentine way was wrong: the zero point almost three months into the first year in the *anno Domini* system is well after the traditional dates for the Annunciation, the Nativity, and the Circumcision.

The point here is that Britain was using the aberrant Florentine system right through until the calendar reform. Paradoxically, this led to a lesser hiccup in the day counting than would otherwise have been the case; if Britain *had* been using the Pisan system, then:

1. Its calendar would have been 354 days ahead of the Gregorian rather than 11 behind, emphasizing the point that somewhere along the line the Roman Church had likewise adopted a count shifted from Dionysius's system (that is, based on the traditional Nativity/Circumcision rather than the Annunciation).

2. Chesterfield would have needed to legislate to shift the year numbers so as to count effectively from nine months later (the start of January in A.D. 1), effectively requiring a year lasting for twenty-one months—which would have been the *annus confusionis* to end them all.

The British Easter *Computus*

Given the demonstrated importance of the Easter *computus* in all previous calendrical questions, it is astonishing that there was so little debate in Britain at this juncture about the consequences in this regard. One of the reasons would certainly have been that so few people understood the *computus* as introduced in the *Inter Gravissimas,* with its complicated rules as described earlier. There is clear and inarguable evidence that this was the case: that the British calendar reformers of the eighteenth century did not understand the *computus.* I have already alluded to this in chapter 8, where the writings of the nineteenth-century English polymath Augustus De Morgan were mentioned.

In 1851, the year of the Great Exhibition in London, De Morgan published *The Book of Almanacs,* a tabulation of the thirty-five possible yearly almanacs corresponding to the thirty-five possible dates of Easter (March 22 through April 25, inclusive). In a prefatory section he discussed the proper definition of

Easter, noting that "the description copied into prayer books from the Act of Parliament for the change of style is incorrect in two points: it substitutes the *day of full moon* for the *fourteenth day*, and the *moon of the heavens* for the *calendrical moon*." That is, the legislators mistakenly equated the fourteenth day of the moon with the full moon (which may in reality be the case, but very often it is not), and they also confused the real astronomical moon with the ecclesiastical or calendrical moon, a difference which I have previously emphasized. De Morgan went on to say: "When it happens that Easter Day falls on the day of real full moon, the apparent contradiction always causes a stir." How this anomaly came about De Morgan makes clear both in the above-cited work, and also in his wonderfully witty collection of essays and reviews entitled *A Budget of Paradoxes* (1872).

The bottom line is simple. The original Easter *computus* was defined in order to avoid the Jewish Passover. The Gregorian Easter *computus* was a second-best kludge adopted in order to head off any counterreform by the Protestants. The British Easter *computus* as enacted by Lord Chesterfield in 1751 was erroneous in that the stated rules in the act do not correspond with the Roman *computus*. If you wrote a computer program using those rules you would get a set of results for the dates of Easter which do not agree with those on which it is actually celebrated within the Western churches (Catholic and Protestant). Nevertheless, the tables giving the dates of Easter which the act caused to be printed in the *Book of Common Prayer* actually agree with the dates as derived from the Gregorian *computus* and set out in the *Compendium* of Clavius. Remember that Great Britain/the United Kingdom is a faith-based association per the national church: the monarch is the *Defender of the Faith,* and acts of Parliament plus royal assent are required to change the organization and conduct of the Church of England and the other Anglican Churches.

So what? The so what is that the whole thing is a fudge. Britain enacted a monstrously silly act to reform the calendar in which the writers bent over backward not to mention the Gregorian calendar, the pope, the Roman Catholic Church, or any such phrase. In doing so they made a huge mistake in describing how the date of Easter is to be calculated, because their algorithm results in dates which differ from the Roman Catholic *computus,* and yet the Church of England's Easter tables coincide with those of the Roman Catholic Church. People write of the High Church pertaining to authority in the Anglican sphere; this was high *farce,* and it's never been put right. It took someone like Augustus De Morgan, an Englishman with high attainments in mathematics and astronomy but no religious axe to grind, to spell out how and why Lord Chesterfield's act was nonsense in this regard, precisely a century after it was passed into law.

Did Britain Really Adopt the Gregorian Calendar?

The above matters have wider repercussions than a quick perusal might suggest. I wrote that there were several things "which led to Britain eventually adopting the Gregorian calendar." But did it? Almost all accounts you read will tell you that the calendar we all work on is the Gregorian. Here I would like to throw some doubt on that supposition.

The crux of my argument is this: The *Inter Gravissimas* of Gregory XIII was a *religious* reform of the calendar, whereas Chesterfield's act of 1751 was a *civil* reform. Yes, part of the latter addresses the question of Easter, but it does so only because it was necessary for the organization of Great Britain plus all of its colonies and dominions, to which the act also applied. The *prime motivation* of Chesterfield's act was not to maintain Easter at some specified time of year within acceptable limits, and in accord with an interpretation of the Bible colored by an imperative to distinguish Gentile Christianity from Judaism (as was the case for the Gregorian reform), but to bring the British calendar into synchroneity with that in use in most other trading nations, for the sake of convenience. In this the act is explicit:

> And whereas it will be of general Convenience to Merchants, and other Persons corresponding with other Nations and Countries, and tend to prevent Mistakes and Disputes in or concerning the Dates of Letters, and Accounts, if the like Correction be received and established in his Majesty's Dominions.

That pertains to the prime motivation; the *effect* of Chesterfield's act was to make the British calendar *coincide* with the Gregorian, but that is not the same as being *identical*. Let me say why the distinction between those points is important. A photocopy is not the same thing as an original: it just looks the same, and it may not be acceptable in a legal context.

As I have shown above, following De Morgan, the statements in the act were a hotchpotch designed to avoid mentioning anything to do with the Roman Catholic Church, and by taking that approach the legislators booped in that their verbal description was wrong, but still the tables of dates for Easter are correct (in that they agree with the Catholic dates). The intention, then, is clear: coincide with the Gregorian calendar, but use an alternative calculation scheme so that the systems are not identical. Differing algorithms can lead to the same result: 64 is 8×8, or 8^2, or 4^3, or 2^6, or 16×4, or $10^2 - 6^2$, or . . . If I asked my six-year-old son what is 8×8, he would likely do it by straightforward addition rather than multiplication, and he would certainly not think about raising some number to a power. There are many ways to bring about a coincidental result, the methods being far from identical. Yes, the British got their algorithm wrong, but as they knew what the answer was supposed to be this didn't make much difference; it just confuses them when there is a full moon in the sky on Easter Sunday.

In fact, there are several features in which the British calendar differs from the Gregorian, with the outcome that they only coincide on a temporary basis (even though the timescale is millennia), thus supporting my contention that they coincide but are not identical. For example, given that the rubric for the Easter *computus* in the act and subsequently the *Book of Common Prayer* is clearly wrong, whereas the tables printed in the *Book of Common Prayer* as a consequence of the act are correct, there is only legislative arrangement for the British calendar for the duration of those tables. While the present table is viable through to 2199, and another table is given for 2200–2299, you would need to do some work to calculate dates for Easter thereafter, using the general table which ends with the century starting in 8500, after which you are on your own. Compare this to the succinct rules set out in the *Inter Gravissimas,* which are stipulated to be obeyed in perpetuity. Finding Easter in the British scheme requires a lot of trivial but error-prone arithmetic following a complicated series of steps through various tables, whereas the Gregorian *computus* can be programmed in about twenty lines of code. Both lead to the same result so long as you don't stumble finding your way through the labyrinthine tables in the *Book of Common Prayer.*

That is a distinction between the British and Gregorian calendars in terms of Easter, and hence the Moon. Let's turn attention to the Sun. Another purported reason for the introduction of the act is given as:

> And whereas a Method of correcting the Calendar in such manner, as that the Equinoxes and Solstices may for the future fall nearly on the same nominal Days, on which the same happened at the Time of the said General Council [of Nicaea], hath been received and established, and is now generally practiced by almost all other Nations of Europe.

Here we see that the British involved in the reform at the middle of the eighteenth century had misunderstood the astronomy, even though John Dee and his cohort had been on the ball 170 years before, as had Omar Khayyam even further back in time. The Gregorian reform had a target of keeping the vernal equinox on more or less the same date, which it does in a second-best way; and the autumnal equinox and the solstices are allowed to wander by greater amounts over long periods of time. The eighteenth-century British, however, seem to have forgotten the distinction (as their modern counterparts have done, along with the U.S. astronomers, through confusing different year lengths). But note that you could turn my argument around here: if Chesterfield's act had an aim of maintaining all of the equinoxes and solstices within a minimized range of dates, then the mean tropical year (rather than the vernal equinox year) is the appropriate target length. If that was the case, then there can be no argument that the British calendar and the Gregorian calendar are indeed different!

What about the United States?

When I write about the calendar in the most widespread use around the world, and say that it may be disputed that this is indeed the Gregorian calendar, one's mind must turn to the United States. Without looking up the legal codes of Spain, Portugal, the eventual Italy, and so on, the places which fell in line with the Gregorian reform over four hundred years ago, one can be fairly confident that indeed their civil calendars were defined as being *identical* with that stipulated by the *Inter Gravissimas.* I've said that the British reform of 1751 led to the calendar in use there from January 1, 1752, *coinciding* with the Gregorian convention. At the time Great Britain was the dominant world power, whereas nowadays it is the United States that fills that role, and the global uniformity of culture (including, of course, matters of time and calendar) is controlled to a very large extent by the influence of that nation. One must ask then, on calendrical matters, what about the United States?

Elsewhere I stated that the United States has no specific legislation defining its calendar, and that in effect it inherited the rules as set down by the act of the British Parliament in 1751. But that is a little simplistic. In fact, one could even argue that it's wrong.

The British act applied not only to the calendar to be used in London and Edinburgh, but also to Boston, Baltimore, and Philadelphia. This was a law for all the British dominions and colonies. But to imagine that this was then directly inherited by all of the United States as currently constituted is to pay attention only to the thirteen red-and-white stripes in the flag, and not the full complement of fifty stars. To see why, let's consider the most recent admissions to membership.

Hawaii became the fiftieth state in 1959, having been administered as a U.S. territory since 1900, previously being a British possession as the Sandwich Islands. The state flag carries a Union Jack at top left, as do the national flags of Australia, New Zealand, and various other old British colonies. The calendrical history of Hawaii over the past few centuries therefore does not need to be spelled out.

But what of Alaska? The forty-ninth state was formerly a possession of the Russian Empire, and was sold to the United States in 1867, when Russia was still on the Julian calendar. In that century there was a difference of twelve days between the dates on the Gregorian and Julian schemes. The jump from one system to another occurred when Thursday, October 6, was followed by Friday, October 18. From then onward the United States administered Alaska using that calendar, although the Russian Orthodox Church, which could claim religious allegiance from the pious in that region at the time, has continued using the Julian calendar.

There is another point one could ask about Alaska in 1867. Which day was it there? Nowadays, of course, Alaska looks eastward toward the contiguous United States for its day reference, but in 1867 it looked westward toward Moscow, which might be thought of as being a day ahead. Did Alaska move a day when its administration was transferred? The point is moot in that the international date line was not introduced until 17 years later—the IDL starts off at the North Pole on the 180 degree meridian, but is then drawn to the east so that it can cut diagonally through the Bering Strait, continuing to the west of that meridian so as to steer around the Aleutian Islands before returning to the 180 degree line—but it is an interesting puzzle. Did an Alaskan frontiersman in 1867 really know which day it was? Did it matter?

Those are the youngest states (in terms of when they joined the United States), but what about the areas first settled by Europeans? You should already have an inkling here of what I am intending to convey from the discussion in chapter 14, when I mentioned Spanish exploration of the regions we call Texas, New Mexico, Florida, and so on in the early sixteenth century. Spain being a Catholic country, one might expect that these regions would have been ad-

ministered under the Gregorian calendar, but the situation was even more explicit than that: the first settlers were Catholic priests, so of course the calendar
they brought with them was the religion-based Gregorian system. When the
first missions were built all across the south, and on the Pacific Coast at the locations which were to grow into the cities of San Diego, Los Angeles, and San
Francisco, naturally the calendar used was that of Pope Gregory XIII.

The French settled thereafter in the area which the United States bought
from Napoléon in 1803—the *Louisiana Purchase,* from the Mississippi in the
east to the Rockies in the west, all the way from the Gulf of Mexico to the
Canadian border. Another vast expanse initially on the Gregorian calendar, and
for three weeks on the French revolutionary calendar (see chapter 19). Similarly, one may wonder about the calendar used in New Amsterdam until the
British seized it from the Dutch in 1664 and renamed it New York.

In view of the above it would be absurd to make any statement along the
lines that the Gregorian calendar was adopted by the United States at some
juncture. It would be wrong on several counts. When the thirteen original
states on or near the northern Atlantic seaboard declared independence from
Britain in 1776, the calendar that they implicitly decided to use in the absence
of any contrary move was the calendar put in place by Lord Chesterfield in
1751. That may be viewed as being the calendar of the Constitution of the
United States, which was framed in 1787, and I would maintain that it is *not*
identical with the Gregorian calendar. Indeed, since the United States is a secular nation, there would surely be difficulties with it adopting a calendar which
is indisputably predicated on religious concerns.

That's one note against the notion of the Gregorian calendar having been
adopted by the United States. Now, what of those regions which later affili-
ated? Clearly the Gregorian calendar was that used in, say, Arizona Territory
from when the first Spanish missionaries arrived, and similarly for the regions
later organized as Texas, New Mexico, California, and so on. When Arizona became the forty-eighth state in 1912, what happened? The answer is that there
was no change in dates necessary, unlike when Alaska was bought from the
Russians, but henceforth—and one could say earlier, too, when the area was
administered as a U.S. territory—Arizona joined with the rest of the country
on the calendrical front. I won't worry about arguing whether one might say
that Arizona at some juncture *gave up* the Gregorian calendar, because there
are more significant things to think about, even calendrically speaking. Let's get
back to Chesterfield's act and see why.

A Peculiarly British Calendar

In chapter 15 it was made clear that in the seventeenth century the annual
round followed in Britain and its colonies was peculiar to the British, with regional differences both within the British Isles (for example, the Scots then, as
now, commemorated the battle of Bannockburn under Robert the Bruce in
1314) and without (for example, in New England the previous holidays of November 5 and 17 became amalgamated to produce Thanksgiving). As a matter

of fact, Chesterfield's act was also explicit in this connection, rendering another reason to argue that the resultant calendar is not identical with the Gregorian.

If one reads through the full act, the impression is strong that this was not a full reform of the calendar at all, but only a half-reform. One set of activities was to be shifted forward in time, with the New Style calendar, while others were to stay on the Old Style (a specifically British, but nevertheless Julian, calendar).

Into the former class fell most legal matters and social activities; for example, the dates of various periodic assizes were to be shifted (it is from these that the circuit courts get their name, since the judge would travel around a circuit from town to town, producing a certain cycle time).

The latter class included economic and agricultural matters. Mentioned way back in chapter 1 was the shift in the British tax year to the presently used end date of April 5; if the date had not been moved by eleven days, then that particular tax year (1752–1753) would have been lessened in duration by that amount, producing an unfair levy and maybe civil unrest. Similarly, all private rents, tithes, and so on were to be delayed on the calendar, nevertheless being due for payment when the requisite number of days had elapsed. Although they got the astronomy wrong, the authors of the act took pains to ensure that the people were not disadvantaged.

Agricultural affairs were another set of events which had to be delayed on the New Style calendar so as to occur at the correct seasonal time of year; the act terms these the same *natural days* (that is, at the same phase in the natural/seasonal cycle). It would be no use scheduling a Harvest Festival for ten or eleven days earlier than usual, when the hay would still be in the fields. Similarly, May Day was a major festivity, and there is usually a marked difference in the weather between April 20 and May 1. The weather on the New Style May 1 caused some discontent, and people wanted the Maypole decoration to be delayed until the season caught up with what they expected for their fairs.

Christmas was another problem. Near where I was born stands the town of Glastonbury, which is thought by many to be a mystic place, and so it is inundated with New Age folk around midsummer each year. In the longer term— and I mean from very early in the history of Christianity—it has indeed been a religious center, and in chapter 11 I mentioned Glastonbury Abbey. Glastonbury stands on the Somerset Levels, an extensive flat area of peat bog which was largely drained in medieval times, and hard beside the town stands a steep hill which almost looks artificial, so round and severe are its lines. It can be seen from many miles away. On top stands a tower, and the whole structure is called Glastonbury Tor. In fact, the tower is the ruin of a medieval church which had been rebuilt several times, but some New Agers like to think of it or nearby Cadbury Castle as the site of King Arthur's Camelot.

There is another interesting thing about the town. There was a tree there called the Glastonbury thorn, supposed to flower at Christmas and Easter each year, and perhaps because of that the Roundheads cut it down during the Interregnum: these Puritans ordained that Christmas was not a time for festivities, so that flowers were regarded as inappropriate. Legend says that the original tree grew from a staff brought to the area from Palestine in the first century A.D. by Joseph of Arimathea, who plunged it into the soil, and that Christ's crown of thorns was made from the same plant.

A type of hawthorn, after it was felled various cuttings were saved and transplanted around the town and elsewhere, and some of these grew and could be displayed openly after the Restoration. It is said that back in 1752 their flowers failed to appear on December 25, putting in a belated appearance about a week and a half later. Obviously the trees had not read the act and changed their behavior accordingly. The upshot was that many locals interpreted the late flowering—late on the new calendar, but at the same seasonal time as previously—as implying that the calendar reform was fundamentally wrong, and an offense to God.

For this and related reasons elsewhere, Christmas was split into two for almost a century in many parts of Britain: the New Style Christmas according to the calendar and the church, but the Old Style Christmas which was recognized and noted by the common people. This can lead to some confusion. In the eighteenth century, December 25 (Julian) was the same day as January 5 (Gregorian), and it was on that date on the New Style British calendar that people celebrated what I have here termed *Old Style Christmas*. In the nineteenth century the calendar differential increased to twelve days, moving the date to January 6. The quirk here is that January 6 is also labeled as *Old Christmas* from way back before this, when the fledgling Church used it as the day (now called *Epiphany*) to mark the birth of Jesus Christ, prior to the time when Christianity was deemed strong enough to wrest the traditional winter solstice celebration date from the control of the Mithraists and other pagans.

During the twentieth century (and continuing into the twenty-first) the calendar differential has been thirteen days. My next-door neighbors in Australia still think of Christmas as being on January 7, but that is because they are Russian; that nation shifted to the Gregorian calendar eighty years back, but the Russian Orthodox Church sticks to the Julian scheme, as do many of the Russian people even if they are not religious. One sees modern-day echoes, then, of what happened in the period following the calendar change in 1752.

These are not the only things which caused an upset in England when the calendar was changed. Another is the orientation of churches. Many English churches dating from several centuries back were built with orientations toward the place on the horizon where the Sun would rise on the day of the dedication saint for that church. Imagine a church named for Saint Mary and built in 1600: it would have been oriented toward a point a long way north of east, because on the Julian calendar the Sun rose due east at the equinox, which was on March 10 then, and by the time of Saint Mary's Day (our favorite March 25), the Sun had moved quite a distance north in its annual round trip up and down the horizon between azimuths of about forty degrees each side of due east. When the calendar was changed the equinox was moved to March 19–21, and in consequence sunrise on the twenty-fifth was thereafter only a few degrees north of due east. Any addition to the church—a new chancel, say—would be oriented toward this new reference point, and so the church would become crooked. This is a well-known phenomenon; there are eighty-one crooked churches in Oxfordshire alone!

Other Features of Chesterfield's Act

There are several other features of the act which it may be of interest to consider, such as the following passage:

> . . . the Years of our Lord 2000, 2400, 2800, and every other fourth hundred Year of our Lord, from the said Year of our Lord 2000 inclusive, and also all other Years of our Lord, which by the present Supputation are esteemed to be Bissextile or Leap Years, shall for the future, and in all Times to come, be esteemed and taken to be Bissextile or Leap Years, consisting of 366 Days, in the same Sort and Manner as is now used with respect to every fourth Year of our Lord.

There are four points I'd like to mention just from that section. The first is that it is quite specific about the year A.D. 2000 being a leap year: there is no excuse for the various writers and programmers who have erred in supposing that 2000 is a common year with only 365 days.

Second, I previously noted that the *Inter Gravissimas* is explicit in providing both a leap-year rule and an Easter *computus* which are applicable in perpetuity (although, of course, the Gregorian calendar may well be abandoned at some stage by church and state alike); here the act at least stipulates a simple and perpetual leap-year rule ("for the future, and in all Times to come"), even if it got the *computus* wrong.

Third, there is no sign of the four-thousand-year rule which is beloved by many interested in calendar myths. This suggestion was not made until almost a century later, by John Herschel; and a suggestion is all that it was, predicated on a false analysis at that.

Fourth, that lovely word *supputation* appears in the text. Remember the *supputatio Romana vetus* in chapter 8? Indeed, Chesterfield's act begins as follows:

> WHEREAS the legal Supputation of the Year of our Lord in that Part of Great Britain called England, according to which the Year beginneth on the 25th Day of March, hath been found by Experience to be attended with divers Inconveniencies, not only as it differs from the Usage of neighbouring Nations, but also from the legal Method of Computation in that Part of Great Britain called Scotland, and from the common Usage throughout the whole Kingdom, and thereby frequent Mistakes are occasioned in the Dates of Deeds, and other Writings, and Disputes arise therefrom.

I'd never seen the word before, so I figured I should look it up in the *Oxford English Dictionary*. The OED is a tremendous resource, giving the first-known usages of all words in order to establish the precedent, in the same way as the British legal system is founded upon the precedent system. The meaning, I found, is quite specific: the act or action of calculating or computing, normally in a calendrical connection. And guess what: the OED gives as an example of the use of *supputation* the selfsame act of Parliament, quoting the initial words above!

The Uptake of the Calendar

Although the framers of the act had been careful in many respects regarding the way in which it was to be implemented, trying to ensure that there was no injustice, there were certainly some things that they did not anticipate. As noted above, the British legal system is based upon precedent, so that there are many thousands of previous acts of Parliament which might be affected, and it was virtually impossible to take them all into account. For example, there might be a four-century-old statute allowing the common people to graze their stock upon some communally owned pasture between certain specified dates, whereas another statute for a neighboring village specified the dates to run for two months abutting the summer solstice; the effect of the act would thus be to shift one, but not the other.

One major problem with the act was that it was vague with regard to what was to be done with events which were due to happen in the missing eleven days. If a contract called for certain items to be delivered on September 6, 1752, when would they be due instead? The people found their way around such hassles by themselves instigating a temporary double-dating system, similar to the previous (say) 1749/50 convention except now applied to days and months only. Such matters were tidied up with later legislation. Another example was the herring fishing season: the fishermen were obliged by the initial act both to put to sea and then return eleven days earlier (in the brief season) than intended, and this required rectification.

There were several other errors and omissions in the act. One is that it said nothing about the days of the week. The most generous interpretation is that it was presumed that as they are not mentioned, they were not to be affected; that is, that the day after Wednesday, September 2, 1752, was assumed to be a Thursday, unaffected by the fact that it was to be counted as September 14. This omission, however, could lead to difficulties. For example, what if a particular fair or mart were scheduled for the first Saturday in July each year? The act said that fairs should remain on their same natural or seasonal days, so that in this case the equivalent would be the first Saturday after July 12. Confusing.

In fact there is another contradiction within the act which stems from this. It is stated that the rules set down are to be followed perpetually ("in all Times to come"). But in various places, to clarify matters, the act states that certain events deemed to occur in the future on the same natural days should be scheduled for dates specifically *eleven* days later on the New Style calendar. Well, that makes sense and is more or less correct, except that after 1800 the difference between the New Style and the Old Style was due to grow to *twelve* days (and *thirteen* after 1900). Perhaps the framers of the act imagined that matters would have settled down by then, and so they felt that they didn't need to be explicit about this obvious evolutionary step. If that were the case, though, why were they worried about spelling out the leap-year rules through to the year 2800 and beyond? (My answer to that: stupidity and sophistry.)

It had been confidently expected that the calendar change would not occasion any great difficulties, but the problems which common people had with following the dates were considerable. To such men and women the only time frames for their lives were the calendar followed by their church and the printed almanacs. Their year proceeded as the work which needed to be done changed, from plowing and sowing in spring, to reaping and storing the grain in late summer and early fall. It was only the gentry, the rich tradesmen and the masters, who could afford clocks, and thus thought to split both the day and the year in any other way. Before 1752, the almanac printers had given two columns for dates, one headed English and another denoted foreign or Roman. In 1753 they thought fit to print only one column giving the New Style calendar, together with a verbal account of the changes, but their customers were so confused (and resistant to having their seasonal fairs and so on shifted) that the printers had to revert to inserting both New and Old Style columns for the following quarter century. The acceptance and usage of the new calendar was far from an overnight thing.

Riots? What Riots?

Pick up most any book about the British reform of their calendar and it will tell you that there were widespread riots, the people rebelling against a change which, various authors state with conviction, they thought was stealing eleven days from their lives. Now, it is true that many people like to think the worst of any government action: for example, I recall that when the United Kingdom decimalized its currency on February 14, 1971, many people seriously argued that the government was thieving 140 pence in every pound from them (there was previously a three-tiered system with 20 shillings each of 12 pence in a pound, making 240 in all; the decimal system is two-tiered with 100 new pence in a pound, each of them worth 2.4 of the old pennies). But the idea that the people thought that they would die eleven days earlier seems far-fetched.

Numerous historical scholars have repeated various claims along these lines, although not always making such extreme statements. For example, G. J. Whitrow, in his book *Time in History* (1988), wrote:

> Some workers actually believed they were going to lose eleven days' pay. So they rioted and demanded "Give us back our eleven days!" (The Act of Parliament had, in fact, been carefully worded so as to prevent any injustice in the payment of rents, interest, etc.) The rioting was worst in Bristol, in those days the second largest city in England, where several people were killed.

Starting from such received wisdom, Robert Poole of Saint Martin's College, Lancaster, England, has looked into what actually happened in 1752, and just published an intriguing book—*Time's Alteration: Calendar Reform in Early Modern England*—which deals with the subject. (Note that this Robert Poole is not related to the deceased Reginald Poole of Oxford University who was mentioned in chapter 11; it seems astonishing that England should throw up two chronologists with the same surname, given that it is a rare field of scholarship.) As our contemporary Poole starts out by saying, "For generations, there has been no better illustration of the collective idiocy of the crowd than the story of the English calendar riots of 1752." Poole then goes on to show that this is just what it is: a *story*. There is simply no evidence to support the belief that workers rampaged in the streets of Bristol, trampling some to death. As Poole continues: "The riots, like the *Snark*, are universally known but defy detection. The riot depositions in the Public Record Office are silent on the subject."

So how did the story begin? I mentioned earlier that Lord Chesterfield published a journal called the *World* in which various writers poked fun at all and sundry. One contributor was Horace Walpole, a prominent author and essayist whose father (Sir Robert Walpole) had been prime minister for more than two decades earlier in the eighteenth century. Walpole wrote a piece mocking those who had predicted that there would be widespread trouble following the reform. While there had indeed been some disruption, the people soon settled down to their business and the circumstances were much more muted than the doomsayers had predicted. One hopes that our passage into the next century and millennium is similarly comfortable, contrary to what many are prone to predict.

This general attitude seems to have been taken up by many of the leisured or prominent class. One was William Hogarth, the engraver whose cartoons of the period give us our best visualization of what England was really like in those days. One of his etchings was published in 1755, and is connected with a parliamentary by-election in Oxford. It shows a drunken brawl in a tavern, with a placard thrown to the floor on which is inscribed: "Give us our eleven days!"

What Hogarth was trying to intimate seems clear: that the political wranglings surrounding this by-election make as much sense as the absurdities spouted a few years before by those prophesying that great civil unrest would attend the calendar reform, an episode fresh in the popular memory. In reality, it never happened.

There were a few hiccups, but it all went through quite smoothly in the end. Whether it deserved to do so—whether Chesterfield and his colleagues should have deserted John Dee's superior calendar, and whether they may be excused for the abomination which is the rubric for the Easter *computus* in the act—is another matter. To paraphrase T. S. Eliot, this is the way the calendar changed, this is the way the calendar changed: not with a bang, but a whimper.

George Washington's Birthday

This book began with a chapter entitled "George Washington's Birthday." It is appropriate that I close this chapter by making yet another point about how Chesterfield's act affected that anniversary.

In that opening chapter I wrote that "throughout his life, Washington celebrated his birthday on February 11." I wrote that he was not mistaken (in that it was the date on which he was born). But was he *wrong* to take that as being the anniversary of his birth?

The answer is *yes*, Washington was wrong. He was born in February of the year which we retrospectively term 1732. He was nineteen years old when Chesterfield's act was passed by the British Parliament. He was twenty years old by the time of the shortened month of September 1752, his twenty-first birthday being due the following February. The final verbiage in the act reads as follows, referring to when a person is recognized as having come of age (reached the age of twenty-one years) or to have completed an apprenticeship or period of indentured service:

> . . . and that no Person or Persons whatsoever shall be deemed or taken to have attained the said Age of one and twenty Years, or any other such Age as aforesaid, or to have completed the Time of such Service as aforesaid, until the full Number of Years and Days shall be elapsed on which such Person or Persons respectively would have attained such Age, or would have completed the Time of such Service as aforesaid, in case this Act had not been made; any Thing herein before contained to the contrary thereof in any wise notwithstanding.

The meaning is clear. Whether George Washington liked it or not, the act proscribed that the mere moving of the dates such that February 11 came eleven days earlier than it would otherwise have done did not affect the time which he needed to wait before he legally attained the age of twenty-one years. It follows that his birthday celebrations in 1753 were eleven days premature. The anniversary of his birth, according to Chesterfield's act, was properly on February 22.

This reconsideration of George Washington's birthday introduces the thought: how did the American colonies respond to the calendar change imposed upon them by their masters in London? In the next chapter we will discuss this question.

Chapter 17

❦

POOR RICHARD'S ALMANACK

A substantial proportion of the precision timekeeping services required by the United States, and indeed the world as a whole, derive from facilities located along the Front Range of the Rocky Mountains in Colorado. Coming in to land at the splendid new Denver International Airport, on the flatland east of the Mile High City, you might be able to glimpse the town of Boulder about thirty miles to the northwest, nestling at the point where the mountains rise precipitously above the slopes and plains.

Boulder's population fluctuates somewhat, depending on whether it is semester time at the University of Colorado. The students (and multifarious staff) of that institution comprise a large fraction of the population of this university town. But Boulder is also the home of various high-tech enterprises, especially computer and aerospace companies, and government laboratories. Driving up the highway from Denver, you enter Boulder on Broadway, passing Table Mesa Drive to the left. If you were to follow that drive toward the mountains, being careful to slow down for the deer which gambol beside the road, at the end you come to the National Center for Atmospheric Research, positioned under the Flatirons, the monolithic slabs which seem to hang over Boulder.

Back on Broadway and proceeding into town, on the left you would notice a building which looks like an oversize beached submarine, a thousand miles or more from the sea. This is part of the National Institute of Standards and Technology (NIST; formerly the National Bureau of Standards), and in particular the Time and Frequency Division. The legal definition of time for the United States is based on a set of atomic clocks and other sophisticated devices at the U.S. Naval Observatory in Washington, D.C., but if you actually need to obtain a precise definition of the time, it is NIST in Boulder which supplies it to you, one way or another, either by phone line or through a radio transmission.

The antennae which broadcast the time signals, however, are not actually located in Boulder. To see where *that* happens, you would need to drive another forty miles north, to Fort Collins. There, using a set of reserved frequencies, radio station WWV provides standard time signals to the world. You can set your watch by phoning the talking clock with sufficient accuracy to ensure that you're on time for your dentist appointment, but if you need to know the time to millisecond accuracy, then WWV has for decades been a convenient source, with the WWVH outstation in Hawaii providing a special service to the Pacific Basin. I remember setting the master clock on a radar system I was building in New Zealand using WWVH, a complication being the fact that for such a long travel distance I had to allow for the finite speed of the radio signal, causing a delay of about a twentieth of a second. One might also wonder whether it is possible to tell between the two, given that they broadcast on the same frequencies. The answer is that for WWV a man announces the time signals, while for WWVH a female voice is used.

But technology marches on, and a more convenient time signal for many requirements may now be obtained using a device no bigger than a mobile phone and costing just a couple of hundred dollars (batteries extra): a GPS receiver. The Global Positioning System is made up of a fleet of satellites in twelve-hour orbits which emit precise time references, allowing a user to switch on a hand-held receiver and quickly be informed of his or her position anywhere in the world with an accuracy equivalent to the area of a typical house plot. Very wonderful, if you are lost on a backwoods hike. GPS makes it easy to know where you are at any instant.

The Global Positioning System is a boon to many areas of civilian life, but it was not built and launched as a direct service to the public. Its major purpose is to provide a precise navigational system for the U.S. Department of Defense, and the GPS satellites are monitored from Falcon Air Force Base. To get there, you would need to drive a hundred miles or so south from Fort Collins, past Denver again, to Colorado Springs; and then ten miles east out onto the plain. But don't expect to be ushered in for a free tour of the GPS facilities. The perimeter fence sign reads, "Deadly force will be used beyond this point," and they mean it.

If you want to discuss GPS in Colorado Springs, the *only* place to do it is in a pizza parlor called *Poor Richard's* on North Tejon. It is frequented by air force staff from Falcon AFB, plus Peterson AFB (which is closer to town), plus various Space Command and NORAD people from Cheyenne Mountain, and cadets from the Air Force Academy nearby. Actually, my discussion above was largely a circuitous introduction to this restaurant. Or at least its name.

The Birth of Poor Richard

The name of the restaurant actually derives from *Poor Richard's Almanack,* which was written and published from 1733 to 1758 by none other than Benjamin Franklin (and until 1787 by others). *Poor Richard's* (the almanac I mean, not the pizza parlor) was a miscellany of sayings and proverbs interspersed with the sorts of material which one would still expect to find nowadays in an astronomical almanac: rising and setting times for the Sun and the Moon, positions of the planets, and predictions of eclipses and other celestial phenomena.

Franklin included other predictions less likely to be correct, like long-term weather reports: farmers and others intimately affected by vagaries of the weather comprised a large segment of his market. By 1753, though, he was writing that

I am particularly pleas'd to understand that my Predictions of the Weather give such general Satisfaction; and indeed, such Care is taken in the Calculations, on which those Predictions are founded, that I could almost venture to say, there's not a single One of them, promising Snow, Rain, Hail, Heat, Frost, Fogs, Wind, or Thunder, but what comes to pass punctually and precisely on the very Day, in some Place or other on this little diminutive Globe of ours.

One would imagine that only the stupidest of his subscribers would have continued to place confidence in his forecasts.

When it came to stars and planets, Franklin did not limit himself to *astronomical* predictions, but instead deviated initially into suggesting the future occurrence of events with a more *astrological* bent. He may have become a prominent scientist, and of course a great statesman, but first and foremost Franklin was a *businessman*. He wanted to sell his products. The first edition of *Poor Richard's Almanack* made its mark by, among other things, predicting on an astrological basis the exact day, hour, and minute of the impending death of a rival publisher, named Titus Leeds, who had cornered a large part of this popular market with *The American Almanac*. It was, of course, a hoax. Franklin had begun this hoax in his *Pennsylvania Gazette*, perhaps copying the idea from Jonathan Swift, who had employed a similar tactic for garnering publicity a quarter century before, and it may even have given him the belief that an almanac with the whimsical, tongue-in-cheek style of *Poor Richard's* could be profitable. To his delight, Leeds rose to the bait, calling Franklin a liar, a fool, a conceited scribbler, and doubtless many other names which never saw the printing press. Sales of *Poor Richard's* boomed, and Franklin continued the joke for years, to the great chagrin of Leeds.

Whence the name *Poor Richard's*? Actually, Benjamin Franklin was not the first almanac maker (or *philomath*) in his family, his elder brother James having published *Poor Robin's Almanack* from 1728, five years ahead of the appearance of *Poor Richard's*. Benjamin seems to have borrowed the form of the name, again a good business practice (even if not likely to bolster fraternal harmony). His brother could hardly complain, however, having stolen the title himself from an English almanac which had been started more than sixty years earlier still.

Why *Richard*? Franklin wrote his almanac under the pseudonym Richard Saunders, this character professing to be a destitute stargazer nagged by his wife into doing something to bring in much-needed cash. In fact, Richard Saunders was the name of a real-life English philomath of the previous century, whose almanac (the *Apollo Anglicanus*) Franklin may have come across while working as a printer in London in 1724–1726.

Benjamin Franklin seems to have used well over a dozen pseudonyms in his writing, a matter to which I will return in chapter 20, where I will suggest another Franklin guise which has so far escaped detection, one directly connected with calendar reform. One of his noms de plume was the notorious Polly Baker of Connecticut who eloquently wrote of the injustice of being fined and whipped for fornication, having borne five illegitimate children, while the fathers, who deserted her in each case, escaped without punishment or responsibility. Laws in those days were different: one could be fined in a civil court for not attending church on Sunday, or for simply traveling on that day. In view of our earlier discussions of when a day starts, it is interesting that legally that period—the Christian Sabbath—began at sunset on the Saturday. Similarly, if one tries to travel in Israel today, say by driving a car, at some time after the start of the traditional Jewish Sabbath (sunset on Friday), then you may find your vehicle the target of stones thrown by Orthodox Jews.

The Sayings of Poor Richard

Franklin was pretty good at borrowing things, and the huge popularity of *Poor Richard's Almanack* over the twenty-six years that he wrote and published it was in the main due to the aphorisms he sprinkled through its pages, adapted for the most part from common sayings and proverbs which he took and gave a little twist. Its circulation reached 10,000, and it was read by an estimated one in a hundred of all those in the American colonies, while word of mouth spread his sayings to a much greater proportion of the populace.

The almanac was also a prolific seller elsewhere, going through fifty-six editions in France alone. This popularity has resurged in recent decades due to republications of the text with modern illustrations by Norman Rockwell. Scholars have almost built careers on investigations of the sources of Franklin's quips and truisms. Perhaps the best known is "Fish and visitors smell in three days," which appeared in the 1736 edition, as did "God helps them that helps themselves." "Holding one's nose to the grindstone" gained frequent usage after Franklin coined a similar phrase in his last edition of *Poor Richard's,* the long introduction to that issue of 1758 having been reprinted many times under the title *The Way to Wealth.* In my opening chapter I used the phrase "Time is Money"; Franklin had his Poor Richard pen that in the prologue to the edition of 1751.

A substantial proportion of the phrases we use without thinking in everyday parlance come, without our recognizing it, from the quill of William Shakespeare or the pages of the Bible. A flick through *Poor Richard's Almanack* would show that Benjamin Franklin also has a certain control over our tongues, even two centuries after his death. That is not to say that Franklin was the *originator* of these sayings, and it seems that only about 5 percent of all the aphorisms he printed in *Poor Richard's* were actually his invention. Many collections of proverbs award him with the credit for "Early to bed, and early to rise, makes a man healthy, wealthy and wise," but that saying was current at least by 1639. Contrary to what many believe, "Many hands make light work" did not appear in *Poor Richard's;* on the other hand, Franklin *did* write about the inevitability of death and taxes, but not until the U.S. Constitution was drawn up in 1787, thirty years after he had ceased publishing his almanac.

About the time that Franklin was telling all that "Time is Money," two major scientific matters were occupying his mind. One was his invention of the lightning rod, stemming from his initial experiments with electricity begun in

1746 and involving his famous investigations of atmospheric discharges using kites. Franklin may have been keen on profiting from his printing, but he did not have the same attitude toward his scientific or technological discoveries. A modern-day patent attorney would lose a lot of sleep (and fees) if Franklin were his client. In *Poor Richard's* for 1753, Franklin gave a description of how houses and barns might be protected against lightning strikes, beginning: "It has pleased God in his Goodness to Mankind, at length to discover to them the Means of securing their Habitations and other Buildings from Mischief by Thunder and Lightning."

The other scientific matter Franklin had in hand was the reform of the calendar, which he was able to view from several perspectives: from the position of scientific expertise; from the point of view of a public personage interested in the future of the American colonies; from a business perspective; and, of course, specifically with regard to being an almanac maker who would need to be on top of calendrical changes, because his readers looked to *Poor Richard's* for advice and definitive knowledge. So what did Franklin tell them?

Calendar Reform in *Poor Richard's Almanack*

Because he is often regarded as being one of the mainstays of the American Revolution and he was central in the drawing up of the Declaration of Independence, one might imagine that Franklin was an avid opponent of calendar reform, perhaps leading an American revolt against the imposition of a new calendar—which we have seen is a powerful tool of subjugation—by the political masters in London. Was there perhaps an early echo of "No taxation without representation," as Franklin exhorted the colonists to man the barricades with a transatlantic equivalent of "Give us back our eleven days"?

Far from it, in fact. While there was a certain amount of unrest on the part of the common people who could not understand what was being done, and how it would (and would *not*) affect them, Franklin was not a part of it. The situation was as muted as in England. Franklin could understand the scientific basis for the calendar reform, and gave an excellent account of the history of the calendar and the reasons for the change in the introduction to *Poor Richard's Almanack* for 1752. Even if he had disagreed with the reform, at least from the scientific perspective, at that time Franklin was still a royalist; it was not until about 1760 that he started to turn against England as a colonial power, seeing independence for the American colonies as being a desirable and necessary target.

That is not to say that he was appreciative of all things English until that time, and he frequently wrote diatribes criticizing how the English ruling class treated subordinates. I have been careful with my words there, deliberately using the term *English* rather than *British,* for Franklin was dismayed in particular with the ways in which such English people treated not only the transatlantic colonists, but also the Welsh and the Scottish, not to mention the Irish, and the English working classes. Some of these diatribes were barbed and serious, while others were a little more in jest (although with a point to make), such as when he suggested that if British convicts were to continue to be sent to the American colonies, then it would be appropriate that a plentiful supply of rattlesnakes be sent back in return, and released into the Royal Parks of London where parliamentarians might encounter them. Not the sort of quid pro quo that the English expected from a subject colony.

Franklin began *Poor Richard's Almanack* for 1752 with the following paragraph:

> *Since the King and Parliament have thought fit to alter our Year, by taking eleven Days out of September, 1752, and directing us to begin our Account for the future on the First of January, some Account of the Changes the Year hath heretofore undergone, and the Reasons of them, may a little gratify thy Curiosity.*

Note that the date of January 1 was mentioned "to begin our Account." People were *already* working on years which ran January–January, as can be seen from the fact that each edition of *Poor Richard's* started with that month, and the first was published in December 1732 (ready for January 1733). What was changed was that the legal year was now to run over that course. Franklin's choice of the word *account* is interesting, however, because the financial or tax year, which we might nowadays call the *accounting year*, was actually left to run from April to April (finishing on a date equivalent to March 25, as discussed in chapter 1).

Franklin continued:

> *The Vicissitude of Seasons seems to have given Occasion to the first Institution of the Year. Man naturally curious to know the Cause of that Diversity, soon found it was the Nearness and Distance of the Sun; and upon this, gave the Name Year to the Space of Time wherein that Luminary, performing his whole Course, returned to the same Point of his Orbit.*

One might here object to the use of "Nearness and Distance of the Sun" on the grounds that it is not the changing rectilinear distance to the Sun which produces our major seasonal variations, of course, but rather the tilt of Earth's spin axis. To give Franklin the benefit of the doubt, however, one could suggest that he meant by "Nearness and Distance" the angle between the Sun and the zenith as it crosses the meridian each day. Outside of the tropics the Sun is nearest the zenith at transit on the day of the summer solstice, and most distant from it at the winter solstice. His words comprise a clumsy explanation, though, for an unlearned reader.

Franklin Knew How Long a Year Should Last

Franklin next gave a historical sketch of how different peoples had arrived at disparate values for the average length of a year (however defined), how the Roman calendar had been developed and then widely adopted, how Pope Gregory XIII had ordered a reform, and how various European nations had at length bowed to the inevitable and come into line with the Gregorian calendar. "And now in 1752, the English follow their Example. . . ," he wrote, noting that "this Act of Parliament . . . extends expressly to all the British Colonies. . . ."

Franklin did realize, though, that this calendar reform led only to a more precise average value for the year, and that eventually the errors would accumulate:

> *Yet is the* Gregorian Year *far from being perfect, for we have shewn, that, in four Centuries, the* Julian Year *gains three Days, one Hour, twenty Minutes: But it is only the three Days are kept out in the* Gregorian Year: *so that here is still an Excess of one Hour, twenty Minutes, in four Centuries: which in 72 Centuries will amount to a whole Day.*

Oh heaven, Oh joy! Read that again, and ponder! Compare with almost everything which has gone before in this book! Franklin knew that the yardstick against which the accuracy of the Gregorian rules (or any other calendar rules based on the vernal equinox) should be judged is not the tropical year, but the time between vernal equinoxes! All praise to Benjamin Franklin!

That point is important, so I must emphasize it. In Britain, the framers of Lord Chesterfield's act had been clumsy and apparently ignorant of the real aim of the Gregorian reform (keeping the equinox on close to the same day, but with the second-best solution) and the astronomical requirements thus imposed. The work of John Dee and colleagues seems to have been forgotten, and British and American astronomers since then, through John Herschel and to the present day, have erred in their statements of how long the year should properly last, given the core desideratum of the Gregorian reform. Franklin saw the difference: he cites "one Hour, twenty Minutes" as the deviation over four hundred years, and not the three hours so erroneously stated by many people since then. So again: all praise to Benjamin Franklin!

Franklin's Seal of Approval

Benjamin Franklin recognized, then, that the wander of the Gregorian calendar would only amount to a single day over seventy-two centuries. Using modern values for the year length, we could quibble over whether it would be closer to eight thousand than seven thousand years before such a deviation accumulates, but we also know that such arguments are specious because there are many other significant variations and uncertainties for which to account.

In the shorter term, of concern to ordinary humans, Franklin could see that the calendar reform was sensible, adequate, and an event of note. He finished by writing:

> *Wishing withal, according to ancient Custom, that this* New Year *(which is indeed a New Year, such an one as we never saw before, and shall never see again) may be a happy Year to all my kind Readers.*

> I am, Your faithful Servant,
>
> R. SAUNDERS.

It seems remarkable that Franklin, in all his voluminous writings, had little else to say about Lord Chesterfield's act, although for the reference of his readers he did give the full verbiage (tedious though it is; be thankful that I only quoted a few sections in the previous chapter). By the time that he was preparing the 1753 edition of *Poor Richard's* he had made the necessary changes to his tables, giving the dates both in New Style and Old Style. This was a feature of the almanacs of the time: the authors had tended to believe that their readers would make an easy transition from the old style to the new, but underestimated how deep would be the desire to maintain the various feasts and fairs—equivalent to our public holidays—at the same places in the seasonal year. Many almanac makers gave just one set of dates in 1752 and/or 1753, but soon discovered that public discontent compelled them to give both dating systems, a practice which needed to continue for a generation or more.

Despite the relative silence of Poor Richard on matters calendrical, it would be remiss of me if I did not mention the one area in which Benjamin Franklin is widely regarded as having priority on the subject of timekeeping. In a whimsical essay written in 1784, the year he invented bifocal spectacles, he suggested the concept which we now call *daylight saving time,* a boon or a bane to civilization, depending upon your opinion.

Later, though, I will be suggesting to you that, some decades before, Franklin was actually the first person to nominate radical reform of the calendar, one far more sweeping than Chesterfield's act and predating it by several years, a contribution hitherto unrecognized as stemming from Franklin's pen. But first we must continue our amble along the historical path by which our calendar was delivered to us.

Chapter 18

❦

PRESIDENT ARTHUR REQUESTS (A.D. 1884)

It was mentioned in preceding chapters that various countries moved over to the Gregorian calendar either immediately after it was introduced in 1582 (the Catholic nations) or else in two waves around 1700 and 1752 (various Protestant states in western and northern Europe, and Britain and its dominions). Many other places still held out, however, either on the Julian calendar, or perhaps some calendar of their own (such as Japan until 1872). Egypt converted in 1875, China in 1911.

Many eastern European nations adopted the Gregorian calendar—or, if you prefer, their own civil calendars declared according to rules which happen to make them coincide with the Gregorian—in the decade of the First World War and the Russian Revolution: Albania, Bulgaria, Estonia, Latvia, Lithuania, Romania, the former Yugoslavia, and Russia itself all capitulated then (calendar-wise, I mean), their procrastination being overcome at last due to religious considerations having been usurped by the new secular (and totalitarian) regimes of government. The power of the Russian Orthodox Church may have waned in those states, but the Greek Orthodox Church managed to keep Greece on the Julian calendar until 1924. In Turkey, December 18, 1926, was followed by January 1, 1927.

Within the Eastern Orthodox churches the Julian calendar persists for liturgical purposes, making Christmas thirteen days later than in the Western churches (the Protestants and Catholics generally agreeing on this, at least). The Eastern Orthodox Easter may coincide with that of the West, but is often one, four, or five weeks later.

This demonstrates the gradual conformity of countries, if not religious sects, to the Gregorian calendar, but there was still a significant difference between nations. Indeed, there was even a significant difference *within* them. What I am referring to is the *time zone* problem. In chapter 16 I wrote that "a particular twenty-four-hour period might be labeled February 1, 1720, in Scotland, February 1, 1719, in England and February 12, 1720, in most of the rest of Europe." But that is not quite true. Even today a particular twenty-four-hour period in Germany, say, only has a twenty-three-hour overlap with a similar period in the United Kingdom, and that is with regularized and quantized time zones, which are a fairly modern concept. In the past each state, each city, each small town even, had its own time system. If you never traveled more than ten miles, and there were no rapid means of transport such as railways, or communications such as telephones or radio, then there was no need for you to keep to the same time frame as a place just twenty miles away. All that mattered was the rising, transit, and setting of the Sun as witnessed from your own little bailiwick.

The major subject of this chapter, then, is the coordinated global time zone system and how it came about. First we should consider some background.

The Time before Time Zones

Gentle reader, again please permit me some indulgence in telling you another story. The town where I was born is skirted by the Fosseway, an old Roman road which begins in central England and then passes southwest through the Cotswolds and by Bath (the Roman *Aquae Sulis*, remember) on its way to the tin mines of Cornwall. In later times this road was used as a coach route, and with the invention of the internal combustion engine it was covered with tarmac, carrying traffic from Bath and beyond towards Exeter and the English Riviera, if you could excuse such a notion.

When I reached the age at which it was legal to enter a public house and drink alcohol—and not before, of course—one of my favorite pubs was called the White Post, just southwest of the boundary of my hometown, and slapbang beside the Fosseway. Traveling a handful of miles towards Bath, one comes to a town called Peasedown St John, and as you drive into it on the Fosseway is another pub called the Red Post. Although these are nowadays solely places to eat, drink, and be merry, their names tell of their earlier history. They were staging posts. In the days of horse-drawn coaches and highwaymen like Dick Turpin, these were combined stables and taverns, where passengers could gain refreshment while the horses were being unbridled and a fresh team buckled in.

The movement of passengers was only one of the purposes of this network of coach routes; another major aim was the delivery of mail, hence the origin of the term *post office*.

But these were not the only things which happened at these staging posts. The area of which I write is two and a half degrees in longitude west of London, and therefore ten minutes away in terms of solar time. These were the days before there was any method for disseminating time signals by telegraph wire or radio pips, let alone Global Positioning System satellites and such ultraprecise modern contrivances. The only time which had any meaning was the time according to the Sun.

For this reason, once mechanical clocks had been invented which could function on a bumpy ride like a coach trip on a muddy or cobbled path (such as John Harrison's famous marine chronometer), these had to be checked against local time at staging posts along the way. At those posts, reasonably accurate pendulum clocks would show the time, recalibrated occasionally by watching the Sun as it crossed the overhead meridian. And to keep correct time according to points along the journey, the small clocks carried on the coaches needed to be adjustable, such that they gained or lost time at the appropriate rate for movement east or west!

For example, traveling at an average speed near 5 or 6 miles per hour (mph), it would take about twenty hours to get from London to Bath; but the clocks in Bath would be about ten minutes behind those in the capital, so that the timepiece carried on the coach would need to lose time at a rate of half a minute per hour, and do the opposite on the trip back. When I went up to study at the University of London as an undergraduate, the first 125-mph trains had just been introduced along the route of Isambard Kingdom Brunel's Great Western Railway; the best time I ever managed from Paddington Station to Bath was sixty-four minutes, a bit quicker than the early eighteenth-century mail coaches, although by the 1780s they were able to boast speeds of 10 mph on the best routes. But fifty years later railways were sprouting up around Britain and the days of the coach system were numbered.

We are habituated to time zones which step up in hours—are *quantized,* rather than smoothly changing with longitude, and, hence, local solar time— but it is quite surprising how recently this innovation came about. Yes, some places are awkward and have quarter- or half-hour time steps, like South Australia where I live—or India, to cite a place with a vast population—but the point is that the changes you need to make to your watch when you land at some airport are discrete, rather than requiring a clock like those on the stage-coaches, with variable rates of time reckoning depending upon the speed and direction of journey.

I wrote that this situation is quite recent. Let me give a couple of examples. If you go to Christchurch College, Oxford, you will see over the main gate a clock tower, which is called Tom Tower. For most of its life, the clock there-upon had not one but *two* minute hands: one showed Oxford time, the other five minutes ahead showed Greenwich time. (Thinking of Big Ben in Westminster, you might imagine that the English give male names to all their clock towers, but that would be based on a fallacy: very few people actually *see* Big Ben, although many *hear* it, because that appellation specifically applies to the great bell inside the tower.) Similarly, two different time systems were used in Holland during the first half of the twentieth century, one being the local solar time and one a standard railway time according to the clock over Amsterdam Station; that nation did not officially fall in with the standard time zone system used in most of the rest of the world until 1956.

So when *did* most nations decide to standardize their time? My mention of railway time gives a clue, but again a short historical sketch might be useful to put the whole matter in perspective.

Railway Time

Not for parochial reasons, but because of historical fact, my story must continue in Britain: this is where the first railways were built, where a standard time system was first introduced, and where, many decades earlier, the observatory was established which was destined to become the reference longitude for the definition of time zones worldwide.

The Royal Greenwich Observatory (RGO) was founded in 1675, and very early on its work was directed toward providing a reference of time—Greenwich time—for navigators. For Britannia to rule the waves, it was necessary that its sea captains, whether military or merchant, should know where they were, and that necessitated excellence in navigation. Early in the history of the RGO the government astronomers favored methods for finding one's position at sea which involved complicated astronomical observations—for example, by timing the eclipses of Jupiter's four large satellites, or by measuring the angular separation between the Moon and certain bright stars. Either of these techniques could provide, in principle, a form of celestial clock, and from the time so determined and an observation of the Sun as it crossed your meridian, the longitude might be calculated.

In the event it proved that the rival system pioneered by John Harrison, which substituted an accurate mechanical clock for the figurative celestial clocks described above, was superior both in terms of precision and convenience. When Captain James Cook set sail in 1772 on his second great voyage of exploration, he took with him one of Harrison's famous marine chronometers, and returned full of praise for its efficacy in matters of longitude determination.

By that time, then, the work of the RGO in this regard was directed more toward providing an accurate time standard for nautical (and indeed astronomical and domestic) applications, rather than in investigating alternative navigational techniques, as had been one of its earlier tasks. For example, if you visit the observatory now, you will see that over the roof towers a pole, upon which is mounted a large sphere. This contraption is the *time ball,* which was raised to the top of the pole just before 1 P.M. every day, and then dropped at the instant of that hour so that the captains of ships laid up on the Thames, which flows past half a mile to the north, would be able to set their chronometers before starting out on a voyage.

The time ball at Greenwich and many other places around the world was an innovation of the 1830s. Thereafter the system was introduced in some locations far from the sea—for example, Denver once had a time ball, and I doubt whether many oceangoing ships passed through that city—wherever a time reference was needed for other purposes, such as coordinating railway service times. In fact, until about that era the common people had hardly any real need for an accurate definition of time, and few possessed a watch.

But things were changing. From around 1830 the first public railways began running, as well as the first transatlantic steamships. Within the next decade or so the railways took over mail delivery, railway timetables were published (and adhered to, making it necessary for people to *know* the time), the first electric-powered clock was constructed, and Sir Charles Wheatstone invented the electric telegraph, making it possible for time signals to be transmitted essentially instantaneously over wide areas.

There was no longer any *need* for Bath to be ten minutes behind London, and in 1840 the Great Western Railway (GWR) decreed that London time should be kept by all of its stations, regardless of the local solar time.

Nevertheless, cities persisted in using their own local time, so that in Bath most clocks would show one value, but those at the railway station would be ten minutes ahead, and woe betide anyone following his own watch and expecting to catch a train to London. It was for such a reason that the Tom Tower clock in Oxford had its pair of minute hands.

Just *why* many people resisted the introduction of standard time throughout Britain is a bit of a mystery. Look at it this way. If you follow the Sun in the sky to give you your time (apparent time), as would rural people, then the lengths of the days and the hours vary during the year. But if you use a clock with equal hours throughout the year, then you have effectively given up on the real sun and are actually following the fictitious mean sun (you are using local mean time). The difference between these systems is the equation of time, as described in appendix A.

At the latitude of Britain, this means that in November the afternoons are half an hour shorter than the mornings, whereas in February the converse is true. This is a much bigger deviation than that caused by moving from solar time to standard time, and in any case the latter produces only a constant off-set. So there was little logic to people's arguments against the system. On the other hand, then as now country people were more affected by the appearance of the Sun than those who lived and worked in the city. If you are habituated to rising with the cock's crow at dawn, it would be annoying to be told that from now on you must use a clock set by the observations of some astronomer in far-off London, an entirely different world than that of a farmworker.

Think not that the question of standard time following mean time was debated only in Britain, though. On the Continent, various *cities* such as Geneva, Berlin, and Paris had decided to use mean time rather than apparent time, in the decades around 1800. The peculiarity about Britain was that the whole nation was to be subject to a standard time following the mean sun, prompted by the railways.

After the GWR had shown the way, most of the major railway companies had adopted Greenwich time by 1847, and at length the cities realized that they were tilting at windmills by refusing to join in. Although one often thinks of Britain as being aligned north to south, in fact Scotland is mostly quite a long way west of London, such that local solar time in Edinburgh is over twelve minutes behind Greenwich, and Glasgow seventeen. In January 1848, those cities went over to Greenwich mean time, helping their residents to avoid missing their trains.

The 1840s saw a boom in public transport. Until then most Britons rarely strayed more than ten miles from their place of birth, but the advent of cheap excursion trains meant that soon they were traveling further afield. London's Great Exhibition in 1851 proved a watershed, with 6 million visitors attending, most getting there by train. This exhibition, at the magnificent Crystal Palace temporarily erected in Hyde Park, also attracted seventy-five thousand visitors from overseas—obviously the more monied and hence influential people—and they went away much impressed with Britain's innovative time standardization, preaching its attractions to their home governments and railway companies.

Although there was yet to be a legal definition of time—that did not come until 1880—Greenwich mean time had become the de facto system by which the British conducted their lives. When the Tower of Westminster was built in the 1860s (and Big Ben installed—a *second* bell, because the first one cracked), the massive clock was sent hourly time signals by telegraph from Greenwich, and its rate adjusted by the operators adding coins as tiny weights to its gigantic pendulum. All manner of organizations depended upon the RGO for their time calibration, from individual watchmakers to the post office, the telegraph companies, and, of course, the railways.

International Time and Longitude

Standard time throughout Britain made a great deal of sense, because as the nineteenth century progressed the various towns and cities were becoming interconnected by the railways and also the telegraph system. But it was only a matter of time before the web spread wider still, and then there would need to be an internationally agreed-upon time standard.

The installation of the first submarine telegraph cable link to France turned this question from being moot into one which clearly needed to be answered. In October 1851 an anonymous correspondent, noting this development and perhaps inspired by the optimistic outlook upon the future which had been promoted by the Great Exhibition, wrote a letter to the *Times* of London suggesting that a prime meridian should be adopted, providing a basic time reference on which other countries could base their own time standards. In the 1850s Britain was telegraphically connected with much of Europe; by the middle of the 1860s the first successful transatlantic cable had been laid. To the British it was obvious that Greenwich should be that prime meridian, because they were by then habituated to using time as defined from that longitude. But not everyone, the French in particular, was likely to agree. And in any case there was already a long history of other reference meridians.

For internal mapmaking practices, most European countries usually adopted some reference point within their borders—maybe the center of the capital city—and measured distances north-south and east-west from there. However, for navigational charts some wider standard system is needed, and the early seagoing civilizations like the Greeks and the Egyptians adopted the furthest western land that they knew—the Canary Islands—as the standard zero point for longitude. The astronomer Ptolemy used Ferro (now Hierro), the western most cape of these islands, and it is interesting to note that nowadays there is a suite of modern astronomical observatories nearby on La Palma.

To the ancients all measurements were then made eastward from Ferro, but the Spanish forays across the Atlantic made it necessary for westward locations to be incorporated in the scheme. Philip II of Spain issued an edict that longitudes should similarly be measured westward with reference to Ferro, a decision which in effect was internationally ratified in 1634 when Cardinal Richelieu called a conference in Paris (attended, of course, only by Catholics), where knowledgeable mathematicians and astronomers considered the question of a prime meridian and agreed to continue with that reference point (which was in Catholic hands, of course), and also the convention of positive numbers eastward, negative numbers for westward. Referring back to the English adventures in Virginia described in chapter 14, one can see that this prime meridian question has repercussions far beyond the grids you place on your maps.

This, then, was the state of the matter until the latter half of the nineteenth century.

The American Railroads

Over its full extent east to west, the longitude spread of Britain amounts to only about thirty minutes of time, so that a single time zone does not cause a great dislocation between clock time and mean solar time. A railway timetable could be printed which used the same time system for London, Plymouth, Cardiff, and Glasgow.

But this was not the case in the United States. From the northeast of Maine to the western tip of California or Washington State represents a full four hours in terms of longitude difference, making it undesirable to have all locations on the same time standard (but note that the former Soviet Union imposed such a system across its own wide longitude band; possible in a totalitarian state, but not a democracy). Nowadays we are familiar with having four time zones across the United States; but soon after the transcontinental railroad system was completed, if you took a train starting in Bangor and traveled through the network to San Francisco, you would have needed to have adjusted your watch no less than twenty times along the way! In Buffalo, New York, four separate time systems might be displayed on the railroad station clocks; in Pittsburgh it was six.

With separate railroad companies using their own time standards, things could be really confusing to a long-distance traveler. There was no need for this mess, since the U.S. Naval Observatory (USNO) in Washington, D.C., had been founded in 1845 to determine the nation's time standard, and was distributing time signals to whoever wanted them, but there was a lack of any central organization or agreement among the railroads.

This salmagundi was brought to a conclusion through the enthusiastic efforts of Professor Charles Ferdinand Dowd of Saratoga Springs, New York. In 1870 he came up with the idea of there being four major time zones across North America, centered on meridians spaced by fifteen degrees and with the prime meridian being the longitude of the USNO in Washington. In 1871 he spoke at the meeting of the North East Railroad Association in Boston, and his persuading the companies to agree to his basic suggestion was crucial for future developments.

It was pointed out to Dowd that the United States had already legally adopted the Greenwich meridian as the prime reference for all its nautical charts, rather than that of its own national observatory in Washington. Because of that Dowd altered his scheme slightly in that his *prime American meridian* was then shifted to be the line of longitude seventy-five degrees west of Greenwich, offset by a couple of degrees from his original conception. As a matter of fact, this shift was quite convenient in that it provided a better fit with where the railroads wanted the time zone dividers to be drawn.

Eleven years of debate followed; finally, at noon on November 18, 1883, all the railroad clocks throughout the United States and Canada were at last set to a standard reference, with four time zones. Thus North America trailed Britain by over thirty years in adopting a standard time system, abandoning local solar time, and you can now see why I wrote above, "Not for parochial reasons, but because of historical fact, my story must continue in Britain." Indeed, the government of the United States, like that of Britain before, was also tardy in legislating to cover this development; it was not until 1918 that a standard time bill was enacted by Congress.

Dowd's vision was not limited to North America. He visualized a standard time zone system around the whole of the globe, all begun from his concern over railroad timetables. I am not sure which word to use to describe the melancholy event—paradoxical? ironical? perhaps inevitable?—but he died in 1904 when he was run over by a steam locomotive.

Which Prime Meridian?

Although many countries were still using Ferro as the reference meridian from which to map their lands, or perhaps the location of their own capital cities or national observatories, for sea charts the use of the Greenwich meridian had spread widely. This was a matter of scientific and technological dominance on the part of Britain: it produced the best marine chronometers, calibrated these at Greenwich, and with those facilities to hand produced nautical maps which could not be rivaled. Most other countries then followed suit.

It was inevitable in this regard that other nations would, during the nineteenth century, gradually come to regard Greenwich as the fiducial point from which all measurements were charted, as foreshadowed above for the case of the United States. In Antwerp in 1871, and in Rome in 1875, International Geographical Congresses were held at which the consensus grew that the Greenwich meridian was the one which should be universally applied.

That is not to say that everyone agreed. For example, Charles Piazzi Smyth, the Astronomer Royal for Scotland and a keen Egyptologist, argued that the Great Pyramid outside of Cairo should provide the prime meridian. Others were a little xenophobic, dismissing the need to consider other nations in their own affairs; Simon Newcomb, a prominent astronomer who was superintendent of the U.S. *Nautical Almanac*, thought that in defining its own geographical reference system the United States should pay as much attention to Europe as to "the inhabitants of the planet Mars." On the other hand, it is surprising, given later events, that France at this stage was agreeable to adopting the Greenwich meridian, although only if Britain and the United States would in return agree to deploy the metric system for their own weights and measures!

There were also alternatives based on Greenwich itself. At this stage the question of when a day should be counted as beginning and ending was still open. Leaving aside various cultural and religious usages of either sunset or sunrise, arguments could be made in favor of either midnight or midday. In the same way, one might choose to make the prime meridian 180 degrees/12 hours *away* from Greenwich; that is, where our international date line's nominal position lies. One can see that this would have had advantages. If longitude were to be stipulated on a scale stretching from 0 to 360 degrees counting eastward, as was proposed, with Greenwich being counted as 0, it would follow that Queen Victoria's residence at Windsor Castle would have been at a longitude of 359 degrees and twenty four minutes, which seemed ungainly. It was two millennia ago that Ferro, at the far west of the known world in the Canary Islands, was chosen as the reference, so that all positions of interest were within tens of degrees east of there; the proposal to put the prime meridian 180 degrees from Greenwich was an attempt to similarly use a reference longitude as far removed as possible from where most of the action occurred.

In the years between 1871 and 1883, several international meetings were organized, attended by diplomats, astronomers, cartographers, and the like. In October of the latter year, Rome was again the location where the conference met and all of the basic ideas were set down. It had already been decided that the various matters would be put up for international ratification in a conference to be held in Washington, D.C., the following year.

Although it was thought that this would merely be a matter of dotting the i's and crossing the t's, in fact two major changes to the proposals were made between Rome in 1883 and Washington in 1884. First, it was decided that the *Universal day* should begin at midnight rather than noon, and with that late change in mind it is hardly surprising that astronomers persisted in using noon as the break between days until 1925, as it still is in the widely used Julian Date system.

Second, the Rome conference had recommended that longitude be measured from 0 to 360 degrees as mentioned above, but in the event it was decided that the count should start at 0 but increase to 180 degrees in either direction, echoing the seventeenth-century convention based on Ferro introduced by Philip II. Note that this means that additional information is required in any statement of location in order to avoid ambiguity; Washington is at a longitude of 77 degrees, but so is New Delhi, one being west and the other east (thus 77° W and 77° E).

The Lead-up to the 1884 Washington Conference

James A. Garfield was elected twentieth president of the United States after a string of unlikely events. At the Republican Convention he had tried to push through the nomination of his friend John Sherman, but failed, and was himself nominated as a dark horse candidate on the thirty-sixth ballot. In the actual election he won by less than ten thousand votes from the Democratic candidate, Winfield S. Hancock, and entered the White House in 1881.

His domicile was destined to be short-lived. On July 2 of that year he was shot by a lawyer who was embittered by Garfield's refusal to appoint him to a consular post. Given the demonstrated connection between railways and questions of time standards, we should note that the assassin's bullet struck while he was standing in a Washington railroad station.

Garfield's death was not immediate, and he lingered on for more than eleven weeks. Doctors were unable to find and extricate the bullet from his body, and so Alexander Graham Bell, the inventor of the telephone, designed a novel type of metal detector with which he tried to identify the projectile's location; the discovery of the penetrating power of X rays was still over a decade away. In the event Bell's attempts failed, largely because Garfield lay on a bed with metal springs and no one thought to move him: the background signals from the springs made it like looking for a snowman in a blizzard. The internal bleeding and infections eventually killed the president.

Garfield was an unlikely president, and his running mate (due to internecine political scrapping) was an unlikely vice president: Chester A. Arthur. When Garfield died on September 19, 1881, Arthur stepped into the presidency. Although he was initially distrusted, he gained wide recognition and respect for the measures he enacted during his tenure. Despite this he was not renominated by the Republican Party in 1884, and he left office in 1885. By then he was already mortally ill with a kidney disease, and he died the following year.

In our present connection, President Arthur had important business to deal with in the period leading up to the election in November 1884, which his Republican replacement, James G. Blaine, lost to the Democratic candidate, Grover Cleveland. As far back as October 1882 the United States had sent out letters to various nations, asking for an indication of whether a conference in Washington to finalize an agreement with regard to matters of a common meridian and time system would be acceptable.

The answers being affirmative, formal letters of invitation were issued from President Arthur's office following the 1883 meeting in Rome, requesting those nations to send delegates to the forthcoming *International Meridian Conference* (IMC) in Washington, D.C., whose deliberations were set to begin on October 1, 1884, and in the event to last for three weeks. The history of the lead-up to the meeting is summarized in the introductory paragraph which begins the *Final Act*, the title given to the formal report of the conference:

> The President of the United States of America, in pursuance of a special provision of Congress, having extended to the Governments of all nations in diplomatic relations with his own, an invitation to send Delegates to meet Delegates from the United States in the city of Washington on the first of October, 1884, for the purpose of discussing, and, if possible fixing upon a meridian proper to be employed as a common zero of longitude and standard of time-reckoning throughout the whole world, this International Meridian Conference assem-

bled at the time and place designated; and, after careful and
patient discussion, has passed the following resolutions. . .

The verbiage is important, as are the various resolutions which were adopted.
Look at it this way: suppose that instead of the Greenwich meridian, the north-
south line through the Paris Observatory had been chosen. If that were the
case, then all geographic longitudes would be 2.3375 degrees different from
the values they are given on modern maps. That, you might say, is no big deal,
but think about the consequences for time standards. When time zones are set
up, decisions have to be made with regard to where the boundaries should lie.
Using Paris as the prime meridian would mean that for most of us, our watches
would be set to a time 9 minutes and 21 seconds different than the time standard we actually employ; but for some, the difference would be 50 minutes
and 39 seconds, because they would be in a shifted time zone.

The Prime Meridian Defined

In all there were forty-one delegates listed as coming from the following
twenty-five nations: Austria-Hungary (as it then was), Brazil, Chili (note
spelling), Colombia, Costa Rica, France, Germany, Great Britain, Guatemala,
Hawaii, Italy, Japan, Liberia, Mexico, the Netherlands, Paraguay, Russia, Sal-
vador (no "El"), San Domingo, Spain, Sweden, Switzerland, Turkey, the
United States, and Venezuela. A few things immediately stick out from that
list, because since 1884 quite a few national boundaries have changed.

Hawaii, obviously, is now a part of the United States, but if one looks at the
delegate list one finds that Colombia was represented by Commodore Franklin
of the U.S. Navy. One might suggest on those grounds that the United States
had sequestered a large number of the votes, whereas the now separated countries of Austria and Hungary (plus some adjacent territory) got only one.

An even greater misjustice, in such an anachronistic view, would be the case
of Great Britain. Four delegates attended, one of them being Professor John
Couch Adams, who, as a student forty years before, had theoretically deduced
the existence of Neptune, his suggestion (along with that of the French astronomer Urbain Leverrier) leading to the telescopic searches which eventually
turned up the planet in September 1846. Another British delegate was Captain
Sir F. J. O. Evans of the Royal Navy. Both nautical and astronomical interests
were represented, then.

The other "British" representatives were Sandford Fleming, a Canadian
who was largely responsible for the impetus which brought about the IMC and
the adoption of the prime meridian, and a Lieutenant-General Strachey, member of the Council for India. The point is that there was no real differentiation
back then between Britain and its dominions. There was no need, say, for separate delegates to be sent by South Africa, Australia, and New Zealand because
the mother country was relied upon to represent the interests of those regions,
which were yet to be constituted as separate nations.

Leaving these matters aside, the actual resolutions of the IMC and who voted *against* them throws up some interesting facts. The first was that the need for a single prime meridian was recognized; unanimously adopted, no surprises there. The second resolution read: "That the Conference proposes to the Governments here represented the adoption of the meridian passing through the centre of the transit instrument at the Observatory of Greenwich as the initial meridian for longitude." Twenty-two nations voted in favor; San Domingo voted against; Brazil and France abstained.

In earlier discussions France had agreed to Greenwich being the prime meridian, but according to its voting stance it had then decided to sit on the fence. Actually, by the time of the IMC, Britain, Sweden, the United States, and Canada were all using a standard time system referenced to Greenwich mean time, and as a result of the IMC's resolutions most other countries followed by 1905.

France, however, was tardy, and it refused to use Greenwich mean time in any explicit way. In 1898 its government passed legislation saying that the standard time in France should be Paris mean time diminished by the nine minutes and twenty-one seconds mentioned earlier, which would make French time coincide with that in Britain without specifically saying so. Even then it took until 1911 before French standard time was put into use. In 1978 France enacted a new law saying that its legal time system was to be determined from coordinated Universal time (UTC). Thus France *never* agreed to use GMT, in so many words!

Some other countries were also somewhat slow in conforming with this time system. Russia did not switch from the Julian to the Gregorian calendar until after the revolution in 1917, but it was another seven years before it went over to standard time based on GMT. One of the nations sending a delegate to the IMC, Liberia, did not move over to such standard time until 1972, which seems strange given its nautical connections (a good part of the world's merchant shipping is registered there), and the fact that back in 1884 it voted in favor of all of the resolutions.

The Netherlands provides another peculiar example, although maybe it has an excuse in that it did not favor all of the IMC resolutions. Earlier it was mentioned that this country used different civil and railway time systems until only some decades ago. In fact *neither* of those was correlated with GMT. Holland was first forced to adopt a time system in step with its neighbors when the Nazis invaded during World War II, all the occupied countries being instructed to use the same time as that in Germany, which was an hour precisely ahead of GMT. After the war the Dutch, in part out of rightful objection to a measure imposed by their former occupiers, reverted to their original out-of-kilter time system, eventually joining most of the rest of the world in a timekeeping sense in 1956.

The Shifted Meridian

The wording of the above resolution has also led to an incorrect understanding of the location of the prime meridian. At the RGO, a transit circle had been installed in 1851 under the supervision of Sir George Biddell Airy, the Astronomer Royal during a large chunk of the nineteenth century, and it was this instrument which was assumed to be the fundamental marker for the meridian. If you visit the observatory now, you will note that the line taken to divide the Eastern and Western Hemispheres of the globe passes through the location of the Airy transit circle.

The problem was that the nautical charts and so on, through which Greenwich gained its precedence, were based upon two previous transit instruments, installed by two earlier Astronomers Royal: James Bradley, in the middle of the eighteenth century, and John Pond, in 1816. These had been located nineteen feet (about six meters) to the west of Airy's telescope, so that, for example, all maps of Great Britain then in hand (in 1884) were offset by that distance from the meridian which the IMC adopted. This was not realized until 1949. Nineteen feet may not seem a lot, but it is equivalent to the Earth's rotation in one-fiftieth of a second, which is easily identifiable through timings of the transits of stars, the very work of the various instruments in question. More important in the modern era, this distance is very significant compared to the precision available using the Global Positioning System. Paradoxically, the resolution which led to the international adoption of the Greenwich meridian as being the standard for all mapping worldwide actually had the effect of making all the British maps inaccurate!

The Other IMC Resolutions

Some of the other resolutions of the IMC have already been mentioned. Longitudes to be measured from 0 to 180 degrees both east and west from Greenwich. A Universal day to be recognized as being a convenient construct, to be counted from 0 to 24 hours, and being defined as the mean solar day beginning for all the world at the moment of mean midnight of the prime meridian. Much against the wishes of various astronomers and nautical types around the world, one resolution (carried without division) stated: "That the Conference expresses the hope that as soon as may be practicable the astronomical and nautical days will be arranged everywhere to begin at midnight." Some arm-twisting was being applied, then, to try to persuade the astronomers and sailors to abandon their practices whereby their days both began at noon, but with a twenty-four-hour difference between them.

Resolution VII of the IMC, the final item, is one of the most intriguing:

> That the Conference expresses the hope that the technical studies designed to regulate and extend the application of the decimal system to the division of angular space and of time shall be resumed, so as to permit the extension of this application to all cases in which it presents real advantages.

Now, let's be clear about what was being recommended here. In effect, it was being advocated that the right angle should contain one hundred rather than ninety degrees, and there is indeed an English word—the *grad*—for such a unit, derived from the French.

That might be livable for most people, but what about a day with ten hours, each hour containing a hundred minutes? The surprising thing is that the resolution was adopted with the vast majority in favor and none against, with only Germany, Guatemala, and Sweden abstaining. One gets the feeling that this was a sop to the French (who have campaigned ever since to get all other nations to recognize the superiority of their metric/decimal system), but one wonders why the British would have voted in support. Is the answer perhaps in the wording? One could simply claim that "all cases in which it presents real advantages" might be interpreted by a suitably determined party as meaning *never.*

Despite this, that final substantive resolution provides some food for thought. If one claims that the world's mapping grid and standard time system derive their authority from the International Meridian Conference held in Washington in 1884, does it not also follow that the same authority says that we should be using decimalized clocks and right angles by now?

Wait a Minute, Where Is Washington?

If you are attentive, your wits may suddenly have perked up when you read above that "Washington is at a longitude of 77 degrees." Now wait a minute, isn't that the important longitude, the quest for which led to the first English attempts to found an American colony back around 1600? Yes, it surely is.

So here's a staggering thing. The devious plan I described in chapter 14 which was apparently the major motivation behind Sir Walter Raleigh's adventures around Roanoke Island, then the Lost Colony and later the Jamestown Island settlement, leading eventually to the opening up of Old Virginny and the formation of the United States as an independent nation, was all based upon trying to occupy the seventy-seventh meridian. And yet in 1884, in the largest city which stands on that line, the capital of the United States, delegates from around the world voted to put the prime meridian for all charts, navigation, and timekeeping back in Greenwich next to the Thames, the very place from which many of the exploratory ships had departed three centuries before!

But the story does not stop there. When John Dee presented his proposed calendar reform to the Privy Council (and thence Queen Elizabeth) in 1583, he prefaced it with a poem. In part this reads as follows:

> *ELIZABETH our Empress bright,*
> *Who in the yere of eighty three,*
> *Thus made the truth to come to light,*
> *And civile yere with heaven agree.*
> *But eighty foure, the Pattern is*
> *Of Christ's birth yere: and so for ay*
> *Eche Bissext shall fall little mys,*
> *To shew the sun of Christ's birth day.*

Three hundred yeres, shall not remove
The sun, one day, from this new match:
Nature, no more shall us reprove
Her golden tyme, so yll to watch.

Now recall what was described in chapter 14. Dee had designed a *perfect Christian* (and distinctly Protestant) calendar, hence "civile yere with heaven agree." In 1584 the phase of the thirty-three-year leap cycle was to be back in line with that in 1 B.C. ("Christ's birth yere"), and 1585 would begin a new thirty-three-year cycle in which the leap years would continue in step with those on the Gregorian calendar, giving the English some decades to grab God's longitude. "Eche Bissext [leap year] shall fall little mys" means that the error (mistake) will not be enough to allow the Sun to reach the solstice on anything other than Christmas ("Christ's birth day"). And how long will this matchup continue to last? "Three hundred yeres, shall not remove/The sun, one day, from this new match."

So here's an even more staggering thing. Dee seems to have thought that his calendar rules would keep Christmas coincidental with the winter solstice for three hundred years, and similarly the vernal equinox occurring on March 21, and that after such a period some accumulated slippage would allow the astronomical events to occur off-date. There are several causes for that slippage; for example, our well-known fact that the year defined by successive winter solstices is not the same as that between vernal equinoxes. The agreement also depends upon whether one is thinking in apparent time (the real position of the Sun in the sky) or mean time (according to the fictitious mean sun), plus the gradual slowing down of Earth's rotation.

But just as Dee's three hundred years were terminating, human civilization got to the point where it needed to make some decisions about what meridian to measure from, and what time system to use. The slippage mentioned in the previous paragraph was such that by the 1880s, the completion of Dee's three hundred years, God's longitude had moved eastward by a couple of degrees. If Raleigh had been sailing across the Atlantic in that later period, he would have been looking to found New Albion somewhere on the seventy-fifth meridian, not the seventy-seventh.

The effect of the railroad companies adopting an eastern time zone in 1883 based on the mean solar time on the seventy-fifth meridian as measured from Greenwich, that decision effectively being supported by the International Meridian Conference in 1884 and eventually passed into law by the United States, was that if a thirty-three-year leap cycle were used along the lines of Omar Khayyam, John Dee, and the present Iranian calendar (to list a few examples), then the vernal equinox would remain on one calendar date within the eastern standard time zone of North America. Those conventions adopted in 1883 and 1884 were just in time: on Dee's calendar the equinox was saved from drifting off of March 21.

Eventually the favored longitude will move so far east that it will depart that time zone, and after some centuries or more it will be the time zone containing Bermuda, which is the one that could keep the equinox on one day. But the question is moot, because we simply do not work on a thirty-three-year leap cycle.

Looking back, if someone had really wanted to fix up the calendar as well as the meridian and time standard questions, then the IMC should have instigated Dee's thirty-three-year calendar. You could cast your eye down the list of nations represented and note that many of them were Catholic, and imagine that this would have made it impossible to get a vote to abandon the Gregorian calendar—but wait. When England was trying to grab part of the seventy-seventh meridian to facilitate a perfect Protestant calendar, there was no United States. But in 1884 the United States existed not only as a major nation, but importantly, as a secular state. Since the United States possessed God's longitude, there would not be any grounds for the various churches to bicker over priority. From this perspective, I would say that a strong case might have been made for choosing the U.S. Naval Observatory in Washington, rather than the Royal Greenwich Observatory in London, to provide the prime meridian. But it was not.

Daylight Saving Time

There seems to be nothing so contentious in the timekeeping sphere than the question of daylight saving time (DST). I have previously noted that it seems to have been Benjamin Franklin who first suggested the idea, back in 1784. He treated the matter quite lightly, obviously not taking it as seriously as many have since.

The idea of daylight saving time is to advance the clocks by an hour during the summer months (or more) so as to minimize the amount of sunlight "lost" prior to the usual rising time of people who work in the industrialized world. That is, if you sleep through the early morning hours when the Sun is above the horizon, no direct use is made of that natural light; then, in the evening, artificial lighting may be required for some pursuits. By shifting the clocks forward one may sleep through an hour less of sunlight, and have that hour naturally lit in the evening.

Note that the argument only works for an industrialized/urban society. For rural dwellers and agricultural workers, their day tends to follow the sunlight anyway, and so again it is they who tend to oppose daylight saving time as an unnecessary imposition. In addition, although the original motivation seems to have been energy saving and added work efficiency, in the present affluent society arguments swing more around considerations of leisure activity: daylight saving time allows added recreational time in the evening for games of tennis and the like. So long as there are more people who enjoy evening activities rather than early birds who jump out of bed at the crack of dawn, daylight saving time will likely continue as a custom in the richer temperate-latitude nations.

Every year people seem to get confused over how to adjust their clocks and watches when the dreaded time comes that means you are going to lose an hour of sleep. In American parlance, it's easy: spring forward, fall back.

The Introduction of Daylight Saving Time

One cannot really say that Benjamin Franklin invented daylight saving time, so who did? The person usually credited with this is William Willett, a British construction manager, who wrote a pamphlet entitled *The Waste of Daylight* in 1907. As a builder he would have wanted to see as much work as possible done during the daylight hours in the summer, but with sunrise as early as 4 A.M. it was impossible to get his workers to start at so early an hour. The answer was simple: shift the hours. Actually, Willett suggested four incremental steps, each consisting of twenty minutes, thus producing an overall change of eighty minutes, rather than the single one-hour jump with which we are familiar, but the underlying concept is the important thing.

Actually the first country to put the daylight saving time concept to work was Germany, as an energy-saving device in 1915 under the stringencies of the war. Britain soon followed, as did the United States. Back then daylight saving time was a temporary measure called *war time*. It was this in part which led to the standard time bill of 1918 in the United States.

After the repeal of nationwide daylight saving time in the United States in 1919, some localities persisted with it while others did not. The entry of the United States into World War II meant that daylight saving time was again instigated, from 1942 to 1945. Thereafter a hodgepodge of daylight saving time utilization (or not) resumed.

From 1966 the dates between which daylight saving time should be used, if at all, were set as the last Sunday in April through the last Sunday in October. Note that these are not symmetric about the trend in sunlight (they are both about five weeks after the equinoxes), but rather they follow more or less the equal-temperature trend, which peaks at the end of July or perhaps early August, depending upon the locality.

The 1970s oil crisis resulted in the start of daylight saving time in the United States being pushed back to January in 1974, and February in 1975, after which there was a reversion to late April. From 1987 daylight saving time has been earlier than in the preceding decades, the first Sunday in April being the onset, while the cessation has remained at the last Sunday of October.

My reason for describing briefly the history of daylight saving time in the United States is just to show that it is a far from concrete matter which is very hard to justify in scientific terms, and that has led to many changes there, and elsewhere. Some countries have occasionally used a *double* daylight saving time, a two-hour jump; this tends to have been the case for places at far northern latitudes where the variation in the duration of sunlight hours during the year is extreme (and there are no Southern Hemisphere nations for which this is the case). On the other hand, some places have never used daylight saving time, for good reasons, while others have it unwillingly imposed upon them. Let us see why.

Why Daylight Saving Time Is Not Used Everywhere

In terms of meridians, going onto daylight saving time means that in effect you are using the mean solar time for the zone fifteen degrees to your east. For reasons which will become apparent, Britain is a good example to use. When Britain moves onto daylight saving time, usually in late March, it is taking up the solar time appropriate for places like Berlin, Prague, and Vienna. The standard time in the United Kingdom is that of the time zone of the Greenwich meridian, which is close to the easternmost coast. The longitudes of the Western Isles of Scotland and parts of Northern Ireland are around half an hour west of Greenwich, though, so that when daylight saving time is instigated each year the effect is that these places have clocks set to a time about an hour and a half ahead of their solar time. For the Republic of Ireland, which for convenience follows the U.K. time, the situation may be even more severe.

Because daylight saving time starts just after the equinox, in the northwest of Scotland the Sun is not rising until around 7:30 A.M. according to their clocks, and so they have reason to be a little annoyed with this convention imposed on them from faraway London. Some people have campaigned for daylight saving time to be permanent throughout the year in the United Kingdom, but although that may make sense in southeast England, a powerful argument against it is that in extreme locations, as above, the Sun would then not rise until near 10 A.M. during the winter.

The bottom line from all of this is that one needs to consider not only the longitude but also the latitude when making decisions about whether various time zones might perhaps shift over to daylight saving time during much of the year.

For example, I live in South Australia, which has the same standard time as the Northern Territory, directly to the north. But when daylight saving time is brought in where I live, the Northern Territory remains on the standard time, because in the tropics daylight saving time is not needed: the length of the sunlit hours varies by comparatively little during the year. Say you live in Seattle, or Montreal, or London, and your summer is not coming up to expectations weatherwise, and so you take a vacation in the Caribbean. In June and July in those northern latitudes you have become used to it remaining light until well after 9 P.M. Even though it may be warmer on your Caribbean island, it is always a disappointment that the evenings are short: by 7 P.M. the Sun has set, and by 8 P.M. it's dark. You've gained in solar intensity, but lost in sunlit hours, by heading south.

Step back to Australia. Like the Northern Territory, the state of Queensland does not employ daylight saving time. In the summer, a clock in Brisbane will show 10:30 when the time in Adelaide is 11:00, despite the fact that in terms of longitude Brisbane is a full hour east, making a discrepancy of an hour and a half. The island state of Tasmania is so far south that its government considers that it should use daylight saving time for longer than the mainland states. The outcome is that at some times of the year there may be five or six different time conventions in use in Australia, and that does not count bickering in Western Australia as to whether the tropical north of the state should follow the same time as temperate Perth, way down in the southwest.

In the United States, similar quandaries turn up, although perhaps not in so complicated a way. Some of the southern states, like Arizona, do not use daylight saving time while others do, and even some southern counties of Indiana exempt themselves from its provisions. Across the contiguous United States, this means that in the summer there are five time zones in operation. Of course Alaska, the Aleutian Islands, Hawaii, Samoa, Guam, and so on are in different time zones. In fact, Guam is kind of interesting, because it's on the opposite side of the international date line, and so it's not just a question of what *time* it is there, it's also a question of what *day* it is.

The International Date Line

Let us start with a puzzle. A heavily pregnant woman, expecting twins, boards a ship in Yokohama, bound for Hawaii. As the ship is approaching the Midway Islands in a storm the bucking of the vessel forces her into early labor. The first twin is born at noon, and the captain diligently records the date as January 1, 1999. Sixty minutes later the second twin arrives, but the time now is two o'clock because the ship has entered a new time zone an hour ahead of that where the first twin saw the light of day. But no, wait. The new time zone is not really one hour ahead; it's actually twenty-three hours behind, because the international date line (IDL) has been crossed. The captain therefore has little choice but to record the birth date of the second twin in the ship's log as December 31, 1998. We therefore have the paradox of twins born an hour apart, and yet their birth certificates will show that the second one (in terms of birth order) was born in the year before the first twin. Pray answer me, then: which twin is the elder?

Such an actuality seems unlikely, but given that there are a lot of people around nowadays with private planes and a lot of money, let me suggest to any of them who really enjoy yuletide festivities a way to experience Christmas on Christmas Island on two days in the same year. There is an Australian territory called Christmas Island, about five hundred miles southwest of Djakarta, Indonesia. There is also an island with that name about fifteen hundred miles south of Hawaii, part of the Line Islands. Thus one could spend much of one Christmas Day on the Indian Ocean's Christmas Island, before jumping into a jet and winging it east across the international date line to the Pacific Christmas Island and there repeat the experience, although without a chance of snow in either case. They are about a hundred degrees apart in longitude, so you might reckon that their Christmas Days overlap by only seven hours. But, let me hasten to add, you should diligently read the rest of this chapter before planning such a trip.

Such flights of whimsy aside, as a matter of fact the crossing of the international date line is indeed a real phenomenon which can cause headaches, or the contrary. When I fly from Sydney to Los Angeles, the plane may depart at 2 P.M., but I arrive in California at perhaps 10 A.M. *on the same date.* Flying back, after the fourteen-hour flight you arrive two days (or, at least, *dates*) later. A modern-day Mr. Scrooge who had saved his money could manage to avoid experiencing Christmas Day altogether by crossing the international date line from east to west in this way, and indeed flights around Christmas tend to be reduced in price because so few people want to be in transit then.

(Historical accuracy compels me to add this note, however, in connection with Mr. Scrooge. Elsewhere I have noted that the festivities associated with Christmas are quite a recent innovation. Santa Claus in his red suit started out as a 1930s advertisement for Coca-Cola. The idea of decorated Christmas trees was brought from Germany to Britain and thence the United States when Prince Albert married Queen Victoria in 1840. And so on. The Puritan ethic of Christmas Day being marked in a somber way persisted well into the Victorian age. Thus, when Charles Dickens wrote in *A Christmas Carol* of Scrooge insisting that his employees turn up for work at Christmas, this was not an unusually harsh business practice. It was the norm back then.)

The international date line is often said to have been defined by the International Meridian Conference in 1884. This is not quite correct. Nowhere is the international date line mentioned in the resolutions of the conference, although the resolution stipulating that longitudes should be measured both east and west up to 180 degrees implicitly defines a junction at the nominal position of the international date line.

I write "nominal" there because the international date line is not a straight line following the 180-degree meridian, but rather it deviates in several places with a path defined for the sake of local convenience. The idea of standard time throughout a region or country came about, as we have seen, in order that all clocks showed the same time within that area. Imagine how difficult it would be if the *dates* within different parts of a country were consistently different. The 180-degree meridian passes between the many islands which comprise Fiji. Rather than have split dates there, the international date line is drawn out to the east.

New Zealand is similarly affected. I lived in Christchurch, on the South Island of that nation, for three years. The seaport for Christchurch is called Lyttelton. One of the constituencies for the New Zealand Parliament consists of the town of Lyttelton plus the Chatham Islands, which lie about five hundred miles east in the Pacific, a single elected member representing both that town and those far-flung islands. This makes sense in that the Chathams have a small population, and are linked to the rest of New Zealand largely by ships sailing in and out of Lyttelton. But the Chathams are on the opposite side of the 180-degree meridian from the rest of the country. Again for convenience, the international date line is drawn as following longitude 172.5 degrees west of Greenwich from Fiji and Tonga until south of the Chathams, putting those islands on the same date as the rest of New Zealand, although they do use their own peculiar time zone.

There is another interesting consequence of this. Imagine that you are writing time-connected software for worldwide use. The standard time zones step from twelve hours behind coordinated Universal time (UTC) to twelve hours ahead. Knowing that daylight saving time can add an hour to standard time, you might put a check in your program to ensure that the computed value is no more than thirteen hours ahead of UTC. The problem is that the people of the Chathams choose to use a time another forty-five minutes ahead of that!

But it doesn't stop there. Perhaps in an attempt to capitalize upon being able to claim the title of the first country to enter each day, Kiribati (which was constituted in 1979 from the previous British protectorates of the Gilbert, Phoenix, and southern Line Islands) unilaterally decided to move the international date line from 1995 such that it passes around the easternmost part of this republic, which as a result is permanently fourteen hours ahead of UTC; straddling the equator, they do not worry about daylight saving time there. This means that the Pacific Christmas Island (Kiritimati in the local lingo) is actually a day ahead of Hawaii, although it is at just the same longitude. Thus, since 1995, my suggestion of enjoying two Christmas Days in one year, as described earlier, has been circumvented, although you could try racing the Sun by flying instead from the Pacific to the Indian Ocean Christmas Island, giving you a Christmas Day lasting for thirty-one hours or so.

The idea of the day change being necessary at some juncture has been around for a long time. The first circumnavigators, such as Ferdinand Magellan around 1520, noted that they had jumped a day in their diaries when sailing westward, and the opposite is also true. For example, Jules Verne, in his *Around the World in Eighty Days* (published in 1873), had his heroes thinking that they had just failed in their race against time, but then realizing that the day they had gained by traveling eastward had enabled them to beat their deadline and win the associated bet.

Nevertheless the international date line, as such, is another quite recent innovation which has been chopped and changed since the 1884 conference which implicitly established it. Prior to that convention, the date used by settlers spreading around the Pacific depended upon the direction from which they had come. Those sailing eastward would use one date (called the *Asian date*) while those traveling westward would have diaries showing one day earlier (the *American date*). We saw before that this caused a hiccup when Alaska changed hands in 1867: not only did people need to change from the Julian to the Gregorian calendar, but they were also on a different day of the week because of this effect. Similarly, if the flight schedules were set appropriately, you could take off from Sydney on a Tuesday and land in Los Angeles on Monday. Somewhere there has to be a junction between the days and dates, and that's the international date line.

Time Signals

It was mentioned above that the introduction of the electric telegraph in England in the 1840s allowed time signals to be disseminated throughout the land, and in the subsequent decades the installation of submarine cables enabled time coordination over much of the developed world, in principle.

That does not mean that this quickly came about. Different countries might decide to employ their own time standard based upon observations made at their own astronomical institutes. The time calibration signals could then be sent out in various ways. In particular, the invention of radio communications allowed a very convenient system to be introduced, and in this France led the way, with time signals being transmitted from the Eiffel Tower from 1910 onward. Although Britain tried some early experiments, it was not until 1927 that a regular radio time service was introduced there, using a complex of antennas near Rugby.

All must be familiar with the time pips still broadcast on various stations. Many people hate to work on a Saturday night, but when I was observing asteroids and comets from Siding Spring Observatory (in northern New South Wales, Australia) I would request telescope time on Saturdays because that gave me the chance to listen to the soccer commentaries from England on the BBC World Service in the middle of the night. On the hour at the start of the program the time pips would sound—*beep, beep, beep, BEEEEEP*—but the announcement of the time by the following plum English accent was spoiled for me by the fact that it was said to be in "GMT." As I explain in appendix A, that acronym is now ambiguous and should not be used; *coordinated Universal time* (UTC) is the correct term.

The time pips are a widely used phenomenon, and inspired my favorite title in a book: *The Beginning of the Long Dash: A History of Time-Keeping in Canada*, by Malcolm M. Thomson. To complete the cycle of this story I should point out that the idea of the broadcast time pips derived from Frank Dyson, the Astronomer Royal in the 1920s.

When different countries began broadcasting their own time signals it was possible to sit down with a variety of radio receivers and compare the times of receipt, and it was soon found that they differed by several seconds. In this connection France took the lead in trying to unify the signals, eventually leading to the Universal time concept. An international conference was called at the Paris Observatory in 1912, and the *International Time Commission* was founded. It was from this that the *Bureau International de l'Heure* resulted, established in Paris in 1920, in order to *define* the time.

The French may actually have refused to *use* GMT explicitly, then, but it was through their efforts that the equivalent which we all use, coordinated Universal time, was both brought into being and defined for worldwide use.

Railroad Time Reprise

The influence of the needs of railway scheduling was what brought about standard time in many parts of the world. You may have thought that I'd overemphasized that above, but let me assure you that it was the action of the railroad companies, rather than the procrastination of governments, which resulted in the time system we are habituated to using, thinking nothing of why we move our watches through a full hour as we move across a state boundary. Let me close this chapter by reiterating my point with a quote from the United States Code which defines standard time:

§ 265. Transfer of certain territory to standard central-time zone

The Panhandle and Plains sections of Texas and Oklahoma are transferred to and placed within the United States standard central-time zone.

The Secretary of Transportation is authorized and directed to issue an order placing the western boundary line of the United States standard central-time zone insofar as the same affect Texas and Oklahoma as follows:

Beginning at a point where such western boundary time zone line crosses the State boundary line between Kansas and Oklahoma; thence westerly along said State boundary line to the northwest corner of the State of Oklahoma; thence in a southerly direction along the west State boundary line of Oklahoma and the west State boundary line of Texas to the southeastern corner of the State of New Mexico; thence in a westerly direction along the State boundary line between the States of Texas and New Mexico to the Rio Grande River; thence down the Rio Grande River as the boundary line between the United States and Mexico: Provided, That the Chicago, Rock Island and Gulf Railway Company and the Chicago, Rock Island and Pacific Railway Company may use Tucumcari, New Mexico, as the point at which they change from central to mountain time and vice versa; the Colorado Southern and Fort Worth and Denver City Railway Companies may use Sixela, New Mexico, as such changing point; the Atchison, Topeka and Santa Fe Railway Company and other branches of the Santa Fe system may use Clovis, New Mexico, as such changing point, and those railways running into or through El Paso may use El Paso as such point: Provided further, That this section shall not, except as herein provided, interfere with the adjustment of time zones as established by the Secretary of Transportation.

Well, that's a tedious amount of verbiage to wade through, so what's my point? Well, first you should note that it is the railway companies which are given priority in fixing the rules. They have the privilege of immunity from the time zone rules which apply to all others; the legal time on board a train may be an hour different than that of the surrounding shops, banks, and other public and private facilities.

Second, you should recognize that in the United States it is the Department of Transportation—and not, say, the Department of Defense through the U.S. Naval Observatory, or perhaps the National Institute of Standards and Technology—which has the authority to define the time zones. As I wrote, the requirements of the railroads resulted in standard time, nothing else, making transportation the relevant governmental portfolio in deciding matters of time.

Chapter 19

❦

MARCHING TO THE SAME DRUMMER?

A lthough for matters of international communication and commerce it is necessary to have a calendar and time system which is universally recognized, it does not follow that all peoples regard the familiar calendar of the West, derived through the tortuous history so far outlined in this book, as being the only—or even the best—system for their domestic affairs, in either a civil or a religious sense.

When immersed in one's own culture, it is easy to be unaware that the concerns of other peoples may be quite different, in the same way that some sports contest takes on an immense perceived significance in one country, but elsewhere it is not newsworthy at all. The most egregious example is the so-called World Series of baseball in North America. One could justifiably say that no one else plays the game (apart from in Cuba, Japan, Taiwan, and a few other outposts of American culture), so that it is indeed the world championship, but the mere fact that few other nations take it seriously should tell you something. The All-Ireland Hurling Final is called just that, even though the winners could claim to be world champions. Where I live many people think that the final of the Australian Rules Football competition is the most important sporting event of the year, but again there's a good reason why so few people elsewhere play the game. In the same way, while the Western calendar may be the standard everywhere for international communication and so on—the equivalent of the Olympic Games, if you like—it does not follow that there are no local calendrical variants.

Ranges of Parameters

In all timekeeping matters one must think of the range of the parameters one has to hand and the implicit assumptions others may be making about their relative significance. A big old mechanical alarm clock is a good picture to keep in mind. The time you can set the alarm to go off is a maximum of twelve hours into the future, and it was reasonable of the manufacturer to design the clock with that built-in limit because people generally do not sleep for more than half a day at a stretch. Given its intended purpose, that alarm clock functions well.

The same is true with calendars, but one must consider the fact that other people may be working on a quite different calendar than your own. For example, your airline schedule may show you arriving at Ben-Gurion Airport at 9 P.M. (21:00) local time, which may be two hours ahead of coordinated Universal time (UTC), on day five of the week (Friday); but that is after sundown, and so by Jewish custom it will be the Sabbath already, so you may have difficulty getting a taxi into Tel Aviv.

That is a general comment, but in order for you to plan your business in Israel you would also need to be aware of the various holidays to which the Hebrew calendar is subject, there being a much greater emphasis upon the phases of the Moon than is the case in the Gregorian calendar. In fact, the length of the Jewish year may have one of six values spread by more than 30 days, because (1) it must contain either just twelve lunar months or else a thirteenth intercalary month; and (2) the lunar synodic period is not an exact number of days, whereas the year must be. Consequently an ordinary year has 353, 354, or 355 days, while a leap year may have 383, 384, or 385 days.

The calendar of a particular people will be shaped by whatever factors they view as being the most important in their lives, contingent upon their history. The Hebrew calendar, although it has evolved in the two and a half millennia since, is founded upon the Mesopotamian system picked up during the Exile, as witnessed trivially by the fact that the months have retained their original Babylonian names. The Mayans, the ancient Chinese, and the Polynesians, who navigated the Pacific a thousand years ago, will all have quite different cultural memories and concerns.

The Beat of the Drum or the Tingle of the Triangle?

In the conclusion to *Walden, or Life in the Woods* (1854), Henry David Thoreau wrote: "If a man does not keep pace with his companions, perhaps it is because he hears a different drummer. Let him step to the music which he hears, however measured or far away."

Let us expand that thought into cultures and their calendars. The calendrical systems employed by disparate civilizations will depend upon the drums which they hear. Some follow the beat of the Sun, whereas others hear the throb of the Moon more strongly. To expand the metaphor to cover the whole percussion section in an orchestra, the kettledrum, the bass, the cymbals, and the triangles may all follow basically the same rhythm, but your ear may pick out one in particular as being noteworthy, depending on your sensitivity, location, and so on. Thus many civilizations may have followed the return of the Sun to a particular constellation each year, or noted the reappearance of the Pleiades, or the heliacal rising of Sirius, and from one of those derived their annual cycle. But only the Egyptians could have been attuned to the flooding of the lower Nile. All of humankind may have been subject to similar experiences in some respects, but in others their annual ebbs and flows would have been unique.

Moon or Sun? To a native living in the middle of Africa the lunar influence extended only to the variation in moonlight each month, whereas to a coastal tribe dependent upon fishing for its subsistence the tides, and therefore the gravitational attraction of the Moon, may have had a much greater importance. In neither case is the Sun of huge significance in that the seasonal cycle is not of great amplitude in near-equatorial regions (although many suffer monsoonal effects). On the other hand, a Caribbean turtle fisherman would know that only once a year does his prey come up onto the beaches, at a time set both by the solar/seasonal cycle and the lunar phase. From such contrasting experiences derive a multiplicity of calendars.

There is one dominant calendar nowadays, but to think that it is the only significant system would be similar to believing that you can understand an ecological network by studying only the top predator. And might is not always right, whether levied by the Catholic or Protestant churches, or the Eastern Orthodox, or the synagogue, or some other religion, or the dominant military power.

It is all the colors in the spectrum which make it so interesting, not just the vivid orange or the aquamarine blue. In this chapter we will consider briefly a few other calendars so as to get some idea of the spectral variety. This is not meant to be a complete list—there are several dozen calendars in widespread use even today—but rather a mere selection of a few postcards from a huge display, one or two mailed home to show what the range of scenery is like.

Some Basic Considerations

Before charging into considerations of different calendars, some thought should be given to the basics, although this can be brief because, if you have read this far, you will recognize that calendars are not always logically founded. For example, we have seen that the design of the Gregorian calendar is definitely suboptimal with regard to astronomical veracity, both in terms of the precision of the leap-year cycle and, even more so, in its description of the motion of the Moon. This provides an illustration of the fact that practical utility, which may encompass intrareligious bickering, is more important than refined astronomical computations: it was more important to the Roman Catholic Church to develop a weapon with which to counter the rise of the Protestants than to bring in a calendar which was more complicated, but scientifically closer to reality. The calendar is also one of the most conservative of all human constructs, and enacting any change is mightily difficult, even if you can put up an inarguable case that your suggestion provides an improvement.

It does not follow that because we use a calendar based largely upon the daily movement of the Sun across the sky, with an additional dependence (especially in fixing various religious celebrations) upon the motion of the Moon, that all human calendars are structured according to those parameters considered independently; that is, a solar day and a lunar month. If that were the case, then the day would always be twenty-four hours long, or very close to it, and the month similarly restricted. A contrary example is mentioned in appendix D: some maritime societies may use a *tidal day* which lasts for about twenty-four hours and fifty minutes, that being the recurrence time for the *Moon* (rather than the *Sun*) passing overhead, high tides occurring twice within that fundamental period. If you lived in a modern seaport where the coastal morphology leads to particularly extreme tidal variations, you would soon notice how the departures and arrivals of ships varied with the lunar day rather than the solar day.

Animals other than humans may be even more greatly affected by such matters. Many crustaceans and shellfish which inhabitat intertidal mudflats live by the lunar day rhythm, and experiments have been conducted which show that it is ingrained: if they are removed to a light-controlled laboratory where no sea tides can act, such shellfish continue to open and shut with a twenty-four-hour-and-fifty-minute cyclicity. The lives of these animals are dominated by the tides, not the Sun.

Numerological Influences

Even if a particular society is not affected by such considerations, so that it is only the solar day and the lunar month which are the short, astronomically imposed units of time, still there is the matter of whether the year should be lunar or solar. A wide variety of human parameters will control the calendar which evolves, logic rarely being one of them. Just to set the ball rolling, let us consider two disparate matters connected with numbers in the calendar. One is thirteen, and the other is seven.

At the conclusion of chapter 7 mention was made of superstitions regarding Friday the thirteenth, and *triskaidekaphobia* itself (the fear of the number 13). There are many explanations (and even justifications) which have been put up for this irrational belief. Some say that it stems from the presence at table of thirteen, including Judas Iscariot, for the Last Supper prior to the Crucifixion, whereas the nonreligious tend to reject this connection. Others see a deep-seated human connection, and claim that it is because a woman menstruates thirteen times during a year (although twelve times is more usual). In discussing the intercalary period inserted into the old Roman republican calendar, I noted that this was considered to be unlucky, as was the whole month of February.

On any lunisolar calendar in which a thirteenth month may be added to every third year, or three in eight, or seven in nineteen, dependent upon the cyclicity used, that intercalary month is often regarded as being a time to be wary by the culture involved. Similarly, a Friday the thirteenth in modern Western society is widely considered unlucky, even by those who claim not to be superstitious, and many hotels do not contain either rooms or floors which are numbered 13 due to customer pressure. Next time you step into the elevator of some skyscraper or tall office block, check for the number 13 button: it may not exist. The root source of all this seems to be the long-term attitude, dating back to ancient Rome, that the thirteenth month in a year is unlucky.

On the other hand, the number 7 is regarded as being lucky by many (but not all: some Chinese peoples revere 8 most highly, while others may put their faith in 9). Seven is the number of days in the week. Since our weekday cycle began over 2,000 years ago, it has not been broken: when ten days were deleted from the Catholic reckoning in 1582, and when Lord Chesterfield's act effectively skipped eleven days in Britain and its dominions in 1752, and when other nations moved onto the Gregorian calendar, the sequence of the days of the week was not interrupted.

On that basis one might consider the seven-day week as providing the most significant calendrical cycle, above all others, but that cannot be the case because it is a slave to the start of the month and the year. That is, neither the months nor the years contain an exact number of weeks. How could one get around this?

The Leap-Week Concept

The answer could be to tailor the year to the week. Instead of having a leap-year cycle in which intercalary *days* or *months* are included every so often, rather have a system incorporating intercalary *weeks*. Iceland once used a calendar with 364 days/52 weeks in a common year, and 371 days/53 weeks in a leap year. Let us look at the possibilities of such a system.

The 400-year cycle of the Gregorian calendar contains 146,097 days, and because that number happens to be divisible by seven there are precisely 20,871 weeks spread over those four centuries. In the leap-*day* system of that calendar the basic cycles are 4, 100, and 400 years, the 100 resulting from those century years which are common. Let us see what could be done over the same gross cycle time using a suitable leap-*week* rule.

If common years were to have 364 days, then there is an excess of about 1.242 days per year which accumulates and needs to be picked up with the leap weeks; I am only giving that quantity to three decimal places because I don't want to worry about which flavor of year should be utilized. If you divide 7 by 1.242, you get the average rate at which intercalary weeks need to be added: once every 5.636 years. Once every 5 years is too often, once every 6 is too infrequent. The thing to do is to mimic the leap-day system we employ, which puts in 1 such day every 4 years (which is too often) and then skips a few so as to bring the average rate down.

The simple answer is this. Have an intercalary week in every fifth year (2005, 2010, 2015, 2020, . . .) unless the year number is divisible by 40 and not by 400. For example, this would make 2000 and 2400 leap years (with a whole week added), but 2040, 2080, 2120, . . . would be common years.

Over the 400-year cycle, 71 years (80 less 9) would have leap weeks inserted. This would render a total number of days equal to (364 × 400) + (71 × 7) = 146,097. And that's precisely the same as the Gregorian system we use. Instead of the (4, 100, 400) leap-day cycles, one would use (5, 40, 400) leap-week cycles.

One could argue in favor of this concept on several grounds. One is that leap years would be divisible by 5, so that they end in either 0 or 5, whereas the four-yearly leap-day system is not so simple (quick: was 1986 a leap year?).

That, though, is a trivial consideration. More important would be the fact that the dates would recur on the same day of the week. For instance, one could arrange for December 25 and January 1 always to be Sundays, or July 4 always a Friday and hence the start of a long weekend. Whatever the decision, one could argue that this sort of calendar could benefit business and industry due to its regularity, with just an extra week occurring every five years (but omitted once or twice in a lifetime).

As a matter of fact this leap-week idea is not a new one, and such calendars *have* been suggested from time to time. When the Eastern Orthodox churches were discussing calendar reform in 1923 they considered a Romanian proposal using leap-weeks, but rejected it largely because it would have meant abandoning the previous Easter *computus*. Again we see how significant Easter has been in calendar history. In 1930 there was another leap-week calendar proposal put forward, this time by a Jesuit, James A. Colligan, but once more the Easter question scuppered it within the Catholic Church.

On secular grounds, though, surely the leap-week system might eventually be adopted if the drift away from organized religion continues in the economic powerhouses of the United States and Europe, the god of mammon being ascendant? Let's look at that leap-week system again. The greatest regularity would result from having not only the year start with the same day of the week each round, but also each month. And regularity, logic argues, leads to greater convenience, heightened efficiency, and better business and productivity. Thus one would arrange the year into months which lasted four weeks, each starting with a Sunday, say. Makes sense?

If that were the case, then each year would have thirteen months; and, for the first of the month being a Sunday, the thirteenth would always be a Friday! By asking that the leap-year cycle be predicated upon our lucky seven days in the week, we have arrived at the unlucky thirteenth month again, and worse.

As I wrote above, I wanted to outline some basic considerations here which are just *examples* of the sorts of things which would affect the choice of any calendar scheme by a particular culture. If you are not superstitious, then you would not be worried about how the numbers 7 and 13 might enter into your calendar. If you are not a Christian, then you might be unconcerned about the Easter question. On the other hand, we have seen that this was actually the major factor which led to the Gregorian calendar. Similarly, the calendars developed by other cultures will surely reflect concerns which would appear ridiculously trivial to an outsider (unless he or she were an anthropologist), but have been the driving forces behind their date-reckoning systems.

The Chinese Calendar

Few will be unaware of the distinct Chinese calendar, with New Year celebrated in late January or in February, and years being accorded names associated with specific animals in a cycle (the Year of the Tiger, Ox, Rat, Dragon, Snake, and so on).

The Chinese calendar, like the Hebrew, is lunisolar, with the time of the year being controlled by the phase of the Moon *and* the movement of the Sun. The Jews employ the autumnal equinox as their marker point with regard to the solar contribution; the Christian churches use the vernal equinox in deciding Easter and thus their liturgical calendar/movable feasts; the Baha'i New Year is at the vernal equinox; and other conventions based on one of the equinoxes could also be mentioned.

The Chinese look rather to the winter solstice. The Chinese New Year is the time of the *second* new moon after the solstice, the approximate range of dates being January 22 through February 19. Note, however, that the new moon in this connection is actually the *dark moon*—the time of conjunction, not the time when the first silvery crescent is spotted—in contradistinction to the usage we have previously assumed. A fifteen-day period of festivity follows; since the count is from conjunction, this means until the full moon.

The ecliptic was an important reference locus to the Chinese. The Moon's path never takes it much more than five degrees away from the ecliptic, which was divided into twenty-eight lunar houses or mansions. Conversely, the Chinese divided the annual path of the Sun into twelve sections which render the twelve zoological names for the years in a cycle. An epicycle was made up of five of those dozen-year units, lasting sixty years.

The origin of this is connected with planetary conjunctions. Jupiter and Saturn come into conjunction about every twenty years, which corresponds fairly closely with two-thirds of a complete solar orbit by the latter, and one and two-thirds orbits by the former. It can be seen that for a repeated conjunction in the same part of the sky (in the same constellation), one needs to multiply by three: five Jovian orbits take as long as two Saturnian orbits, and the required duration is sixty years. Other quite distinct civilizations recognize the same cycle. For example, the Dogons of Africa view this as being a full human lifetime, the first score years being youth, the second adulthood, and the third seniority.

The start of the Chinese calendar is often allocated to a planetary conjunction in March of 1953 B.C., although the history is vague. Some trace the thread back only to the fourteenth century B.C., while there is a legend that Emperor Huangdi introduced such a calendar in 2637 B.C.

Even over quite brief spans of time it is erroneous to assume that nations have been homogeneous and unchanged, so that describing a specific "Chinese" calendar lasting for more than four millennia seems fraught with danger. Historical studies show that calendar changes have been very frequent in China, especially in times of social and regnal upheaval. Between the fourth century B.C. and the eighteenth century A.D. there were more than a hundred distinct calendars introduced by royal edict, so that on average they each lasted for only twenty years or so, giving the lie to any notion that calendars are difficult to reform in *all* cultures. In this situation calendar change was invoked more as a weapon of control, the fact that date changes perturb people being used as a mechanism of subjugation. The contemporary Chinese calendar is that which has survived through all this confusion.

From the sixty-year cycle the Chinese also developed a sexagesimal day-counting system similar to the Babylonians, providing for a sixty-day week. Although the Chinese have clearly used a seven-day week for some time, it is not clear the extent to which this may have been self-derived, diffused through India from Babylon in time immemorial, or been brought in by early Christian missionaries. Perhaps a mixture of all three.

If one studies a table of the Gregorian dates for the Chinese New Year, one sees the date regressing in eleven-day steps and then jumping forward (later in the year) by a twenty-nine- or thirty-day step, behavior reminiscent of the epacts described in chapter 13. What is happening, of course, is that the eleven-day difference between the solar year and twelve lunar months causes the New Year to arrive sooner two or three years in a row, but then an intercalary month must be inserted.

This is done again according to the Metonic cycle of nineteen years, the intercalary month normally appearing after the seventh month of the year. Every nineteenth year, however, in completing the cycle the rule is that it should be inserted after the eighth month, being called *ren ba yue,* or "double August." This is viewed as being an especially unlucky time, with disasters to be expected. Many Chinese ascribed the great Tangshan earthquake of 1976, which killed over two hundred thousand people, to that being a double August year, and they breathed a collective sigh of relief when no such calamity befell them in 1995 (although that *was* the year of the Kobe earthquake, which left nearly six thousand Japanese dead, and also a major tremor on Sakhalin Island just to the north, which killed a lesser number).

The nineteen-year cycle is not necessarily perpetual, however, and Chinese calendar development has followed a trend similar to that described earlier in connection with the calendrical advances of the ancient Greeks, whereby cycles of seventy-six years, and then longer cycles still, were tried. In some ways, though, the idea of using such extended cycles is misleading when one considers the frequent calendar reforms which have occurred in China. It has more been a case of realizing that the nineteen-year cycle followed had led to a discrepancy of a whole day in the phase of the Moon, and spasmodically correcting it, rather than adhering to some long-term calendrical plan like the epact table of the Gregorian system.

The Japanese Calendar

The lunisolar calendar used in Japan is similar to the Chinese, although with some significant differences. In particular, improved astronomical knowledge in China—of the date of the winter solstice and the prediction of the lunar phase—meant that the Japanese tended to lag behind developments on the main Asian landmass. This was not just the simple matter of, say, a one-day adjustment after seventy-six years in the tabulated dates of the new moon; larger gaps were possible. Say that one's tabulated date for the solstice was incorrect by a day or two; this could lead to the New Year being celebrated either a month too soon, or too late, according to the simple rule of "second dark moon after the solstice."

Similarly, the fact that the actual synodic period for a particular lunar orbit may vary between about 29.2 and 29.8 days could lead to the calculated dark moon being at a time which is too early or too late if it is presumed that the 29.53-day mean period is adequate. Another contribution to discrepancies would derive from Tokyo being at a different longitude than Beijing, sometimes allowing the Chinese to witness the crescent new moon twenty-two hours earlier than the Japanese. One sees here shades of the Easter *computus* problem in the Christian churches, the inexact assumptions made in framing the rules leading to Easter being marked sometimes on the "wrong" date.

Over the past several centuries much of the astronomical development in China, followed by Japan, was driven largely by these calendrical considerations. There was an impetus to obtain a far better understanding of the clockworks of the heavens, especially when Jesuit missionaries arrived carrying obviously superior astronomical knowledge.

It was mentioned above that while a cycle involving the number 60 was used in China to name both years and days, the seven-day week cycle also reached there quite early. In the case of Japan, that cycle seems to have arrived at some stage between A.D. 800 and 1000. Presumably the later arrival of Christian missionaries, from the sixteenth century, led to its ascendancy and the eventual naming of the days according to the planets.

While the Chinese and Japanese calendars have similar bases, different dates within them are associated with specific celebrations or events to be marked, in the same way as the United States marks Independence Day, France sings on Bastille Day, the English wear a rose on Saint George's Day, and the Irish wear green on Saint Patrick's Day. For example, in Japan the equinoxes are important: a seven-day period called a *higan* ("equinoctial week") occurs when they traditionally visit the graves of their ancestors. In China the same sentiment is manifested through each family's "Tomb Sweeping Day," which occurs around *ching ming,* one of the traditional divisions of the solar motion along the zodiac, this occurring in early April.

When did Japan give up on the Julian calendar and adopt the Gregorian? The answer is: never, because the Julian was not used at any stage. As a result of the Meiji Restoration in 1868 the Gregorian system was adopted from January 1, 1873, following a government decree proclaimed just twenty-three days earlier. Thus, in Japan, the calendar used throughout most of the world is officially employed for convenience, but the old lunisolar calendar still governs many public holidays and local observances.

An Example of a Recent New Era Definition

The idea that a new era—a new year-numbering convention—may be suddenly declared might seem a little strange, but it does happen often, at least somewhere in the world. Sometimes it is retrospective, as in the gradual uptake of the *anno Domini* system. The accession of a monarch in Britain leads to all parliamentary papers being dated according to the year of the new reign, but that is little used in common life. Japan provides a contrary example. There, the regnal year-counting system persists as a parallel to the Gregorian tally, the Showa era ending in its sixty-fourth year when the Emperor Hirohito died in 1989, and that year is also the first of the Heisei era.

It is also interesting to note in this connection that while China abandoned its dynastic year count after the revolution of 1911, following which it adopted the Gregorian calendar for official business, in North Korea the opposite has just occurred. In 1997 that Stalinist state instigated a year numbering it calls the *Juche era*. *Juche* is a Korean word meaning "people create their own destiny." The following is the official press announcement from the state-run Korean Central News Agency:

> *Pyongyang, September 10 (KCNA)—Juche era is available in Korea from September 9. The whole country was yesterday in deep emotion and joy at this great national event which records the Juche era with 1912 when the respected President Kim Il Sung was born as the sun of humankind, as the first year. The Korean people courteously printed the first Juche era on telegrams, documents and information addressed to Secretary Kim Jong Il, with deep gratitude, unfailing loyalty and devotion to him who ensured the continuation of President Kim Il Sung's history. On Tuesday national news agency and radios started broadcasting with Juche era and capital and local press organs published newspapers on which the Juche era was printed. All transport is operated according to this era. People began to use it in communications and other daily routine. In maternity hospitals across the country babies were given the date of birth with the Juche era.*

In the Democratic People's Republic of Korea, then, calendar reform is alive and well. What to many people will be A.D. 2000 will be counted as Juche 89 by North Koreans.

The Soviet Calendar

The fact that the Russian "October Revolution" in 1917 actually occurred in November (on the Gregorian calendar) has been alluded to elsewhere. I am writing this on December 16, and the local newspaper tells me that the "mad monk" Grigory Rasputin was murdered on this day in 1916, and again a date shift is really required. But even given that shift, the system later employed was anomalous. The whole history of the calendar used in Russia is a remarkable story, in particular during the postrevolutionary phase, when that nation was the mainspring of the former Soviet Union. They gave up the seven-day week!

Until about the fifteenth century the Russians, through the Orthodox Church, had persisted in using an *anno mundi* system (a count from the sixth millennium B.C.) with March 1 as New Year. Thereafter September 1 (the indiction date) was used to begin the year, and that calendar is still used in some places; it is called the *Byzantine calendar*, counting from 5509 B.C. but with New Year's Day now being September 14 due to the thirteen-day Julian-Gregorian difference. In Russia that system was employed until 1700, when

Peter the Great instigated the usage of January 1. At that time many other non-Catholic European nations were switching to the Gregorian calendar, but Russia remained on the Julian, largely through the influence of the Orthodox Church.

From time to time attempts were made to convert Russia to the Gregorian system, but without success—the conservative czars resisting any reform, although considerations of trade and commerce made it inevitable that change must come. Leo Tolstoy (1828–1910) wrote that "it is obvious that the eloquent demand of our century and of modern life will not stop before any difficulties, and therefore undoubtedly a practical solution of the calendar question is near." But he was long-since dead when, soon after the revolution, Lenin directed that the Gregorian calendar should be immediately implemented.

In its rudimentary form this did not last for long. While the Russian Orthodox Church persisted on the Julian, and the Soviet Union used the Gregorian dates for external communications and the like, internally the state restructured its calendar in various ways from 1923 to produce what was called the *eternal calendar.* This never really caught on, and by 1929 there was a new reform which revolved around the encouragement of industrial productivity.

The seven-day week was abolished, and instead a five-day cycle was substituted such that each worker had one day off during that time. This meant that 20 percent of each week was a rest period, as opposed to the previous two days in seven, or 28.6 percent. The aim was clearly one of greater efficiency. On top of that, these rest days were staggered: each worker was given a colored card showing which was his or her day off. This meant that there was no weekend, as such, so that the factories could keep functioning, with various economic advantages resulting. Keeping the lines running in the Soviet Union might have led to more efficient use of resources, but the people had reasons to be unhappy.

Sets of six of these weeks were arranged into months, there being twelve of those 30-day months in each year. In ancient Egypt such a 360-day calendar was used, with a 5-day epagomenal period not within any month being a time for festivities. Russia brought in a similar system, except that the 5 days were not contiguous: these were used to commemorate various facets of the revolution, and their Gregorian dates were January 22, May 1 and 2, and November 7 and 8.

After a few years of trial (and that word may be interpreted in several ways here), the five-day week with staggered rest days was abandoned in 1932, and instead a six-day week with all workers having the sixth, twelfth, eighteenth, twenty-fourth, and thirtieth days of the month off was introduced. This meant that the ratio of rest against workdays that they enjoyed was further reduced.

In its initial form that revision lasted only a couple of years. From 1934 on the month lengths have conformed to the Julian/Gregorian convention, but the seven-day week was not reintroduced until 1940. This meant that in the period 1934–1940 there was some confusion, and much of the aim of added industrial efficiency was lost: because of the six-day cycle, no one really knew what to do on the thirty-first day of the longer months, so that the comrades were either given extra pay or an additional day off, while the lack of a February 30 meant that many took March 1 off instead.

More than two decades earlier, Lenin had published a decree legislating for the Gregorian calendar to be used "for the purpose of being in harmony with all the civilized countries of the world." After 1940 the Soviet Union reverted to that scheme, complete with the seven-day week. "Eternal" was hardly the best adjective to have been applied to the short-lived Soviet calendar experiments.

The French Revolutionary Calendar

Prior to the Soviet fracas, there had been another attempt to introduce a de-Christianized calendar along revolutionary lines. Following the French Revolution, which began in 1789, a novel calendar was declared to count from September 22, 1792, onward, a new era dating from then being employed. The numbers of the years did not grow greatly, however: after Napoléon gained power in 1799, the revolutionary calendar went out of favor, and the French senate proclaimed toward the end of year 13 (during A.D. 1805) that the Gregorian calendar would be resumed from the start of 1806.

My terminology there had to be a little convoluted because the French revolutionary calendar had abandoned January 1 as the beginning of the year, instead employing the autumnal equinox. The start was also declared post hoc; it was actually introduced on November 24, 1793, in year 2 of the new era count.

As in the ancient Egyptian calendar, the 5 days prior to the start of a year were a holiday period, but unlike the Egyptian year of exactly 365 days leading to the 1,460-year Sothic period, the French allowed for leap years and thus a 6-day holiday from time to time, although the way this was accomplished during the revolutionary calendar's short lifetime was haphazard.

The division of the other 360 days again echoes the Egyptian practice, with twelve months each of 30 days, subdivided into three 10-day weeks. The Egyptians called these the decans, but clearly here there is another implication: the French have long been keen on decimalization. This led to discontent among the people, though; one reason that they disliked the new calendar was that now there was only 1 rest day in 10, rather than 1 in 7 as before. As in the case of the Soviet calendar, this new system introduced supposedly for the benefit of the masses actually caused them to have fewer rest and recreation days each year.

We saw earlier that the French have tried for many decades to get the rest of the world to use a decimalized time system, and after the Revolution they got the chance to try it out at home. Clocks were manufactured displaying ten hours in a day, each split into a hundred minutes, but this system was rejected by the people even more quickly than the radical calendar.

The twelve months were given novel names. The fifth was called *pluviôse*, and obviously that was a time (straddling mid-January to mid-February) when it was expected to rain a great deal. The last month of the year, mid-August to mid-September, was called *fructidor*, for the fruit ripening then. One of the month names which has continued in common usage in English is that of the eleventh, named *thermidor* because it was the hottest time of year; the culinary creation known as *lobster thermidor* got its label because it was first served to Napoléon in that month.

Both the five end-year days of celebration and the festivities themselves were known as the *sans-culottides*. The origin of that term may be perceived by an anglophone with some knowledge of women's fashions during the past few decades. In some phases among the fads of oscillating hemlines leading to abominations like the miniskirt and hot pants, a rather more attractive style—I cannot call it an innovation, for reasons which will become apparent—has gained favor. This style is called *culottes*, the word being derived from the French term for breeches. Now, in French *sans* means "without," but that does not necessarily mean that the *sans-culottides* festivities consisted of a mass orgy, people running around without their breeches. Rather, the origin of the term comes from the prerevolutionary peasants who could not afford breeches/culottes like those worn by the rich men who had been the first to meet Madame la Guillotine. The name of the five-day festival was chosen to honor those peasants.

Since its abolition there have been various attempts to revive the spirit of the French revolutionary calendar, such as during the Paris Commune in 1871 (which lasted for just ten weeks), and also during the Spanish civil war in 1936–1939. It remains an interesting oxbow of history, a lesson in how resistant people are toward anyone trying to reorganize the temporal bedrock on which they build their lives.

The Coptic and Ethiopian Calendars

The Coptic Church is the major Christian sect in Egypt (the adjective derives from the Coptic language, a derivative of ancient Egyptian which is now extinct except in the liturgy of this church). The important points to remember are that our calendar traces a history back through ancient Egyptian learning, through the Roman era in Egypt, and later through the Council of Nicaea effectively delegating responsibility for the *computus* to the Alexandrian Church. Dionysius Exiguus recognized this when he used the lunar phase rules from Alexandria rather than from Rome, where he lived.

Now, Dionysius took a step which eventually brought about the *anno Domini* year-numbering system when he decided to abandon the *anni Diocletiani* which counted from August 29, A.D. 284, and was used by the Church of Alexandria. But that church (the Coptic, in its present form) persists in using that era, and the leap-year rule of the Julian calendar. Because of the thirteen-day difference between Julian and Gregorian, the Coptic New Year is now on September 11 (the twelfth in a leap year), and that date in A.D. 2001 is the first day of *anno Diocletiani* 1718. Actually, the Coptic Church refers to this system as being the *Era of Martyrs*, Emperor Diocletian being remembered as a persecutor of Christians, Jews, and any others who failed to follow the

Roman state religion; it was for precisely this reason that Dionysius wanted to drop the Diocletian era.

I wrote that the Coptic Church uses the Julian leap-year rule (intercalary day *every* four years), but that does not mean that it uses the Julian calendar. Rather, a system based on that ancient Egyptian calendar is employed, as described above for the French revolutionary system: twelve months each of thirty days followed by five (six in a leap year) epagomenal days. Those are September 6–10 on the Gregorian calendar, at least until 2099 (because the following year is a Gregorian common year, increasing the difference to fourteen days). There are nonetheless vestiges of the Julian calendar as used in the other Eastern Orthodox churches; for example, Christmas is celebrated on January 7, equivalent to December 25, Julian.

Heading south from Egypt to the source of the Blue Nile one comes to Ethiopia. That this nation has a long-term link with the Judeo-Christian tradition is clear; for example, there are many black-skinned Jews living there, and some have moved to Israel. There is a unique Ethiopian calendar, denoted by E.C. This calendar is the same as the Coptic, except that the start of the era is different by precisely 276 years. For some reason the Ethiopians count from A.D. 8, which means that the aforementioned date of September 11, A.D. 2001, is the first day of 1994 E.C. To look at it another way, 2000 E.C. begins in September of A.D. 2007, and 2001 E.C. in September of A.D. 2008. One might imagine that the Ethiopian government will be anticipating a tourist boom, as people take advantage in those years of the opportunity to celebrate the start of *another* millennium.

The Islamic Calendar

The calendars described above were either (1) lunisolar, the number of *months* being adjusted by some cycle like the Metonic so that the duration of a particular year may be thirty days or more different from the adjacent year; or else (2) purely solar in that some leap-year cycle is followed with single intercalary *days* being inserted. The Gregorian calendar is of the latter type, with the Moon affecting only its internal structure (when Easter will occur, etc.), and the lengths of the calendar months being divorced from the lunar synodic period.

It is possible, though, to have a calendar which exclusively follows the Moon. Such a calendar is the Islamic, or Muslim. In this system a year consists of twelve observed lunar cycles, so that it is almost 11 days less than a solar year. On average the Islamic year lasts for $12 \times 29.5306 = 354.367$ days, but be clear that this is an *average* length. First, the new month cannot be proclaimed until the new moon is actually sighted, giving a clue as to why the crescent appears on the flags of so many Islamic states, and particular months may deviate some way from the average. Second, it follows from this that the year must be quantized, containing an integer number of days, but that number can vary from one year to the next.

The difference between the average length of the Islamic year and that according to the Sun is near 10.875 days, implying that the seasons slip by that amount from one year to the next, and go through a complete cycle in about 33.6 solar years. Thus the month of Ramadan, during which there is a rigid fast from dawn to dusk observed by Muslims, may be in midsummer in one year, but will have shifted to midwinter 17 years later. In the Islamic system the day begins at sunset (as it must do because the spotting of the new moon controls all such things).

In the previously discussed calendars, converting from a year number in one era system to that in another is quite straightforward, merely involving a regular offset, even if they use different start-of-year dates. The number of years in the Jewish *anno mundi* era starting near the autumnal equinox in 3761 B.C. can easily be converted to an *anno Domini* count (but be warned that the calculation of the first day of Tishri, Rosh Hashanah, is very complicated, although basically it is near the new moon occurring close to the equinox). In fact, in expressing that matter I did the sums back to front as I have throughout this book: I used the familiar B.C./A.D. system as the framework. Rather, one should say that the Jewish year 5761 begins in A.D. 2000. The point, though, is that the yardstick being used is of uniform length; the average length of the Jewish year is the same as the average length of the solar year.

This is not the case for the Islamic year, because it is almost eleven days shorter. Converting from an Islamic date to a Gregorian calendar date is therefore a headache. In investigating this problem I came across a lovely little book entitled *The Islamic and Christian Calendars, A.D. 622–2222 (A.H. 1–1650)*, by G. S. P. Freeman-Grenville. This version was published in 1995, but there were earlier editions which carried an introduction which began as follows: "At the twenty-fourth annual meeting of the East African Swahili Committee, held in Zanzibar in September 1960, I was told that the absence of any comprehensive work dealing with the Muslim and Christian Calendars was causing difficulties."

This opening sentence is one of the most delicious I have ever come across in a book, being redolent of English expatriates discussing colonial matters over postprandial port and cigars after an excellent dinner of roast beef and Yorkshire pudding, no matter that the temperature outside is over the century mark. I likely do them an injustice, but what a way to start a book! I'd put it right up there with "It was the best of times, it was the worst of times . . ." (Charles Dickens, *A Tale of Two Cities*) and "It is a truth universally acknowledged, that a single man in possession of a good fortune, must be in want of a wife" (Jane Austen, *Pride and Prejudice*).

Unfortunately, Freeman-Grenville then spoils it with his beginning of the book proper, at least in those early editions: "The moon revolves round the Earth in 29⅓ days. But because the Earth is itself in motion, this revolution in fact takes approximately 29½ days. The Earth itself performs a complete revolution once in twenty-four hours, and at the same time revolves round the Sun in slightly less than 365¼ days." Of course, it actually takes the Moon a little over 27, not 29, days to orbit Earth, the terrestrial motion indeed leading to a synodic period near 29½ days. Also, Earth does not spin once in 24 hours, but rather in a length of time somewhat less than this: it revolves on its axis about 366¼ times during a year, not 365¼.

No matter, Freeman-Grenville gives a nice introduction to the Islamic calendar. In the title of his book given above, the abbreviation A.H. appears. It is usual to employ this Latinized version when writing in English; it stands for the *anno Hegirae,* meaning the year counted since the prophet Muhammad departed Mecca. The book's title further informs one that A.H. 1 began in A.D. 622, although not on January 1, of course; the start of the first month, Muharram 1, was calculated to be July 16 in that year, although this was done post hoc, 17 years later by the caliph Umar, and in any case it is acknowledged that Muhammad likely left Mecca a couple of months later. The book title further indicates the relative lengths of the Islamic and Christian years: about 1,649 of the former must be equal to 1,600 of the latter, a difference of 1 part in 33.6, as indicated earlier.

Several times I have alluded to the fact that the declaration of a new moon requires the actual sighting of the lunar crescent just after sundown, and that will depend upon various factors such as the observer's latitude and longitude, and the weather conditions. In connection with the latter, sometimes cloudiness intervenes, and so rules exist for declaring a new moon even though it is impossible to have seen it. A week of solid cloud cannot be allowed to postpone the start of the month for such a duration.

That is a meteorological matter, whereas lunar visibility depends upon your geographical location, and that is connected with astronomy. The position of the Moon at any near-future time (within decades) can be computed with utmost precision, but as to whether it may be seen is subject to various criteria such as the elevation above the horizon, the angle which it lies north or south of the ecliptic, the sky lighting conditions which depend upon your latitude, and so on. Numerous attempts have been made by astronomers, both Muslim and Western, to quantify all these effects. One outcome is the concept of the *international lunar date line.*

The international date line discussed in the previous chapter is based on the Sun, giving the location where the solar day jumps from one value to the next. Imagine applying the same basic concept to the Moon. A mullah in Mecca, say, does not catch a glimpse of the Moon before it sets soon after the Sun, and so the declaration of the new month has to wait until the next evening. But by ten hours later the Moon has moved about five degrees further from the Sun in the sky, and so an Islamic cleric at a mosque in Los Angeles is able to spot it and call that the new month has begun.

For such an episode to occur, clearly there is some dividing line between the two locations, but unlike the well-known IDL, the international *lunar* date line is not just a simple line of longitude (admittedly with a few kinks). Rather, because lunar visibility depends upon both longitude *and* latitude, it is a curve which is symmetric north-south about the declination of the Moon. Not only that, but because the synodic month is not an integer number of days, the curve will move east-west from lunation to lunation. Places within the curve can witness the new moon the evening before those to the east of it. Thus Muslims living in different parts of the world may follow different calendars: if one lives in North America, say, your new month may start the evening before that of a Muslim in the Middle East.

The Sikh Calendar

The Sikh calendar is another lunisolar calendar, with twelve lunar months in the majority of years and thirteen in leap years, but there are two peculiar factors which make it worthy of mention.

The first is that the solar-year duration employed, termed the *Bikrami* year, is not the same as that used in most other calendars. We are habituated to a year which is slightly less than 365¼ days long, whereas the Bikrami solar year is actually a little *longer* than 365¼ days. It is actually close to the period between perihelion passages of Earth (see appendix B). Because of this discrepancy the Sikh solar year drifts away from the Gregorian dates at a rate of about 1 day every 70 years.

The second thing to note is that on the Sikh calendar the anniversary dates are recognized according to the lunar year, which may last for either twelve or thirteen months, rather than the solar year. This is best understood with an example. One of the holy men of the Sikh sect is Guru Gobind Singh, who was born on the twenty-second of December in 1666, on the Gregorian calendar. The one-day-in-seventy-years shift meant that his birthday was to be celebrated on December 25 in 1998; one might have expected a date one or even two days later than that, but recall that this is another calendar in which the Moon is followed rather precisely in terms of its *sighting*, with the result that theoretical calculations may be a little out. However, the Gregorian year of 1998 contained only twelve lunar months on the Sikh calendar, and so the birthday of Guru Gobind Singh was also noted on January 5 in that year: two birthdays in one year, according to the calendar of the West. But there is no such thing as a free lunch, and the following Sikh year (straddling A.D. 1999) contains thirteen lunar months, such that there is no birthday for Guru Gobind Singh in 1999, the next such event occurring on the fourteenth of January in A.D. 2000.

Within the Sikh community two different dating schemes are used, but both employ an era-numbering system which is fifty-seven years in advance of the *anno Domini* convention. There is current discussion concerning a possible reform of the Sikh calendar to try to bring about agreement between that and a related calendar, the Khalsa.

The Hindu Calendar

There are several distinct branches of Hinduism. The Sikhs of the Punjab broke away in about A.D. 1500, being unhappy with the caste system and the assumed supremacy of the Brahman priests, and the era of the Sikh calendar counts from 57 B.C., as implied above. One of the two main Hindu eras (*Vikram Samvant*) reckons from the same epoch, while the other (*Saka*) counts from 78 B.C. Also because of unhappiness over the caste system, the Jain sect separated much earlier, in the sixth century B.C., and as a result the Jains count from 527 B.C. Like the Sikhs, the Jains follow a lunisolar calendar.

The mainstream Hindus, however, have a quite different calendrical system. The Hindu calendar is purely solar in that the year is made up of twelve months, five containing thirty-one days and seven with thirty apiece, plus an occasional intercalary day, so that the Moon does not affect the length of any particular year.

Nevertheless, the phases of the Moon are followed for other purposes. The lunar brightness cycle is divided into waxing and waning portions called *paksha,* within both of which each of the nominal 15 days has a name counting up to the full moon and then repeating in the second paksha up to the new moon. Those days, though, are *lunar* days of a different form than the tidal days of twenty-four hours and fifty minutes described earlier. Each of these lunar days is called a *tithi* and lasts for one-thirtieth of the 29.53-day lunar synodic period, or about 22½ minutes short of 24 hours.

One recognizes, then, a lunar dating system internal to the solar year, and religious festivals are mostly kept on regular dates according to that system, in the same way as the Gregorian calendar maintains a solar year but the church has an internal lunar calendar which controls the dates of the movable feasts. For example, like Sikhs, mainstream Hindus mark the birthday of Guru Gobind Singh.

About a week after that celebration is a quite remarkable event, called *Thaipusam.* (*Thai* is the tenth month of the Tamil calendar, while *pusam* means "star.") A decade or so back I flew into Singapore for a stopover on my way back to Australia from Sweden. Unable to sleep with jet lag, at about 3 A.M. I went out for a walk through the center of the city, and unwittingly I came across a bizarre spectacle, consisting of men and youths struggling along the street, each carrying elaborate and heavy decorated structures called *kavadi,* which are supposed to mimic the appearance of a peacock. These were supported on metal hooks and spears which were dug into their bodies. Some had nails piercing one cheek before passing *through* their tongues and then out of the other cheek. The rib cages of some were lined with sharp hooks of a size suitable to catch a shark, from which limes hung on silvery metal chains. Thus adorned, and with blood seeping from every self-inflicted wound, these penitents were staggering the several miles from one major temple to another, while their friends and families accompanied them chanting words of support, and the police stopped the traffic to let them pass (which, one has to say, is a remarkable witness to the interracial and interreligious tolerance in Singapore). This display of masochism in devotion to Lord Subramaniam or Murugan is performed by Hindu men either out of contrition for some wrong done, or to seek heavenly favor—for example, in economic activity, in forthcoming examinations, or in their wives having safe and successful pregnancies. It has been officially banned in India. As pastimes go, I'd recommend that you restrict your participation to spectating.

As in many belief systems, Hindus also refer to ages of the world like the Christian and Jewish *anni mundi.* A distinction is that Hindus recognize three earlier ages (*Yuga*) of Earth prior to the present age, this being the *Kali Yuga,* which is said to have begun in 3102 B.C.

The Zoroastrian Calendars

Zoroaster was a spiritual teacher in the area now called Iran, who founded the religion named for him around 600 B.C. The *Magi,* or Three Wise Men, in the story of the Nativity of Jesus are thought by many scholars to have been Zoroastrian priests. In the modern world there remain two main groups of adherents to the sect: the Parsis of India and the Gabars of Iran.

The Zoroastrian dating system is interesting in that even within such a restricted set of people united in religion, there are three separate calendars in use. One, favored largely by the Parsis, is the *Shahanshahi*, in which an intercalary month is added once every 120 years. This averages one extra day every 4 years (and indeed the addition of a little under a quarter of a day per year, as required astronomically, could be accomplished by making some of the intercalated months only twenty-nine days long).

Over the past two millennia the devotees have failed to maintain the system rigorously, and there is currently a proposal to put it straight. In 1995 New Year's Day (*Naw Ruz*) was August 23, but that will gradually move earlier on the Gregorian calendar, by one day every four years, until the next extra month is added, whereupon it will jump to a later date again.

In part because of the failure to keep to the once-per-120-years rule, the Zoroastrians living in the region of Surat in western India in 1746 decided to switch to the *Qadimi* ("ancient") system from Iran. The Qadimi is simply a month ahead of the Shahanshahi.

Those two conventions employ intercalary months, meaning that over a century or so the seasons slip away from the dates by up to thirty days before being reregistered. The third Zoroastrian calendar, which is clearly based on the Julian/Gregorian tradition, is the *Fasli*. Here harmony between the seasons and the religious holidays is maintained by intercalating one day every four years, with Naw Ruz being marked on March 21, the fixed equinox of the Gregorian reform.

The Voodoo Calendar

The *Voodoo* (or more correctly *Vodoun*) religion has been much misrepresented in popular culture (Hollywood movies, I mean) as being a black-magic superstition followed only by a few crazy Caribbeans. In fact, it is a widespread religion, with over 50 million adherents. Voodoo is founded on the traditional West African Yoruba religions, which were brought to the Caribbean by slaves on French ships. The cult thrives most in Haiti, although there are also strong communities in Jamaica, Trinidad, Cuba, and in various U.S. cities such as Miami and New Orleans.

In the present context it is of interest because its existence is owed largely to the former French colonial presence in the Caribbean, the Catholic plantation masters trying to suppress Voodoo among their slaves and thus driving it underground. Consequently the Voodoo calendar follows the same basic conventions as the Gregorian, except with distinct folk memories being recalled on different days during the year. For example, the few days leading up to the vernal equinox are marked by various ceremonies, including the eating of a black goat.

Another example is Halloween. In many places such as the United States, October 31 is a day to dress in weird and perhaps scary costumes, a hollowed-out pumpkin carved into a horrible face to scare away bad spirits being placed on the front steps with a lighted candle inside. In the Catholic Church, the next day is All Saints' Day, and celebrated as such. In Voodoo the events occur in the opposite order. On October 31 mass is sung by Voodoo adherents in the Roman Catholic churches, and on November 1 reincarnated corpses are represented as rising from the cemeteries and congregating at the Voodoo meeting places, the *hounforts*.

The Iranian/Persian Calendar

Which is the best calendar in use, according to astronomical veracity? The vote must surely go to the Iranian/Persian solar calendar, which has been mentioned elsewhere as following a thirty-three-year leap cycle, producing an average year length very close to the actual time between passages of the vernal equinox. It seems that Iranian scholars may have been misled by erroneous statements in the official publications of U.S. and U.K. governmental observatories, making them think that a correction of their calendar is necessary, but one would hope that they will realize that no such alteration is needed before the deed is actually done.

Not only is the Persian calendar an excellent approximation to the length of the year, but also the distribution of the months within it leads to the seasons being nicely split up. The year starts near the vernal equinox, as might be expected, and the first six months are each allotted 31 days, giving a total of 186 days for the first half of the year. This puts the end of the third month close to the summer solstice and the end of the sixth month near the autumnal equinox; that is, these six months, which occur while Earth is further than average from the Sun and thus moving more slowly, aphelion passage being on about July 4, have durations which reflect the astronomical and thus seasonal reality. The other six months are each 30 days long except for the last one, which is only 29 days in a common year, and this is as it should be because the final season (winter) should be the shortest because perihelion passage in early January is contained within it.

One would have to admit, then, that the Persian calendar does rather well in reflecting Earth's orbit about the Sun.

Changing the Tune

In this chapter we have considered just a handful of the many contrasting calendars which are either still in use today, or else have been tried and rejected over the past couple of centuries.

It may well be that until now you had never given much thought to the possibility that other cultures might march to a quite different drummer, just finding it a little peculiar that Jews send each other Hanukkah cards and light the menorah candles around some date which is in late November or early December, but which seems to skip around from year to year. On the Hebrew calendar they are keeping to the same date—the eight days of the Hanukkah festival start with the twenty-fifth day of the month of Kislev—but that calendar is predicated more on the Moon than the Sun. Similarly, others may find it bizarre that Christmas is celebrated on December 25 each year regardless of the phase of the Moon.

These have all been calendars which have been tried by human societies, and in some cases found to be wanting. But let me assure you that there are plenty of people around who would like to make major changes to the calendar which is the norm for international trade and communication. They want to change the tune which our lead drummer is following. In the next chapter we look at proposals for calendar reform.

Chapter 20

∽

CALENDAR REFORM

Many people will proudly tell you that they studied the liberal arts, and yet few know which subjects are strictly encompassed by the term. Traditionally the seven liberal arts are split into two divisions, the lower of which (consisting of logic, grammar, and rhetoric) is called the *trivium*, with an etymology similar to that of *trivial*. The more advanced division (consisting of arithmetic, geometry, astronomy, and music) is called the *quadrivium*. Clearly anyone wishing to devise an alternative calendar must have a command of the quadrivium: it all has to add up, be appropriately arranged and proportioned, follow the basic rules dictated by astronomical considerations, and yet be harmonious and conducive to the rhythms of human life. But, for successful calendar reform, the trivium must also be mastered: the logic to weigh up the pros and cons of the proposed reform, the grammar to describe it succinctly and clearly, and the rhetoric to promote it orally.

The fact that much of the world uses a calendar which has not been altered substantially over two millennia, only tweaks and minor adjustments having taken place, seems to show that no one person or group has had a sufficient command of the liberal arts to get the calendar grossly revised, despite many attempts over the past couple of centuries. In this chapter we will take a passing glance at some of the reforms which have been proposed from time to time.

The Dewey Decimal System

Many people know the second name of Melvil Dewey, though few would be able to cite his unusually spelled initial moniker. He lived from 1851 through 1931, so perhaps his parents were influenced by the name of Herman Melville, the American author whose famous *Moby Dick* was published in the year Dewey was born. Indeed, the spelling he chose to use was a diminution of the full *Melville* with which he was christened.

We know Dewey's name through the eponymous *Dewey decimal* system of book classification, so that it would be entirely appropriate if he were indeed named for an author. Dewey was an educator and administrator, and at the age of twenty-five he invented the bibliographic system which is unwittingly used by millions of library visitors every day. The Dewey decimal classification owes

its preeminence not only to its logical design and utility, but also to the pushi-
ness of its inventor; in describing the personal nature of Dewey, the *Concise
Dictionary of American Biography* states that "abounding in energy and self-
confidence, he was tactless and indiscreet."

If you want to take a trip to the library and find out about the Dewey decimal system itself, look up books classified under 025.431. Therein you will
find the following broad classes:

500   Pure Sciences

510   Mathematics

520   Astronomy

530   Physics

Astronomy, then, occupies the ground between mathematics and physics,
which seems appropriate. I have often had cause to peruse books and journals
cataloged under 520 onward, sometimes overshooting and thus accidentally
coming upon some interesting volumes. You could go to an academic library
to see what I mean, or we can just take a closer look at the Dewey decimal sub-
division of the 520s:

520   Astronomy and allied sciences

521   Theoretical astronomy and celestial mechanics

522   Practical and spherical astronomy

523   Descriptive astronomy (specific celestial objects)

525   Earth (astronomical geography: tides, seasons, solar orbit)

526   Mathematical geography (cartography, surveying, geodesy)

527   Celestial navigation and nautical astronomy

528   Ephemerides (astronomical and nautical almanacs)

529   Chronology (time and calendars)

One might note that there is no number 524. You might imagine that perhaps
Dewey was anticipating the beginning of the space age almost a century ahead
of time and so leaving a gap, but space research actually appears under 629, and
he also failed to foresee the development of computer science, a major section
in any academic library nowadays.

My intention in discussing the Dewey decimal system, though, should be
apparent from the above listing. As one steps through the classes from pure, abstract astronomy one comes to ever more practical matters. A surveyor in train-
ing would have reason to visit the 526 section frequently, while books about
the calendar and timekeeping may appear under 529 (but not all of them:
clocks may be under 529.78 but equally well may be classified under horolog-
ical science at 681.11, while time systems and standards occur at 389.17). In

walking the library stacks I have often tarried awhile at 529.05 J8; for this is the wonderful *Journal for Calendar Reform*.

Elsewhere in this book I have had cause to write that only a brief summary of a specific subject will be given, because a full treatment would need an entire volume. Calendar reform represents an extreme example. The *Journal for Calendar Reform* occupies not just a few hundred pages, but near half a yard of shelf space. Although it ceased publication back in 1956, a few minutes spent inspecting its pages will quickly convince one of the enthusiasm and vigor with which some determined (but ultimately unsuccessful) people have argued for an amendment of the calendar. With the availability of the Internet, you don't even need to visit the library to obtain this impression. A short while surfing the Net will turn up many examples of suggested revisions to the calendar of everyday use, many of them radical. Let us consider just a few from the past.

The Blank Day Concept

In chapter 19 we mooted the possibility of having a leap-week system, most years containing 364 days/52 weeks while sporadic leap years contain 371 days/53 weeks. That protects the 7-day week, which has persisted for so long and has such a powerful influence upon our lives. The years so produced would vary in length by 7 days, but so what? We have seen that there are many widely used calendars (such as the Hebrew and the Chinese) with lengths varying by a full month, and the Islamic year is always 10 or 11 days less than a solar year.

Nevertheless, the basis of the Julian and the Gregorian systems has been to maintain the calendar year within a day of the solar year, and so most advocates of calendar reform aim for a novel system which is merely a tweak of that in effect now, but with added regularity in terms of month lengths and the date/day-of-week correlation. For example, they might want to achieve a perpetual calendar in which January 1 is a Sunday, there are fifty-two weeks in the year, and yet the year is 365 days long (366 in a leap year). Sounds impossible.

There is a solution, though. Blank days. Imagine a calendar with thirteen months each of four weeks. Each month starts with a Sunday, and the extra month you can name for your pet cat or world peace. That makes 364 days. The 365th day is tacked onto the end of the year as a universal holiday, being a part of no month and (importantly) no week. It is a true epagomenal day in the spirit of the ancient Egyptian calendar. In a leap year the 366th day may be added either coupled with the 365th at year-end, or perhaps in midyear.

These blank days would indeed regularize the calendar. You would know that the second of each month is always a Monday (making the identification of lemons among automobiles easier, the dates on the production record quickly showing the Monday cars). On the other hand, how would a businessman be able to compare quarterly sales figures? The thirteen months divide evenly into neither halves nor quarters. An alternative solution would be to have twelve months split evenly into four quarters, each containing three months made up of thirty-one, thirty, and thirty days. That would make sales comparisons much easier, every quarter containing ninety-one days rather than the present split of ninety, ninety-one, ninety-two, and ninety-two days.

Then again, there would be more Sundays in some months than in others, and in many places the shops are shut then (or on whichever is the Sabbath for the dominant local religion). Regularity would be achieved according to the months and the quarters, but not the weeks. Is that surmountable in this context? Again, the answer is *yes*. Employ months which consist of either four weeks precisely, or five. Each quarter could consist of three months lasting for twenty-eight, twenty-eight, and thirty-five days. The first day of each month could be a Sunday, as would the eighth, fifteenth, and twenty-second days, and the twenty-ninth in the long months.

All of this depends upon the acceptability of having either one or two blank days in each year—epagomenal days which are not counted as part of any week. There would be great reluctance on the part of the populace to abandon the week in that way, and one can imagine that the church would be even more dead set against it given that the Easter *computus* is based in part on the sanctity of the week cycle. Nevertheless, enthusiasts have campaigned vigorously for a calendar reform based on the utilization of blank days, as we will see below. Let us identify the reprobates who have suggested such a heinous notion.

The Thirteen-Month Calendar

The beginning, history-wise, may be a very good time to start, but it can complicate a story. So I am going to leave the first real post-Gregorian suggestion for a radical calendar reform until the end of this chapter, and instead start in the nineteenth century.

Again the Eternal City enters into our account. Marco Mastrofini (1763–1845) was an ecclesiastical philosopher who lived in Rome, serving also as a professor of mathematics at the College of Frascati. He was therefore well equipped to balance both the scientific and the religious aspects of the calendar question, and in 1834 he published a report recommending a perpetual revised Gregorian calendar in which there were blank days inserted at year-end.

This basic idea was seized upon fifteen years later by the famous French philosopher Auguste Comte (1798–1857), the founder of the secular religion of positivism. For this reason it has become known as the *Positivist calendar*. Comte had thirteen months each of four weeks, surmounted by the one (or two) blank days. The main reason that his suggestion failed to find favor with many people seems to have been that he insisted on naming the months for various notable persons from historical to modern times, such as Moses, Aristotle, Archimedes, Gutenberg, and Descartes. One must admit that it would seem strange to give the date as the third day of Homer, and with a month named for the bard a reference to "Shakespeare's Twelfth Night" would be ambiguous.

The plan of Mastrofini and Comte was revived by the Englishman Moses B. Cotsworth, who was born two years after the latter's death. Cotsworth was employed in calculating various rates and charges for the railway companies in Britain, and his masters were unhappy with the fact that their monthly income and expenditure figures fluctuated due to the inequalities of the Gregorian calendar. This set Cotsworth to thinking about how the calendar might be better fitted to the requirements of business, and he published a summary of his conclusions in 1899. He planned to reduce the currently named months to four weeks each, and call the month added (to the middle of the year) *Sol*, reflecting the fact that sunlight persists longer at that time of the year in the Northern Hemisphere—even in Britain, with its wet summers. Four years later he followed his initial suggestion with a 573-page tome entitled *The Rational Almanac*.

At this juncture Sir Sandford Fleming reenters our story. Fleming appeared in chapter 18, as the Canadian whose lobbying brought about the International Meridian Conference in 1884. Fleming was much impressed by Cotsworth's ideas, and invited him to visit Canada to make presentations to various scientific bodies and the dominion government. Their endorsement led to the British government being petitioned to approach all other nations with a view to holding some form of international calendar conference.

In his home country, though, Cotsworth met with little success. It was on the western rim of the Atlantic that his arguments were best received. The League of Nations (the forerunner of the United Nations) formed a committee of inquiry on the calendar question in 1922, and Cotsworth's plan was adopted as the best of the 130 put forward, whereupon he spent most of the rest of his life lobbying among government and industry for its official adoption. At this stage an important convert was gained: George Eastman, the founder of the Eastman Kodak camera company and a generous philanthropist.

Cotsworth and his supporters formed the *International Fixed Calendar League,* and Eastman played a prominent part in its activities; so much so that the *Cotsworth calendar* became known, in the United States, as the *Eastman plan,* although Eastman himself gave the credit to the Englishman. Eastman argued strongly and persuasively among his many contacts in government and business, writing numerous magazine articles during the 1920s espousing the cause of calendar reform.

For a whole host of reasons the Cotsworth calendar did not find favor. For example, Americans objected to the fact that the Fourth of July would suddenly be renamed Sol 17, if it were to keep to its familiar seasonal location a couple of weeks after the summer solstice, or else get shifted to late May, seasonally speaking, if the label July 4 were to be kept instead. The initial wave of enthusiasm soon waned, and both Cotsworth and Eastman were not young men, the latter dying in 1932, with which the organization lost its prime benefactor. By 1937 the offices of the International Fixed Calendar League in London and Rochester, New York (the home of the Eastman Kodak Company), had shut up shop.

The World Calendar Association

There was another factor which stymied the advance of the thirteen-month calendar discussed above. This was the formation of the *World Calendar Association* in 1930.

This calendar is based on one of the notions described above when we discussed the possibility of blank days. Four identical quarters would be employed, each containing three months made up of thirty-one, thirty, and thirty days. The epagomenal day appears at the end of the year, with an extra such day between June and July in leap years. This perennial calendar is shown in figure 4, on page 311, with a flyer summing up the arguments of its proponents comprising figure 5, on page 312.

Going through the months one sees that, apart from the added blank days, the changes from the present system dating back to Julius Caesar are simply that March, May, August, and December lose a day each, while April gains one and February two. In terms of secular matters this does not seem a great drama; the problem, of course, is with ecclesiastical considerations like Easter, and the breaking of the seven-day cycle.

As with the Cotsworth-Eastman campaign, the prominence gained for the World Calendar derived largely through the efforts of a few people, and in this case the major figure was Elisabeth Achelis (1880–1973). Achelis was a lady of independent means, her father having made a fortune through rubber manufacturing. In 1929 she was inspired by a lecture by the one-and-only Melvil Dewey, who spoke of the need to simplify modern life decades before "small is beautiful" became a catchphrase.

Serving as president of the association from its inception, Achelis funded most of its activities herself, setting up an office in New York City from which to campaign. The *Journal for Calendar Reform* soon started to appear, carrying articles lauding the great benefits which the World Calendar would bring if only the nations of the world could get together and see the sense of what was proposed.

In 1931 the League of Nations was holding meetings in Geneva to discuss the question of calendar reform, and Achelis and her cohort presented their case so well that the previously favored thirteen-month calendar was dropped and the World Calendar adopted as the leading candidate. Eastman and Cotsworth would have died disappointed men.

By 1937, when the League of Nations next considered this question, the World Calendar Association had affiliates in thirty-two countries. The United States was resistant to the idea of calendar reform, and so the association persuaded Chile to present its case formally to the membership of the Council of the League. In the event, after some time for discussion at home, fourteen separate nations approved the concept of adopting the World Calendar as the international standard.

In 1939, though, the beginning of World War II put a stop to all advances on the calendar reform front. Although it stuttered on, the momentum which had built up during the 1930s was lost.

The formation of the United Nations after the war, with its headquarters in New York, presented the opportunity to keep the ideals of the World Calendar Association before the ambassadors and diplomats, who might then persuade their home governments of the sense of what was being discussed, but Achelis and her colleagues were blocked by the intransigence of the United States in this regard.

The World Calendar
(Symmetrical and Invariable)

First Quarter

January

S	M	T	W	T	F	S
1	2	3	4	5	6	7
8	9	10	11	12	13	14
15	16	17	18	19	20	21
22	23	24	25	26	27	28
29	30	31				

February

S	M	T	W	T	F	S
			1	2	3	4
5	6	7	8	9	10	11
12	13	14	15	16	17	18
19	20	21	22	23	24	25
26	27	28	29	30		

March

S	M	T	W	T	F	S
					1	2
3	4	5	6	7	8	9
10	11	12	13	14	15	16
17	18	19	20	21	22	23
24	25	26	27	28	29	30

Second Quarter

April

S	M	T	W	T	F	S
1	2	3	4	5	6	7
8	9	10	11	12	13	14
15	16	17	18	19	20	21
22	23	24	25	26	27	28
29	30	31				

May

S	M	T	W	T	F	S
			1	2	3	4
5	6	7	8	9	10	11
12	13	14	15	16	17	18
19	20	21	22	23	24	25
26	27	28	29	30		

June

S	M	T	W	T	F	S
					1	2
3	4	5	6	7	8	9
10	11	12	13	14	15	16
17	18	19	20	21	22	23
24	25	26	27	28	29	30

Leap Day*

Third Quarter

July

S	M	T	W	T	F	S
1	2	3	4	5	6	7
8	9	10	11	12	13	14
15	16	17	18	19	20	21
22	23	24	25	26	27	28
29	30	31				

August

S	M	T	W	T	F	S
			1	2	3	4
5	6	7	8	9	10	11
12	13	14	15	16	17	18
19	20	21	22	23	24	25
26	27	28	29	30		

September

S	M	T	W	T	F	S
					1	2
3	4	5	6	7	8	9
10	11	12	13	14	15	16
17	18	19	20	21	22	23
24	25	26	27	28	29	30

Fourth Quarter

October

S	M	T	W	T	F	S
1	2	3	4	5	6	7
8	9	10	11	12	13	14
15	16	17	18	19	20	21
22	23	24	25	26	27	28
29	30	31				

November

S	M	T	W	T	F	S
			1	2	3	4
5	6	7	8	9	10	11
12	13	14	15	16	17	18
19	20	21	22	23	24	25
26	27	28	29	30		

December

S	M	T	W	T	F	S
					1	2
3	4	5	6	7	8	9
10	11	12	13	14	15	16
17	18	19	20	21	22	23
24	25	26	27	28	29	30

Worldsday**

* The Leap-Year World Holiday (366th day), falls outside any week or month.
** The Year-End World Holiday (365th day), falls outside any week or month.

Fig. 4. The perennial calendar advocated by the World Calendar Association. Each quarter would be identical, and the same every year. The months would last for thirty-one, thirty, and thirty days, beginning with a Sunday, then a Wednesday, then a Friday. At the end of each year would be an epagomenal day which is not part of either a week or a month. In leap years another such blank day would appear in the middle of the year.

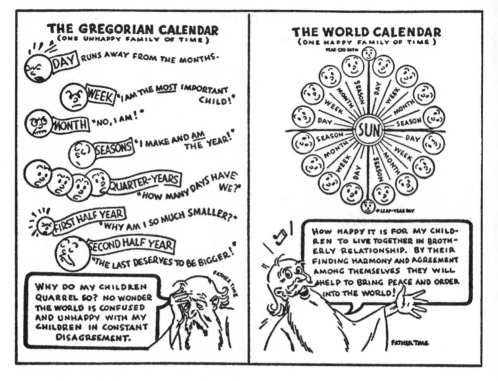

Fig. 5. An example of the propaganda of the World Calendar Association, from *The Calendar for Everybody,* by Elisabeth Achelis (1943).

The UN secretary-general, Dag Hammarskjöld, wrote to all member governments in October 1954 asking for their views on calendar reform. The following March the United States, through its representative to the United Nations, Henry Cabot Lodge Jr., made its views known. It was felt that no such calendar reform could be countenanced unless it was favored by a large majority of the inhabitants of the United States, and that the opposite was the case with much opposition "based on religious grounds, since the introduction of a 'blank day' at the end of each year would disrupt the seven-day sabbatical cycle." That being the case, the opinion of the U.S. government was that "any additional study of the subject at this time would serve no useful purpose."

With that, Elisabeth Achelis resigned the presidency of the World Calendar Association. At the age of seventy-five, she felt that she had done all that she could, although she continued to support the cause, and the association still exists, a few diehards persistently arguing the case. After a quarter of a century of publication, the *Journal for Calendar Reform* ceased publication. But you may find it in a well-stocked library. Melvil Dewey not only brought about the classification label of 529.05 J8 under which the journal is shelved, but he also inspired the movement which led to its publication in the first place.

The Question of Easter

Quite apart from adjusting the lengths of the months and perhaps using blank days, the civil year formatted as it is suffers some disruption each year due to the thirty-five-day range of dates on which Easter may occur. What I mean here is that the interests of business and industry are certainly not served, even within a monoreligious modern state, by having Easter move from one year to the next. For international commerce the situation is much more severe, with different Christian sects celebrating Easter on separate dates, and those of other religions being perturbed by the thirty-five-day wander of Easter.

One solution would be to fix the date of Easter, perhaps to the second Sunday in April, and forget the phase of the Moon. For example, in his thirteen-month calendar Moses Cotsworth planned to fix Easter on the date equivalent to our April 9.

It is surprising how close this reform—a fixed Easter—has come to adoption in various spheres. In 1928 the British Parliament passed the *United Kingdom Easter Act,* which would have put Easter on a date to be agreed upon internationally and by the different Christian churches. The proposal was accepted by the archbishop of Canterbury for the Church of England, and later by Pope Pius XII; the Second Vatican Council (1963) approved the possibility of such a move. Other churches, however, refused to agree to the concept either of a fixed Easter or to one calculated according to other than the Julian calendar.

The dissidents in this connection were largely the various Eastern Orthodox churches, but it is important to realize that even within that grouping there is considerable schismatic debate on the calendar question. Elsewhere I have mentioned that, although the Orthodox churches had persisted with the Julian calendar post-1582, rejecting the Gregorian reform, by the 1920s some parties within the Eastern sects advocated a change so as to make their calendar more astronomically precise, but with an imperative of avoiding the system of the Roman Church. This led to the misguided suggestion of the *Revised Julian calendar* in 1923—I write *misguided* because although it is claimed (still, by some) to have a more accurate leap-year rule, in fact it is inferior when judged against the proper yardstick—that calendar having been temporarily adopted by a few of the dozen or so Eastern churches, but rejected by others. Within the Orthodox community there is an ongoing dispute along these lines, the two warring parties being termed the *New Calendrists* and the *Old Calendrists*. Be clear that neither of those support the Gregorian calendar. Confused? One should be.

Present Deliberations on Changing the Easter *Computus*

Back in chapter 13 it was mentioned in passing that there are current discussions within the World Council of Churches (WCC) aimed at reaching an agreement to alter the Easter *computus*. In March 1997 the WCC and the Middle East Council of Churches had a joint consultation in Aleppo, Syria. The fact that two distinct groups are mentioned there demonstrates that unity is far from being achieved within the Christian faith, but it is more complicated than that. The WCC consists of 330 separate churches from more than a hundred countries, but the Roman Catholic Church is not a member, although it is said to work cooperatively with the WCC.

This meeting in 1997 was therefore not attended either explicitly or implicitly by all churches, meaning that its resolutions cannot be judged to be pan-Christian. Nevertheless, the outcome with regard to the Easter *computus* is of interest because it demonstrates the line of thinking within a major part of the Christian faith. So let's see what they said.

A press release issued after the consultation reads, in part:

> *Churches in the East and West calculate the date of Easter by the same principle. This is the one given by the Council of Nicea in 325 which determined that Easter should be celebrated on the Sunday following the first full moon after the March equinox.*
>
> *Differences in dating occur because churches follow different calculations of the equinox and full moon.*
>
> *The recent Aleppo gathering has proposed that, from the year 2001 the date of Easter should be calculated using precise astronomical determinations.*

The first paragraph contains a clear error of fact, a matter which has been emphasized elsewhere. Again and again people ascribe the principles for the Easter *computus* to the Nicene fathers, and that is simply incorrect. They did not take that responsibility upon themselves, but effectively delegated it to someone else. It might, of course, be an uncomfortable truth for the modern Christian hierarchy to admit that it was not the Council of Nicaea, to which all churches would cede some authority in the early years of Christianity, but rather the Alexandrian Church alone which decided upon the Easter rules, with some later help from Dionysius Exiguus.

The second paragraph is also misleading on several grounds. There is a difference in dating because different dating schemes are used, period. And in the Gregorian scheme there is no *calculation* of the equinox; it is *defined* to be March 21, period.

The third paragraph quoted makes sense, however. When the calendars we use and the Easter *computuses* themselves (plural because paragraph two is based on fact) were adopted, it was appropriate to work with single-day precision, not the split-second accuracy we can bring to bear in the modern era.

All manner of peculiarities could be hypothesized under the present system, and they do happen in reality, leading to confusion. If the real astronomical equinox were on March 19, and the astronomical full moon on the twenty-first, the latter might occur two full days after the former and yet not be admissible under the ecclesiastical rules at present, because the ecclesiastical equinox is March 21 and the ecclesiastical full moon might be calculated as occurring that day or earlier, postponing Easter by a month over the date which would result if astronomical reality were followed instead.

Starting with the advances in analytical celestial mechanics and the theory of the motion of the Moon which came about in the late nineteenth century, it has been possible in principle to calculate the times of full moon with some precision, but it has only been in the computer age that such an innovation has really been practical vis-à-vis the Easter *computus*. We are now able to define the instants of the equinox and full moon to a split second some way ahead of time.

Will the Easter *Computus* Be Changed?

The proposal to apply scientific rigor to the calculation of the dates for Easter which were announced in 1997 have not been taken further at the time of my writing a couple of years later. Will this proposal be adopted soon? There is a time constraint in that the reform of Easter would best begin with a year in which it has no effect.

What do I mean by that? The point is that we are here talking about determining Easter through three different methods. One uses the Gregorian calendar, as is the case for the Western churches. Another employs the Julian calendar, the situation for most Eastern churches. The third, the new method, would be the astronomical reckoning whereby one directly compares the actual instants of equinox and full moon in order to determine the date for Easter. It happens that all three coincide for A.D. 2001, so that the reform would have no outward effect in that year.

All is not lost, however, if that opportunity is missed. Coincidence recurs in 2004 and 2007. Given that one might imagine this potential reform will necessitate at least a decade of negotiation, a good target date might be 2010, because that year *and* 2011 have Easter on coincidental dates for all three computation methods.

If (a big *if*) this reform of the Easter *computus* could be put into effect throughout Christendom, then at last the resolution of the Council of Arles in A.D. 314 (*not* the Council of Nicaea in 325, as is often misstated) will be achieved, that all Christians should mark Easter on the same day. There are two additional consequences to note, though. One is that Easter Sunday will never again suffer the apparent anomaly of being a day of full moon, which has occurred in the past due to the Easter rules applying to the ecclesiastical rather than the astronomical equinox and moon. The other is that Easter would henceforth perennially avoid the Jewish Passover, which was the only real edict connected with this matter to arise from the Council of Nicaea, but which the Gentile churches have failed to ensure over the seventeen centuries since then.

The proposed reform of the Easter *computus* would thus lead to the intentions of the councils of Arles and Nicaea being fulfilled at last, but there is still a Jewish connection. The astronomical method for Easter makes use of two instants of time—the passage of Earth through the equator and hence the time of the equinox, and the arrival of the Moon at superior conjunction and hence the time of the full moon—which are independent of one's position on the surface of the planet. Those instants are the same for Christians everywhere.

A problem arises, though, because Easter must be a Sunday and postdate the equinox and full moon, and that requires a local definition of time and hence a longitude. The WCC proposal calls for the meridian of Jerusalem (35.2

degrees east of Greenwich) to provide that standard, that being the place of Christ's death on the Cross and the Resurrection. If one is really looking for astronomical veracity, then it would be necessary to define Sunday in Jerusalem according to the local (not mean) solar time. That would be out of step with the local clocks, and would vary from year to year with the date because of the equation of time.

An alternative would be to use the Israeli local time and date, and count the days from midnight, but then another problem rears its head. One could use as the underlying basis for the date the time zone two hours ahead of coordinated Universal time, but then what happens when a switch to daylight saving time (DST) occurs?

Most of the predominantly Christian states in the northern temperate zone adjust their clocks on a date within the thirty-five-day range of possible dates for Easter, the norm being in the early morning hours of a Sunday. Obviously there is the potential for conflict. Despite being close to the tropics, Israel does use DST, but without any specific rule governing the dates when it is to be invoked each year. The reason for this is that while the secular politicians might like to have DST introduced on regularized dates, the religious (Jewish, of course) parties object to DST because it interferes with the morning prayer ritual, which must be conducted after sunrise but before work.

The upshot is that the exact period in which daylight saving time is in effect in Israel varies according to the state of the political parties in the Knesset (the Israeli parliament) and also the phase of the Moon, and is usually only pronounced a few weeks in advance. With this situation in mind you might consider that DST in Israel (and thus when Sunday starts, and thus potentially when Easter is to fall) is partly under the authority of the chief rabbi of Jerusalem, which the Nicene fathers clearly would not appreciate. I will leave you to ponder the various permutations which could arise, and their effect upon Easter.

As an atheistic astronomer you might imagine that I would support the notion of calculating Easter by a method based upon the real positions of the Sun and the Moon. But I am not so sure. The preceding paragraphs indicate some of the factors which could confuse what on the surface might appear to be a straightforward matter. I tend to support the great astronomer Johannes Kepler who, four hundred years ago, opined that "Easter is a feast and not a planet. You do not determine it to hours, minutes and seconds."

Will the Calendar Ever Be Reformed?

Obviously the question posed in the heading above is a silly one: "ever" is a long time, and in chapter 22 we will see that if humanity manages to survive for long enough, the natural lengths of the days and the months will alter to such an extent that our present calendar simply will not serve. But that is looking some millennia into the future, at least.

There have been many, many proposals for calendar reform, each with its own devotees (in some cases, though, just the lonely originator riding his or her favorite hobbyhorse). Some have quite a lot going for them, logic-wise, but that is simply not enough. Others are just too complicated; their proponents forgot the KISS rule—*Keep It Short and Simple* (or *Keep It Simple, Stupid*, if you want to be less charitable).

So far as I can see, the advent of the computer age has obviated any slim chance that existed for immediate calendar reform. Now that software is written to handle all our dating needs, and it would cost such a huge amount to make any change, it seems most unlikely that the calendar will be altered in the foreseeable future. Any reform of the calendar would need to be scheduled for the reasonably distant future; say to begin operation in A.D. 2100, allowing a century for the gradual change of software through redundancy.

In the meantime, I anticipate that we are going to continue with our year numbering according to the Christian era, the planetary weekday names taken largely from Norse mythology, the Roman names for unequal-length months, and our Arabic day numbers.

Benjamin Franklin Rides Again

As foreshadowed above, I have chosen to leave a rather juicy morsel of calendar reform history until last, because I think that it's a nice little story to be contemplated in isolation.

At the conclusion of chapter 17 it was mentioned that Benjamin Franklin may have been involved in the matter of a suggested calendar reform, and here I am going to argue that this is the case. "Wouldn't it be obvious?" you might ask, and the answer is *no*. In that chapter I said that Franklin used many noms de plume, and specified the infamous but imaginary Polly Baker of Connecticut as an example; Franklin admitted that authorship in 1777, thirty years after her plaintive letter was published in the *Gentleman's Magazine,* decrying her unfair treatment for bearing bastard children.

One should be clear here that there is a distinction between the various nicknames or epithets that were applied to Franklin, and his self-assumed pen names. Examples of the former are the American Socrates (or Newton), the Tamer of Lightning, the Sage of America, and less generously the Jolly Imbiber. That's a fairly short list. Here are some of the self-imposed aliases under which he published:

Polly Baker	Richard Saunders (Poor Richard)
Father Abraham	Bridget Saunders
Old Master Janus	Mrs. Silence Dogood
Timothy Wagstaff	Anthony Afterwit
The Busybody	Proteus Echo
Historicus	A. B.
The Left Hand	Medius

Cilia Single   A Tradesman of Philadelphia

Alice Addertongue   Philomath

Do not imagine that Franklin could claim the record for the number of pseudonyms used by one person. During the eighteenth century he was totally outstripped by two others, both of whom are best known by their most common pen names. Daniel Defoe (just Daniel Foe, 1660–1731, in real life) used at least 198, including Count Kidney Face, Obadiah Blue Hat, and the Man in the Moon. Not far behind him was Voltaire (really François-Marie Arouet, 1694–1778), who employed 173 aliases, maybe more.

In the present context the last-listed of Franklin's assumed appellations is of interest. A *philomath* is an almanac maker, which Franklin clearly was in the guise of Richard Saunders, but the term can have a wider meaning as shown by its etymology, implying a lover of mathematics and hence the computation of calendars. I argue here that Franklin wrote advocating calendar reform under a pseudonym which has hitherto escaped detection.

The Georgian Calendar

The *Gentleman's Magazine* (published in London) printed the purported letter from Polly Baker in April 1747. Almost two years before, in the issue for July 1745, the editor published a letter dated February 2 and headed "The Georgian Calendar," from a correspondent in Maryland who signed himself as Hirossa Ap-Iccim. If ever a pseudonym were used, this seems to be the case.

The title of the letter stems from the monarch then reigning over Britain and the colonies, George II, who Ap-Iccim proposed to honor with the naming of a new calendar to be used throughout the British dominions. Such sycophancy was not unusual. When William Herschel discovered the planet Uranus in 1781, he and the rest of the British astronomical contingent decided they would call it the *Georgium Sidus* ("George's Star") in honor of King George III, and even a half century later most references to the planet in British journals persisted with that name. Continental opposition, largely from the French, eventually won out in favor of Uranus, but in the meantime Herschel received a knighthood, a generous annual monetary allowance, and the title Royal Astronomer (there already being an Astronomer Royal). Astronomers know that it is not only stars which glitter.

Hirossa Ap-Iccim enjoyed no such reward, which is hardly surprising since he was anonymous. Let me summarize his calendar reform suggestions. Recall that this was in the era when Britain was still working on the Julian calendar with New Year on March 25, although change was afoot.

In the Georgian calendar there were to be thirteen months each of twenty-eight days and four weeks, all starting on a Sunday. The months were either to be numbered, as is the Quaker practice, or else the new month was to be called *Georgy* (sycophancy again). The blank day in common years was to come as the last of the year, and be celebrated as Christmas Day. In leap years the additional blank day was to come after Christmas and be called the *British Lustrum* (the latter word has a calendrical meaning as a periodical time of cleansing, once every 5 years in the epoch of the Roman Empire). Leap days were to be inserted every 4 years, except that every 132 years the extra day was to be skipped; this was in accord with the requirement based on the best estimate for the tropical year length then available. All feasts such as Easter were to be arranged on regular dates within the year. New Year's Day was to be moved to the winter solstice (which was then on December 11, Julian) in the belief that Julius Caesar had intended, but failed, to start his calendar with the Sun reaching its most southerly point. This would mean that Christmas Day (as above) would be marked on or adjacent to the day of the solstice, in accord with the origin of that Christian feast as an appropriation of the festival from the pagan sun-worshiping religions.

The proposed Georgian calendar, then, had nothing to do with the jump from the Julian to the Gregorian system which Britain was to undertake a few years later, but was rather a quite radical deviation from previous calendars, using a framework similar to the thirteen-month system of Cotsworth and Eastman. Most significantly, the revolutionary concept of a blank day, no part of any week or month, was put forward for the first time since the artificial seven-day week had taken a hold so many centuries before.

The Case against Benjamin Franklin

Why do I think that this letter was actually written by Benjamin Franklin? There are several grounds on which I base the suggestion. Of course, one is that this just *looks* like the sort of mischief that Franklin would get up to: it couples a detailed knowledge of the astronomy, mathematics, history, and religion of the calendar with a piece of lateral thinking a century ahead of its time, but a perusal of the phrasing of the letter also gives a hint of a tongue-in-cheek aspect. That name, for example, looks bizarre, hardly the sort of handle that one would expect a loyal subject of His Brittanic Majesty to rejoice under.

I suggested to various of my calendrical correspondents that Hirossa Ap-Iccim might actually have been Franklin. Two answered that the name looks like an anagram, which is perhaps a fruitful idea. One can get anagram-generating programs which allow you to enter a name and all sorts of variants will then be spat out, using the same set of letters. Here are a few of the results for Hirossa Ap-Iccim:

a cosmic air ship	(But even the hot-air balloon of the Montgolfier brothers was still nearly forty years in the future)
impair chaos sic	(Which could mean that the previously disorderly calendar was being rectified)
pharisaic osmic	(The adjective might refer to the early Jewish sect who lived around the start of the Christian era, or perhaps an insincere or sanctimonious outlook on the part of the writer. The second word would really need

an *s* on the end to produce *osmics,* the
science of smells, again giving a hint of a
possible devilish streak on the part of the
author. Or were those two words to be
conjoined, producing *cosmic* again?)

Those are just some of the possibilities. Having chewed them over, I don't
find any of them convincing, and I would guess that you would agree.

The reality, I believe, is that Hirossa Ap-Iccim would have given his name
just once, at the end of the letter (which is where it appeared when printed),
and it would have been handwritten. Especially because the name is so strange,
the possibility that the copy editor or compositor misread the handwriting rears
its head.

Rearranging the letters as published yet again, one can derive *Isaac Isomr-
phic;* and with one extra *o* you get *Isomorphic.* That word means "of the same
form," hinting at it being a pseudonym. The resultant alliteration also looks
like Franklin's frequent practice (Alice Addertongue, Anthony Afterwit). Similarly, among the list of aliases I gave earlier, one could imagine that with Cilia
Single (again with alliteration in the pronunciation if not the spelling) Franklin
had actually meant to appear as either Celia or Cecilia, both common girl's
names, whereas *cilia* is a biological term for something hairlike, such as an eye-
lash. The compositor's eyes or fingers may have erred again.

Thirty years after assuming the Polly Baker alias, and causing a storm of in-
dignation, Benjamin Franklin at last admitted responsibility. Why, then, did he
not own up to being Hirossa Ap-Iccim, if it were indeed he who wrote that letter? My answer lies in the sycophancy emphasized above. In the 1740s Franklin
was strongly pro-British, so that the letter would not have been considered an
unusual offering from a loyal subject of the king living in the American
colonies. By the 1760s Franklin's attitude had changed markedly, and of course
he is now remembered as a strong advocate of the American independence
movement in that era. Polly Baker, an instrument for bringing attention to
social misjustice, he could claim in later life as a credit to himself; but to have
admitted to having invented the Georgian calendar would have made him a
target for derision and ridicule in the newly formed United States.

Hirossa Ap-Iccim, unsuccessful pioneer of calendar reform, I therefore hy-
pothesize to have been Benjamin Franklin. All scientific hypotheses should be
testable, and I'll leave it to some Franklin scholar to try to tie this one down,
or disprove it. For example, was Benjamin Franklin perhaps in Maryland on
February 2, 1745?

I have not, however, yet finished with the calendar reforms promulgated in
the letter of Hirossa Ap-Iccim (or Benjamin Franklin, if it were he). Part of the
text read as follows:

> *Even the present commencement of our Æra is known to be erro-
> neous, Christ being born 4 years before the date we refer to; for his
> nativity was in the 4710th year of the Julian period; whereas
> Dionysius Exiguus (who first brought this Æra into use so late as
> the 6th century) thro' mistake placed it in the 4714th. . . .*

Suppose the 4 defective years were added to our date, so that next year (for instance) was called 1750: since in reality 'tis not the 1746th, but the 1750th year from the birth of Christ.

The intention, then, was to skip four years in the *anno Domini* numbering scheme in order to compensate for a perceived error made by Dionysius Exiguus, the year 1746 instead being labeled 1750 in Britain and her dominions (as the *anno Georgii* or *anno Regis*, one would surmise). That correction was based upon the realization that the Nativity was indeed some years earlier than the zero point at which we choose to begin our era numbering. In the next chapter we will consider whether it is possible to ascertain when Jesus Christ was born.

Chapter 21

❧

THE COMET OF BETHLEHEM

The Hirossa Ap-Iccim letter nominating the Georgian calendar with its revised chronology shows clearly that it was common knowledge, at least by the 1740s, that the *anno Domini* year-counting system is wrong, in that Jesus' birth actually occurred some years before the traditional date of December 25, 1 B.C. One may wonder, then, when Jesus was born in reality.

That letter seems quite definite in showing the firm belief of the writer that the Nativity was actually four years earlier than the traditional era, although the matter of the date within the year is not addressed. That is, the possibility that Jesus was born on a day other than December 25 is not discussed. This does not imply that the writer was uncritically accepting the date; quite to the contrary, it is clear that he knew enough of the origin of Christmas to recognize that it is celebrated on a date assumed from the pagan winter solstice festival, in total disregard for any real coincidence with the birth of Jesus. That writer was concerned only with the year numbering being out by four, and did not address the question of when the Nativity took place *within* the annual round.

Given that the count is wrong by four years (or maybe more, as we will discuss below), it would seem to be an impossible task to try to ascertain an actual birth *date,* and hence a historically accurate anniversary on which to celebrate Christmas. That may well have been part of the thinking of Hirossa Ap-Iccim, whoever he may have been. In this chapter we will discuss some considerations which *might* lead to the date of the Nativity being identified to within a month or so, based largely on interpretations of the star of Bethlehem story. These matters deal with astronomy, of course, and as we are considering the astronomical constraints upon the dating of the start of the life of Jesus, it is natural then to examine whether that science can also give some indication of the timing of the end of his mortal life, with the Crucifixion.

A Four-Year Error?

The fact that the chronology of Dionysius Exiguus is in error was discussed in chapter 8, and possible origins for the discrepancy summarized. Let us quickly recap those matters.

The source of Dionysius's mistake appears to have been through his interpretation of when the reign of Augustus began, taking this to begin from when the emperor assumed that name in 27 B.C. rather than when, under the appellation Octavian, he became the supreme Roman ruler in 31 B.C. following the battle of Actium. This may be the *source* of a four-year discrepancy; but how do we know that it exists in the context of the year of the Nativity? That is, how do we know the limits upon the possible dates for the birth of Jesus?

The Bible says that Jesus was born during the reign of Herod the Great, and history books tell us that Herod died in the spring of 4 B.C., apparently pushing the Nativity into 5 B.C. or earlier, a shift of at least four years backward from the zero point we use in the *anno Domini* convention. On the other hand, this widely accepted date for Herod's death is based on rather tenuous evidence, particularly the later account of the Jewish writer Josephus, who was not born until about A.D. 37. Josephus is well known to have tailored his histories so as to fit the image desired by his masters, the Romans, and he is the only early writer to assign a year to the death of Herod. Some modern-day historians have argued for Herod dying as late as 1 B.C. or even A.D. 1, although this is a very contentious issue. The fact remains that we have no verified death certificate for Herod, the sole written source being Josephus, who is well known to have been inaccurate in many respects, and in any case he was working many decades later.

Nevertheless, the evidence seems strong that Jesus was born a handful of years before 1 B.C./A.D. 1. The dissidents *may* be correct, but most biblical scholars would put the year of the Nativity between 8 and 4 B.C.

The Star of Bethlehem

Recognition of the fact that the era originated by Dionysius was aberrant by at least four years was a long time coming. Although the matter had been addressed occasionally by scholars during the Middle Ages (Bede had questioned Dionysius's dating, as had a ninth-century German monk known as Regino of Prum,) the first clear statement along these lines did not come until about 1605 when Laurentius Suslyga, a Polish historian, published a tract in which he argued that Dionysius was in error by at least four years.

We have previously seen that this period was one in which there were very active disputes between the Catholic and Protestant spheres of the Western Church, and that astronomers were playing a central role in one of the major fighting grounds: the calendar. Thus it is not unsurprising that Suslyga's book did not pass unnoticed (to coin a sentence with a quadruple negative). Not so far away, in Prague, Johannes Kepler was in the employ of Emperor Rudolph II, working on the celestial observations of the recently deceased Tycho Brahe from which he eventually produced the Rudolphine tables and deduced his eponymous laws of planetary motion.

Fate, or simple coincidence, often plays a hand in prompting human endeavors, and this is a case in point. Late in 1603 Kepler witnessed a conjunction of Jupiter and Saturn in Pisces, and he knew that Jewish tradition paid special attention to such planetary comings-together in that constellation. Next, in 1604, there was a massing of Jupiter, Saturn, and Mars in the constellation Ophiuchus (which, by the way, is the thirteenth constellation on the ecliptic and therefore the thirteenth sign of the zodiac; if some horoscope bore ever pesters you for your star sign, tell him or her that you are an Ophiuchid). The planets were spread by some degrees, so perhaps it would not quite be correct to describe this as a mutual conjunction, but nevertheless it was a remarkable sight to Kepler. He was subsequently astonished when a galactic supernova occurred nearby in the same year.

A supernova produces a burst of light, persisting for some months at least, caused by the explosion of a star when the fuel of lighter elements in its core, which undergo nuclear fusion to power the star, is exhausted. As the star collapses on itself, the crush in the middle can synthesize the heavier atoms (those above iron in the periodic table of the elements) during that brief implosion stage only, the rebound from the center then throwing these atoms out into space, providing the building blocks from which planets and people may be formed. In its death glory the supernova makes possible life itself.

Supernovas are vivid events, perhaps being bright enough to be seen during the daytime. But they do not occur very often within any random galaxy, and of course the really bright ones are those within our own galaxy, making them close to home. A couple of eleventh-century supernovas were mentioned in chapter 12. The expected rate is one every few centuries, and since the seventeenth-century, astronomers have waited in vain for a chance to turn their telescopes upon a Milky Way supernova, although there was some excitement in 1987 when one occurred in the Large Magellanic Cloud, a satellite to our own galaxy, about 170,000 light-years away (and hence in our backyard, cosmically speaking).

Of course Kepler, an avid sky watcher, was much impressed by the supernova of 1604, no doubt having been informed about the one which had been seen by Tycho a few decades earlier. The proximity to the planetary massing had a great impact upon him, making Kepler wonder what the events might portend. Do not think that scientists then (or even now) were not superstitious, or perhaps believers in astrological prophecies. A century later Isaac Newton believed in alchemy, that he would be able to transmute base metals into gold if only the secret could be teased out.

Planetary Conjunction, Supernova, or Neither?

In this case, Kepler knew that mutual conjunctions of the three known outer planets recurred every 805 years, and he suggested that these coincided with great events in history: the culmination of the Reformation in his own time, the ascendancy of Charlemagne in A.D. 800, and the appearances of Isaiah in 812 B.C. and Moses in 1617 B.C. (note that present chronological ideas put the latter events in rather different eras). The missing link in the chain was the birth of Jesus Christ, and on this basis Kepler was led to the idea that the Nativity must have occurred around 6 B.C., when he knew that the three planets had massed in the sky. His theoretical calculations were confirmed by Babylonian almanacs from that epoch.

The grand planetary conjunction provided Kepler with a possible explanation for the star of Bethlehem, and this hypothesis has been taken up by many others over the centuries since. In fact, the concept was by no means new when Kepler described it, the eighth-century Arab scholar Masha'Allah having suggested it many centuries earlier. With modern computational techniques and knowledge of the planetary orbits we are able to visualize much more accurately the relative positions of the planets back in the era in question. There were also other conjunctions of pairs of planets which were worthy of note: Jupiter and Saturn in 7 B.C. (repeated three times during the year, in fact, due to the peculiarities of the motions of those planets and Earth), and Jupiter and Venus in 3 B.C.

Actually, even in his own time the conjunction idea was not original to Kepler, because Michael Mästlin, a professor at the University of Tübingen where Kepler had been a student, had made similar statements in his lectures, and one presumes that this is where Kepler garnered the basic concept. But it was Kepler's computations which indicated the three-way massing around that time, which he recognized from his reading of Suslyga's tract was about the right epoch for the Nativity.

On this basis, as Christmas nears each year you will find magazine and newspaper stories which say that Kepler thought that a planetary conjunction at that time was interpreted as a star for the three Magi to follow, eventually to Bethlehem. In fact Kepler rejected this simplistic notion. Remember that in his day the nature of stars and their behavior was largely unknown. What Kepler argued was that the great conjunction of 1604 had *caused* the nearby supernova, portending great things. Similarly, he saw the thrice-repeated conjunction of Jupiter and Saturn in 7 B.C., followed by the three-way massing of Jupiter, Saturn, and Mars in 6 B.C., as the herald of a new star—*de stella nova*—which must have appeared soon thereafter, in accord with Suslyga's suggestion of the Nativity occurring in 5 B.C. To Kepler, then, the star of Bethlehem was a supernova, a brilliant star which flared temporarily to show all watchers that the Messiah was soon to walk the earth.

This idea that the star of Bethlehem was a supernova is another conjecture which has been revisited many times since by others. The fact is, though, that if a bright galactic supernova had appeared around the time in question, say between 10 and 1 B.C., then we would expect it to be recorded in independent accounts of celestial activity, such as the detailed annals of the ancient Chinese, where comets, meteor showers, supernovas, and other transient phenomena are all diligently entered. And no such supernova is recorded.

Less spectacular than supernovas, some stars go through repeated nova stages in which their brightness increases for a while and then dims again. Such events in our galaxy are much more frequent than supernovas, but again the Chinese records are silent for the period of interest on this score.

Three possible basic identities for the star of Bethlehem have been given above, then: (1) a supernova: vivid, unexpected, infrequent, and temporary, but not evidenced in scientific records from the time; (2) a nova: an unexpected flaring star, again not evidenced in the Chinese records but in any case rather humdrum and lacking in the spectacular qualities one might expect to explain the travels of the Magi; and (3) a planetary conjunction: traceable through modern methods, evidenced in the records, but perhaps *too* temporary to explain a pilgrimage like that of the Magi on its own.

The literature in the field is huge, and several other explanations have been put forward. Of course if you believe that the star of Bethlehem story is a myth, or that it was a divine sign with no physical cause, a miracle, then no further explanation is required. I do not want to be dismissive here of such beliefs. Many theologians and laypeople view this and many other biblical events as being legendary rather than historical, metaphorical tales portraying underlying human truths. In the discussion I present here of the stories of the star of Bethlehem and the Crucifixion I can do no other than take the received reports on face value, as historical records.

The biblical account has the star moving, and so it is an obvious step to invoke Mercury or the bright, easily seen, and fast-moving Venus. But the ancients knew all about these planets, seen in the sky every year, so they are not plausible candidates.

What about other planets? Although we consider Uranus to have been discovered in 1781, it is actually visible to the naked eye, just. Perhaps the Magi were following Uranus? This seems a dubious idea because there are many hundreds of brighter celestial objects, and although that planet moves against the stellar background, it does so rather slowly, being so distant from Earth. As a matter of fact, when William Herschel spotted Uranus those two centuries ago, he initially thought that he had found a new comet, and it was some time before it was recognized as being a planet.

Leaving planets and stars aside, then, what about comets? These may move swiftly across the sky, several degrees from one night to the next, and they may be bright compared to other celestial objects, some even being visible in daylight. So, was the star of Bethlehem actually a comet? For various reasons this is the explanation which I happen to favor. Colin Humphreys of the University of Cambridge has developed an explanation based on a specific bright comet witnessed by the Chinese in 5 B.C., the details being consistent with what we know of the star of Bethlehem story. This he published in the *Quarterly Journal of the Royal Astronomical Society* in 1991. I will summarize the main points below.

The Comet of Bethlehem

Surely one would be able to tell from the records of the times whether the star of Bethlehem was a comet rather than a star? Actually, the biblical description of the phenomenon is quite vague, with only the Gospel of Matthew (2:1–12) providing an account, and that is fairly ambiguous. Although not pointlike as are stars, comets do not always display the long tails which we usually associate with them (the word *comet* derives from the Greek for "long-haired"), and so may appear simply as small, fuzzy blobs crossing the sky. But they *do* cross the sky, moving relative to the background stars, and that is consistent with the reported wandering star but *not* some stellar phenomenon. (And in any case, we will see below that Chinese records indicate a tailed comet at the requisite time.)

Humphreys pointed out that there are three basic astronomical characteristics of the star of Bethlehem, as stated in Matthew's Gospel (the received text of which was finished by about A.D. 85–90). First, it was indeed a wandering star, moving through the sky against the far celestial background; this indicates that it was a member of the solar system, rather than part of the much more distant panoply of stars and nebulae. Second, the star had newly appeared, which seems to rule out all the naked-eye planets.

Those two considerations alone indicate rather strongly that a comet was the culprit, but Humphreys goes further. He notes that Matthew (2:9) states that the star "went ahead of them and stood over the place where the child was." Others have placed perhaps too much emphasis on this, wishing to believe that the star pointed directly to the actual stable where Jesus lay in a manger. In reality it seems that from Jesusalem, where the Magi had arrived and met with Herod, the star appeared immediately over Bethlehem, which is about five miles away, slightly to the west of due south, in which direction Herod headed them. The Magi could have carried on that path to Hebron, but in Jerusalem it had been decided that Bethlehem was their destination.

An essential point here is that the star is said to have "stood over" Bethlehem. Such a term, or alternatively "hung over," is used exclusively in ancient literature describing celestial sightings in connection with comets. For example, the Roman historian Dion Cassius (ca. 150–235) described one seen in 12 B.C. this way: "The star called comet stood for several days over the city." Some centuries later the writer Marcellinus described how "a sign appeared in the sky hanging like a column and blazing for 30 days," this being the great comet of A.D. 390. One reason for the ancients connecting comets with great events like the birth or death of kings was that they looked like mighty swords brandished in the sky. (And note that I implied there that comets were superstitiously regarded as being harbingers of both evil *and* good, not just the former: "The king is dead; long live the king!")

The Magi who followed this star from the East are often thought of as being kings themselves, but this is an embellishment which was not added to the story until about the sixth century, perhaps introduced to add to the perceived importance of the birth of Jesus, the Messiah, or Savior-King, whose arrival had long been anticipated by the Jews. The Magi were not kings as such, but priests learned in astronomy and astrology, perhaps members of the Zoroastrian sect which had spread from Persia in the preceding centuries, officiating at religious ceremonies and interpreting portents. The particular trio who traveled to visit Jesus seem to have originated in Mesopotamia, where a strong Jewish presence had existed since the Exile following the sack of Jerusalem by the Babylonians in 586 B.C., or perhaps in related lands in the northern part of what is now Saudi Arabia, closer to Palestine/Judaea. Whichever was the case, throughout this region there had been a long-held Jewish doctrine of a forthcoming Messiah, an expectation in which the Magi would have been well versed.

It was described earlier how Kepler, in the seventeenth century, interpreted the thrice-repeated conjunctions of Jupiter and Saturn in 7 B.C., followed by the three-way massing of Jupiter, Saturn, and Mars in 6 B.C., as the heralds of a new star which would proclaim the birth of the Messiah. It is not unreasonable to think that the Magi and others, at the actual time, would have thought likewise, especially because something repeated three times would have been viewed as especially auspicious. Such a belief has continued through to the present era: as Lewis Carroll wrote in *The Hunting of the Snark* (1876), "What I tell you three times is true."

The triple conjunction of the two giant planets in Pisces, approaching each other in May, October, and December of 7 B.C., was a once every nine hundred years event; the three-way conjunction of 6 B.C. a once every eight hundred years event. These alerted the Magi to something important being in the offing. Obviously the third celestial portent, confidently expected to follow soon thereafter, was to be a harbinger of some great happening, the birth of the Messiah awaited for so many centuries. One can almost imagine the Magi packing their bags in readiness, and jumping on their camels when the wandering star indeed appeared in 5 B.C.

The Comet of 5 B.C.

The idea that the star of Bethlehem was a comet is not a new one. The Alexandrian theologian Origen, who lived between about A.D. 185 and 254, suggested as much. He wrote:

> *The star that was seen in the East we consider to be a new star . . . partaking of the nature of those celestial bodies which appear at times such as comets. . . . If then at the commencement of new dynasties or on the occasion of other important events there arises a comet . . . why should it be a matter of wonder that at the birth of Him who was to introduce a new doctrine . . . a star should have arisen?*

The idea of a cometary identity for Bethlehem's star has persisted through the ages in ecclesiastical circles. When the European Space Agency sent a probe to investigate Halley's comet in 1986, it was named Giotto for the Italian painter who, in the early fourteenth century, painted a fresco of the Nativity in the Arena Chapel in Padua, in which he depicted a comet hanging in the sky above the head of the infant. The cometary hypothesis has been revived occasionally over the past several decades, but it is Humphreys who has gone further in tracking down an actual apparition at the required time, then seeing how it would fit in with other aspects of the story.

The Chinese annals for the three decades 20 B.C. through A.D. 10 do not contain records of any novas (or supernovas), but three comets are listed. The first of those is the aforementioned comet of 12 B.C., which was actually Halley's comet. That famous celestial visitor returns to the inner solar system about every seventy-six years, and various records exist for its apparitions in 240, 164, 87, and 12 B.C., plus A.D. 66, 141, 218, and so on right through to 1835 (when Mark Twain/Samuel Langhorne Clemens was born), 1910 (when he died), and 1986. We may confidently expect its return in 2062.

Could comet Halley have been the star of Bethlehem? This has been suggested by several authors, but it seems that 12 B.C. is too early. In the Gospel of Luke it is written that Jesus began his ministry when he was "about thirty," the time of his baptism by John the Baptist. As this is stated elsewhere in Luke as being in the fifteenth year of the reign of Tiberius, making the autumn of A.D. 28 the earliest possible time that could be referred to (the start of the regnal year), and allowing the longest possible interpretation of "about thirty" as being thirty-four years old, one derives an earliest possible year for the Nativity of 7 B.C. (thirty-four years before A.D. 28), five years after Halley's comet had passed by.

The third of the three comets in the Chinese records is that of 4 B.C. Unlike comet Halley, but similar to most observed comets, this one does not seem to have been periodic, and it has not been possible to identify it with any later-observed comet. This appeared in April of that year, very close to the generally accepted time of Herod's death and therefore too late to fit in with that ruler meeting the Magi and seeing the star of Bethlehem, and subsequently ordering the slaughter of the infant boys. Indeed, the Chinese made little of this third comet, so that it would seem to have been rather dim and short-lived.

It is the second of the comets recorded by the Chinese in this era which is the prime candidate. The official history of the Han dynasty (206 B.C.–A.D. 9), written at the time, says this: "Second year of the Chien-p'ing reign period, second month, a *sui-hsing* [tailed comet] appeared at *Ch'ien-niu* for over seventy days." The month stated converts to March 9 through April 6 of 5 B.C. on the Julian calendar, while *Ch'ien-niu* means the part of the eastern horizon where the Sun rises at the winter solstice, which at that time was in the constellation Capricorn. In the West, that is how the tropic of Capricorn gets its name—the furthest south latitude at which the Sun passes overhead on the solstice—and similarly for the tropic of Cancer; the fact that precession has pushed the winter solstice into Sagittarius since about A.D. 1000 has not led to a change in terminology.

Skipping momentarily to the origins of Christianity in Judaism and Mithraism, one can see that the comet rising in Capricorn in that era would also be viewed as being significant. The winter solstice contained the festival of the unconquered sun, so that Capricorn was the place where the Sun was reborn. Precisely the place where one would anticipate a divine sign of the birth of the Son of God to appear.

In particular because this comet rose in a part of the sky which they considered extremely important, the Chinese regarded it as a harbinger of change, that concern being exacerbated by the extended time over which it was visible: "The appearance of the *sui* was a sign of change; the long period of its appearance was due to the importance of the incident [to follow]."

Other investigators—starting with J. F. Foucquet, the French expert in Chinese history, in 1729—have noted the record of an apparition in the sky in 5 B.C., but have misinterpreted it as having been either a nova or supernova. In fact the use of *sui*, meaning a "tail," in the Chinese accounts make it quite explicit that the object was a comet. Just to reiterate that comets are viewed as portents of good as well as bad in many cultures, let me note that the ancient Chinese called comets *broom-stars*, and around 300 B.C. the writer Tsochhiu wrote that "a comet is like a broom, it signals the sweeping away of evil."

Clearly the comet of 5 B.C. was an impressive sight from Asia, and there is no reason to suspect that it was otherwise elsewhere in the Northern Hemisphere. Although the Protestant Bible itself may not give an extensive account of its appearance, other Christian literature from about the time is not so reticent. For example, the Apocryphal New Testament, which was written down in about A.D. 150, quotes the Magi as saying that "we saw how an indescribably greater star shone among these stars and dimmed them." It would seem that with the comet of 5 B.C. we are moving closer to the identification of the real star of Bethlehem.

The Interpretation of the Comet of 5 B.C.

In the text above I have alluded to various explanations one could put forward for the comet of 5 B.C. having been regarded as being a divine messenger, telling believers of the arrival of the Messiah. Another factor is that the object was seen in the east, rather than the west; the astrological credo of the era was for this signifying a rapidly approaching event rather than one for which a long wait should be anticipated.

Let us next consider some modern-day interpretations of what may have transpired, in terms of this conjectured cometary phenomenon.

Matthew (2:9) says, "The star, which they saw in the East, went before them." This refers to the final leg of the journey of the Magi, from Jerusalem to Bethlehem. Since they had started out on their pilgrimage, the comet had gradually moved across the sky against the background of stars, starting in the east and inching about ninety degrees across the sky until it was in the south. Note that this is quite independent of the diurnal motion of all things in the heavens, objects rising in the east and setting in the west due to the rotation of our planet. Here we are addressing the movement of the comet against the background of stars, and since such objects typically move at a rate of one or two degrees per day, over the seventy days' visibility reported by the Chinese, transit across ninety degrees of sky is entirely consonant with the overall picture being considered.

That is the motion of the comet in the sky. What about the movement of the Magi across the ground? The furthest one might think that the Magi had traveled would be from Babylon to Jerusalem, which is about five hundred miles directly across the desert regions in the north of present-day Saudi Arabia, but as much as nine hundred miles if the friendlier route arcing through the Fertile Crescent were taken, perhaps northwest following the Euphrates to Aleppo and then south through Syria and Lebanon to Palestine. One would be pressed to average more than fifty miles a day by camel, and if the Magi started out very soon after the earliest date on which the Chinese may have seen the comet (March 9), then they would have arrived in Jerusalem in early to mid-April, although arrival as late as mid-June is possible if the comet first appeared late in the Chinese month in question, and the Magi did not get to Palestine until the terminal part of the seventy-day apparition.

On the other hand, as a schoolboy I sang the following Christmas carol (although not always with the correct words):

While shepherds watched their flocks by night,
All seated on the ground,
The Angel of the Lord came down
And Glory shone around.

The identification of angel myths with comets, complete with halos and wings, has been argued by many other authors, and I won't dabble with that here ("Fools rush in where angels fear to tread" seems apposite), except to note that the last two lines quoted seem to fit a bright comet rather well. Turning to the first two lines, all should know that lambs are born in the spring, and that's when shepherds need to be vigilant in wolf-infested country.

I could give other arguments, but it seems clear that the evidence of both the Magi and the shepherds is consistent with the time of year being spring and not midwinter, to which Christmas later migrated. Let us try to refine this crude estimate of the date of Jesus' birth.

Why Were the Inns All Full?

Countless Nativity plays acted out in elementary schools each December portray Mary and Joseph arriving in Bethlehem (known as the City of David, his family origin) from Nazareth (where they lived) in order to fulfill their obligation in registering for a census. This is what is stated by Luke (2:1–5). The problem is that there is no record of a census being taken at that time, and the censuses carried out on behalf of Augustus in 28 B.C., 8 B.C., and A.D. 14 were for Roman citizens only. Not only that, but such a census was conducted quite differently from the modern mode whereby all the forms are supposed to be filled in on one specific day, for consistency. Such a census in the ancient world continued for a protracted period. That contradicts the notion that all the inns of Bethlehem were full because everyone had arrived at the same time for a census count.

An alternative is that Joseph and Mary went to Bethlehem for a quite different type of registration. Rebellion in Judaea and surrounding areas was rife, resulting in the Jewish wars some decades later in which Josephus (the aberrant historian) fought first on on side, and then the other. Josephus and other writers recorded that at about the time in question, a few years B.C., a form of census was called, but not to count the Roman citizens (such a thing being an ongoing task as the number could vary significantly, because the majority of the citizens of the empire never visited the Eternal City: the citizenship could be earned by foreigners through service, such as fighting as a mercenary in the Roman armies). Nor was the call to the people in about 5 B.C. to go back to their places of birth made in order to levy taxes, as was the case at other times. Rather, it was to extract from the indigents *an oath of allegiance to Emperor Augustus*. Given that we are here discussing the birth of He who would be regarded as the King of the Jews, sent to lead His people from oppression, the real reason for the purported census is hugely significant.

This explains why Joseph and Mary needed to travel to Bethlehem, but not why the inns were all full. It simply makes no sense, especially to a bureaucracy (which is one of the least sensible of organizations), to have myriad men all try to swear oaths in one day, one week, or even one month. The way this would have been handled would have been to stagger the registrations: "Turn up some time over the next half year" would have been the instruction.

Let me backstep here and revisit my comment that the inns were all full. The common belief that there were inns around in those days with private rooms is a fallacy. In fact, practically all travelers stayed overnight in what we would think of as stables, with hay to sleep on in the company of the oxen and the horses. What was lacking for Joseph and Mary was not room to stay in such an establishment, but *privacy*. This misunderstanding has come about through the translation of the Bible through different languages, the nuance being lost along the way. In the scenario whereby the couple arrived in Bethlehem with Mary soon to deliver the child, what they needed was privacy, but the barns where everyone stayed were crowded. There was room to sleep, but no private space.

So, why was Bethlehem crowded? The answer is Passover. Pilgrims would come to Jerusalem, the site of the rebuilt Great Temple (commenced in about 20 B.C.), and that city would soon have filled up. Outlying towns and villages like Bethlehem would soon be occupied, too, being only a couple of hours by foot from the Temple. Living in Nazareth, about sixty miles to the north, Joseph and Mary may well have decided to combine two trips into one: coming to Jerusalem for Passover, and registering at Bethlehem at the same time. But the latter was full, with the overflow from the city, and so they ended up in the overcrowded stable.

This gives us another date. Passover was on April 20 in 5 B.C., the arrivals and departures of the pilgrims and the associated ceremonies lasting from about a week before until a week after, say April 13 through the twenty-seventh.

Narrowing Down the Feasible Dates

But there is more. Luke (2:22) says that Jesus was taken by his parents to the Temple in Jerusalem. But this could not have been until after the purification time for the mother (forty days, per Leviticus 12:2–4), and the Magi are said to have visited him *after* this trip to Jerusalem. This means that they could not have seen him until he was at least forty-two days (six weeks) old. The latest date that could have been can easily be ascertained. The Magi saw the comet over Bethlehem, and that cannot have been later than seventy days after April 6, the last day of the Chinese month in which the comet appeared, provided that one assumes that what was visible in Judaea was the same as that in Cathay. That renders June 15 for the latest day on which the Magi could have brought their gold, frankincense, and myrrh to Jesus; taking six weeks off of that, the latest feasible date for his birth becomes May 4.

The *earliest* feasible date comes from a consideration of the shepherds. Although they are often represented in Nativity plays as walking in with the Magi, in fact they seem to have visited the newborn Jesus very soon after the birth, and certainly before Mary and Joseph took him to the Temple in Jerusalem.

Sheep were generally penned under shelter during the winter and put out to grass in March, consistent with when the comet was first seen (within the four weeks following March 9). One must take that to be the earliest date for the birth, although a month or more later is preferred on account of the Passover considerations.

Sheep enter our deliberations in another way. Jesus is often referred to as the Lamb of God (for example, John 1:29), which one must admit is a strange epithet in an abstract sense. But does it have some specific meaning? The lambs which are slaughtered for the Passover meal a few days later are selected on Nisan 10. If Jesus had been born on that express day, then being described as the Lamb of God, who was expected to die later for the benefit of all humankind, would make some sense. For the record, the Jewish day Nisan 10 in 5 B.C. lasted from nightfall on Sunday, April 14, until nightfall on Monday, April 15.

The interpretation of the comet of 5 B.C. as being the star of Bethlehem leads to a possible range of dates for the Nativity, then, of March 9 through May 4 in that year. The inns being fully occupied is consistent with many people having arrived for Passover, narrowing the range of dates to April 13–27. An arrival of Joseph and Mary close to the beginning of that range of dates, and some imperative to find shelter for the imminent delivery of the baby, fits in with the birth occurring on April 14 or 15.

The Massacre of the Infants

The above scenario is also consistent with other things we know about events at that time, if we proceed as above under the assumption that the story we have received is based on historical facts rather than being a parable confected much later.

Herod was called *the Great*, but he doesn't seem to have been too wonderful a person. By all accounts his family was a malfunctional affliction upon society, as unlikable a bunch as one could wish to meet, and he seems to have had few friends either within or without his immediate circle. Much of the year he lived in Caesarea, down on the Mediterranean coast, and one of his least favorite times of year was Passover, when he was obliged to travel to hot and dusty Jerusalem merely to ensure that the crowds of Jews were kept under control by his troops. In 5 B.C. he would already have been suffering from the cancer which would kill him about twelve months later, and it seems that he had also been instructed to extract from the residents of Judaea an oath of allegiance to Augustus, which would have kept him in Jerusalem longer than usual.

The Magi arriving, perhaps in late May, wanting to know the location of the newborn King of the Jews who was to lead them to salvation, would not have put Herod in an improved temper. Herod asked the Magi "the exact time the star had appeared" (Matthew 2:7). While one might jump to the presumption that only the star of Bethlehem was being considered, in the context of the present interpretation one may conjecture instead that the Magi would tell Herod not only of the presently visible comet, but also of the planetary conjunctions of 7 and 6 B.C. which had forewarned them of the Messiah's coming birth.

This provides an explanation for Herod deciding to massacre all Jewish boys under two years of age (but note that historical evidence for such an edict or massacre is lacking). To be conservative, he would have counted back to the first celestial events described by the Magi, which occurred precisely two years earlier (in May of 7 B.C.). Thus Matthew (2:16) says that Herod set the age limit "according with the time he had learned from the Magi." That way he would have felt secure that the baby Jesus had been killed, and the possible focus for some future Jewish uprising was removed.

In the event, Joseph was warned of what was to happen. Soon after the visit of the Magi, he, Mary, and Jesus quickly left for Egypt, staying there for two years. Why two years in that case? Because it was one year before Herod died, and then to be sure they waited another year to see which way the wind blew with Herod's successor, his son Herod Antipas, before feeling that the time was right to return safely to their homeland. Thus the chronology of their enforced stay in Egypt is again consistent with the star of Bethlehem being identified with the comet of 5 B.C., and Jesus having been born near the middle of April in that year.

The Calendrical Significance of the Crucifixion and the Resurrection

I wrote at the start of this chapter that astronomy can give some clues about not only when Jesus was born, but also when he died on the Cross. Now that we have dealt rather exhaustively with the Nativity, let us turn to the Crucifixion. In fact, the aforementioned Herod Antipas was still alive and in power when Jesus died, and officiated at His trial, accomplishing what his father Herod the Great had failed to do: kill Jesus. As this Herod died in A.D. 39, his son Herod Agrippa taking over, we know that the Crucifixion occurred well before then.

Another thing which has been discussed rather exhaustively in this book has been the significance of Easter in the Christian calendars, the calculation of its date each year based upon the biblical accounts of the Crucifixion and the Resurrection being of overriding importance, connected with all form of religious schism and resulting in the calendrical system used as the worldwide norm. Knowing when these events took place, then, might be viewed as being a major quest.

On the other hand, we have seen above that the birth of Jesus Christ occurred neither on the feast day at which it has been commemorated since the end of the fourth century, nor in the year which would be appropriate if our *anno Domini* year-counting convention were to be correctly zeroed. It is the later mythology and its effect upon the evolution of the calendar, Christian and civil, which led to the system we use today, and the same may be true for Easter and the Crucifixion/Resurrection: it is the tenets of the Nicene fathers and their contemporaries, for example, which led eventually to the Easter *computus,* and if those tenets bear little resemblance to actual history it matters little.

Early Accounts of the Crucifixion

Notwithstanding all of the above, it might be interesting to know when the Crucifixion did occur, in history. We must begin with some basics.

The four Gospels all agree that the Crucifixion of Jesus occurred on the day before the Jewish Sabbath, which begins at nightfall on Friday, and other early sources tally with this (e.g., the writings of Saint Justin Martyr, A.D. ca. 100–ca. 165, and the Roman historian Tacitus, A.D. ca. 56–ca. 120). These and other authors agree that Jesus died on the Cross in the few hours prior to the onset of that Sabbath. This implies that His death took place between about 2 and 5 P.M. on a Friday.

The Synoptists Matthew, Mark, and Luke, however, imply that the Crucifixion was on Nisan 15, based upon a supposition (by us, as later interpreters) that the Last Supper they describe was the Passover meal held at the rise of the full moon at the beginning of that day. The Gospel of John, and the Jewish Talmud, differ from that position; they agree with each other in placing the Crucifixion on Nisan 14, the day of preparation near the close of which the paschal lambs are slain, between about 3 and 5 P.M. (that is, a few hours before the meal held that evening, the start of Nisan 15).

None of the various sources specify the year, but because they all agree that Pontius Pilate was procurator of Judaea at the time, it must have been between A.D. 26 and 36.

From that information one wants to find a Nisan 14 or 15, depending upon which you prefer, which ended on a Friday evening some time between those years. That would be an easy task if the current rather-specific rules for defining the Hebrew calendar had been followed during the first century, but actually the calendar at that time was quite haphazard. Nowadays the Jewish year is stipulated using a set of rules which disregards such things as the variation in time between new moons due to the eccentricity of the lunar orbit. Back then, the onset of each month was defined in much the same way as is the case now with the Islamic calendar, with the *sighting* of the new moon being essential.

Unfortunately, records of the numbers of days counted in each month resulting from those sightings, and the numbers of months in the year, have not survived. Thus we cannot be sure of the date on the Julian calendar when Passover was called each year. In discussing the Nativity above, the Passover date of April 20 in 5 B.C. results from an assumption that our modern computations of the lunar visibility give a fair representation of when Nisan would have occurred in that year. Similarly, one can make an educated guess at when Nisan was counted in each of the years A.D. 26–36. Given that Jesus was not baptized until at least the autumn of A.D. 28, and that he is supposed to have preached his ministry for about two years, and that he died well before the age of forty, there are reasons to confine one's search to the middle years of that range, A.D. 29–33.

Nisan 14 or Nisan 15?

Scholars have debated the relative importance of the above constraints upon the date of the Crucifixion for many centuries. Their opinions fall into three main categories:

1. The Last Supper was a straightforward Passover meal eaten on the evening at the start of Nisan 15, the trial and execution of Jesus occurring later on the same Hebrew day, between 6 A.M. and 6 P.M. This implies that the reading of the Synoptists as above is correct, and John's Gospel is wrong.

2. Jesus knew of his impending arrest and so convened the Last Supper as an impromptu pseudo-Passover meal a day early, in the evening starting Nisan 14, and the trial and execution occurred after the following sunrise. This agrees with John, who states that the meal preceded Passover, and the fact that the Synoptists do not mention eating Passover lamb. In this interpretation all four Gospels would be in agreement, and the Crucifixion was on Nisan 14.

3. The calendar that was being used to frame these reports was not the majority Jewish system, but rather that of the Khirbat Qumran community, living fifteen miles east of Jerusalem, where the Dead Sea Scrolls were found a half century ago. In this interpretation the Last Supper was indeed a true Passover meal, but at the Passover time according to the Qumran sect's calendar. This would have put it at the start of the Jewish Wednesday (Tuesday evening). This interpretation has appeal in that it provides a rather less frenetic schedule for the events: the trial could have been on the Thursday, and then the Crucifixion on the Friday, before the Sabbath and the official (non-Qumran) Passover on the Saturday. This would not contradict any of the accounts, and again one arrives at a conclusion that the Crucifixion was on Nisan 14.

Both the second and third opinions provide an explanation which is in accord with feasible interpretations of all of the accounts, and lead to a date of Nisan 14 for the Crucifixion. On this basis alone that is the date of a Friday between A.D. 29 and 33 which we are looking for, although consideration of the lunar phases also provide support for Nisan 14, as we will see below.

The Lunar Evidence

At this juncture Colin Humphreys again enters the picture. Working with Graeme Waddington (when both were at the University of Oxford), Humphreys investigated the lunar visibility from Judaea throughout the period of interest, and from that attempted to reconstruct the Hebrew calendar during that time. Their results were published in the journal *Nature* in 1983.

Allowing some flexibility in the years, they computed the instants of lunar conjunction throughout the era when Pontius Pilate was procurator of Judaea, from which the times at which the faint lunar crescent (new moon) could have been seen were deduced, and thus the Julian calendar dates and days of the week for Nisan 14 and 15 could be found. There will always be difficulties with such chains of reasoning—for example, cloudy weather could have delayed by a day the actual declaration of the start of the Hebrew month, and it only takes a very slight haze to make a crescent of otherwise borderline visibility actually undetectable. But their results throw up some interesting facts.

In A.D. 27, Nisan 15 is indicated to be a Friday. That year, however, seems too early by far, and we favor Nisan 14. On the other hand, if the atmospheric transparency two weeks earlier had been poor, then this Friday may have been Nisan 14. Nevertheless, A.D. 27 can be rejected as too soon, being even before the year in which it is thought that Jesus was baptized.

In A.D. 30, Nisan 14 is indicated to be a Friday (April 7), although it is just feasible that the new moon might have been spotted a day earlier than the nominal model suggests, in which case Nisan 14 would have been on the Thursday.

In A.D. 33, Nisan 14 is indicated to be a Friday (April 3), without any great doubt.

The only other possibility thrown up is in A.D. 34, although that year seems too late according to our A.D. 29–33 restriction. In A.D. 34, Nisan 14 would have been on Wednesday, March 24, according to the computations of Humphreys and Waddington, but they also point out that if the weather had been exceptionally severe, then Nisan might have been delayed through the insertion of an intercalary month, the purpose of which would have been to allow the paschal lambs more time to grow, and also the ripening of the fruit which is presented on Nisan 16. In that scenario Nisan 14 would have been on Friday, April 23. As a matter of fact, Isaac Newton proposed that date for the Crucifixion, although on an erroneous basis; he may have advocated it in part for jingoistic reasons, that being Saint George's Day!

The two dates one is left with are April 7, A.D. 30, and April 3, A.D. 33. These are both Nisan 14 dates; as I presaged above, the calculated lunar phases provide support for Nisan 14 over Nisan 15, and therefore the interpretation that the Last Supper was a strict Passover meal seems to be wrong. On the other hand, this would mean that the accounts given in all four Gospels are in agreement, and there is other biblical support. For example, this chronology would have Jesus dying in the mid- to late afternoon on Nisan 14, at the same time as the slaughter of the paschal lambs, consistent with the many New Testament references to Jesus being the Lamb of God who was sacrificed for the benefit of humankind.

Other Constraints

How may we differentiate between the A.D. 30 and 33 candidates? We have previously rejected from consideration any of the first several years in which Pontius Pilate was the local civil authority on the grounds that Jesus was not baptized until at least the autumn of A.D. 28, perhaps later, and he preached his ministry for two years following that. The Gospel of John states that three different Passovers occurred during that time, culminating in that of the Crucifixion. If one takes the autumn of A.D. 28 as the actual start time, then the Passovers involved would be those of A.D. 29 and A.D. 30, the Passover of A.D. 31 being the earliest feasible date for the Crucifixion. This would exclude the candidature of A.D. 30 for the Crucifixion.

Further, the erection of the Great Temple in Jerusalem—the third to be built there in biblical times, the first by Solomon, and the second by Zerubba-bel—was finished in about 16 B.C., and John (2:20) states that Jesus was told by the Jews during the first Passover of his ministry that the Temple had taken forty-six years to complete. That Passover would therefore be that of A.D. 30 or 31, again counting against a Crucifixion in A.D. 30, handing the baton to A.D. 33.

A Lunar Eclipse?

I have yet to mention the major step forward taken by Humphreys and Waddington in 1983. In various places in the Bible, and other written records, allusions are made to the Moon being "turned to blood" at the time of the Crucifixion. For example, in Acts (2:14–21), an account is given of the apostles gathering at Pentecost (the seventh Sunday after Easter, hence the fiftieth day, counting inclusively) and addressing a hectoring crowd. Peter says: "No, this was what was spoken by the prophet Joel: In the last days, God says, I will pour out my spirit on all people. . . . I will show wonders in the heavens above. . . . The Sun will be turned to darkness and the Moon to blood before that great and glorious day of the Lord shall come." This may be interpreted as Peter implying that the prophecy of Joel 2:31 had recently been fulfilled. The "great and glorious day" on which wonders had been shown was clearly the day of the Resurrection. So what happened briefly before?

That the Sun had darkened on the day of the Crucifixion is attested to by various sources such as Matthew (27:45). This may well have been due to a dust storm caused by the khamsin, a hot wind from the south which blows through the region for about fifty days commencing around the middle of March, in accord with the chronological framework being reconstructed here.

But what of the Moon turning to the color of blood? This is a phenomenon which is well known to occur during a lunar eclipse. The shadow of Earth stops most sunlight reaching the Moon, but some leaks through by traversing the terrestrial atmosphere and thus illuminating the lunar surface to a muted extent. As the sunlight passes through the atmosphere, though, the blue end of the spectrum is preferentially scattered by the atoms of our air (which is why the sky appears blue), and red light is transmitted on the way to the Moon with greater efficiency. The full moon—a lunar eclipse can only occur then, at superior conjunction—is therefore darkened and reddened in eclipse. In short, it is turned to blood.

With that in mind, Humphreys and Waddington computed the dates of all lunar eclipses visible from Jerusalem between A.D. 26 and 36. And lo and behold, they found one on April 3, A.D. 33.

Dating the Crucifixion

In all, Humphreys and Waddington identified a dozen lunar eclipses spread over this eleven years, but only one (mentioned above) was near the time of Passover. More than that, the eclipse was in progress as the Moon rose on that Friday evening, within hours of Jesus expiring on the Cross. Although the eclipse was not total—only about 20 percent of the lunar disk would have been within the terrestrial shadow at moonrise at 6:20 P.M. local time, the start of Passover—this blood-red apparition rising so soon after the death of Jesus must have had a profound effect upon his disciples.

The 20 percent obscuration diminished over the next half hour, the eclipse terminating at about 6:50 P.M. While this is not a major fraction of the disk, the portion within the direct shadow (called the *umbra*) was toward the top of the image of the Moon rising above the eastern horizon, with the result that as it crept upward the majority of the first-seen section would have been dark and red. The rest of the disk would also have been colored a yellow-orange by the effects of sunlight being scattered and refracted as it passed through the atmosphere on the way to the Moon.

The atmospheric atoms themselves contribute to the reddening of the rising moon, compared to its usual bluish-white color when overhead, due to their scattering properties, just the same as the sunset is always red compared to the midday sun.

On top of that, the evidence that the Sun had been darkened earlier in the day by a dust storm leads to the expectation that there would still be dust suspended in the air, producing a separate darkening and reddening of the Moon on its own account, especially with the long, slanted path through the atmosphere to the observer when the Moon is just above the horizon.

The evidence of the eclipse, then, leads to a dating of the Crucifixion to April 3, A.D. 33. The rising of a temporarily blood-red moon immediately after the death of Jesus would have produced a strong impression in the minds of all witnesses, perhaps changing the attitude of both Pontius Pilate and the Jews toward the body of the dead man, resulting in the guard being posted on His tomb.

Through all these aspects of the historical and biblical accounts of the Crucifixion, and the application of modern astronomical knowledge, Humphreys and Waddington managed to tease out a pretty definitive time for the death of Jesus Christ on the Cross: the afternoon of Friday, April 3, A.D. 33, when He was within a few weeks of becoming thirty-seven years old. It may seem remarkable that no one had managed to come to this conclusion earlier, but as the pair conclude their paper by saying, "The gospel writers were not primarily interested in providing clues for chronologists."

The Fly in the Anointment

The above account may well have led you to a secure belief that the date of the Crucifixion has been established beyond all doubt. But in science there is no such thing as proof: there are only working hypotheses which are subject to disproof. Thus one cannot *prove* that the law of the conservation of momentum is true; all one can say is that no experiment has ever been done which has indicated that it is false.

With that background, one has to lament the fact that Humphreys and Waddington decided to say in their paper that "the accurate calculations presented here prove that this eclipse was visible." In 1990 another astronomer,

Bradley Schaefer (now at Yale University), presented a refined set of computations in which he showed that using an alternative set of values for the orbits of the Moon about Earth, Earth about the Sun, and the spin phase of Earth, the eclipse could have been practically over by the time that the Moon rose as seen from Jerusalem. Schaefer's analysis, published in the *Quarterly Journal of the Royal Astronomical Society,* indicates that only 1 percent of the lunar disk would have been within the umbra at moonrise, in which case there would have been no discernible change in the Moon's appearance except to someone specifically looking for this minor coloring.

This means that doubt has been poured onto the definite identification of A.D. 33 as the year of the Crucifixion. Schaefer's work does not imply that it must have been A.D. 30, only that the lunar eclipse concept may not provide such strong support for A.D. 33 as might have appeared to be the case previously.

My own view is that the work of Humphreys and Waddington will eventually be vindicated, that April 3, A.D. 33, will finally be recognized as far and away the best candidate for the date of the Crucifixion, and that it will be shown that a partial lunar eclipse *was* visible as the Moon rose that evening, a few hours after Jesus had died. (And of course I may well be wrong.)

One of the reasons I hold that opinion is that our evaluations of the local times at which phenomena such as moonrise took place all those years ago depend critically upon our knowledge of the history of the spin rate of Earth over the past several millennia. Because of tidal drag preponderantly due to the Moon, the day is lengthening—but it does so at an inconstant rate controlled by a wide variety of factors such as climatic variations, the continuing rebound of the continents after they were compressed by an overburden of ice during the last Ice Ages, and other geophysical and astronomical phenomena. In the next chapter we will consider these questions.

Chapter 22

❦

HOW MANY DAYS IN A DINOSAUR YEAR?

Throughout this book we have been looking at humankind's attempts to design calendars which follow some regularity based upon the cycles of the heavens (the lengths of the day, the month, and the year), in many societies the culturally imposed seven-day week or some other unnatural cycle of days also being a consideration.

Claims have been made for early humans keeping tabs on the lunar phases as far back as thirty thousand years, although this is a contentious matter. Calendar design as we would now understand it seems to have been going on for about five or six millennia, although the major developments have occurred during only the latter half of that period. The question arises: have any of the natural cycles themselves altered significantly during this era—short compared to the age of Earth—over which human civilization has thrived?

It is immediately possible to give an affirmative answer in that we know that leap seconds are inserted into many years. To be clear, these are nothing to do with leap years, and the name they are given is unfortunately misleading. Leap seconds are made necessary by the fact that the atomic second is defined according to the length of the day in the year 1900, and tidal drag is slowly decreasing the spin rate of the planet and therefore lengthening the day, by an amount imperceptible in everyday life but measurable using various high-technology techniques (radio astronomical tracking of distant quasars and so on). Leap seconds are used to keep the time according to the rotation of Earth within 0.9 seconds of the time shown by an atomic clock, which is not affected by our slowdown in spin rate.

There are various interconnected effects which stem from the spin rate of Earth dropping, and we will consider those at length. First, though, we need to consider the longest of the heavenly cycles contributing to the annual round: the length of the year. The question is one of whether the absolute length of the year has changed over geological and astronomical time, counted in billions of years.

Has the Size of Earth's Orbit Changed?

For detailed calendrical purposes it is necessary to define rather specifically what is meant by a *year*. The time for the stars to return to their same positions in

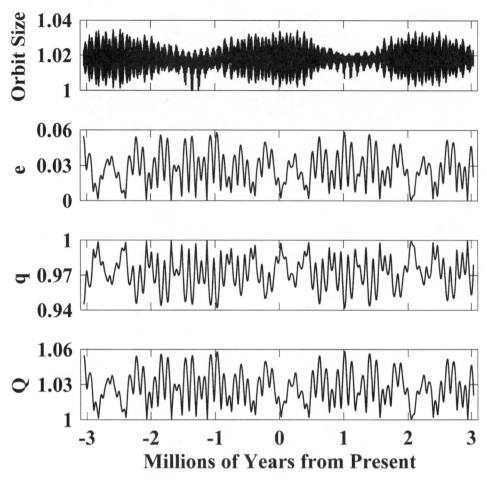

Fig. 6. Earth's orbit about the Sun varies significantly over timescales longer than a few millennia. Shown here, from top to bottom, are the *orbit size* (see below), the eccentricity of the orbit e, the perihelion distance q, and the aphelion distance Q, the latter two parameters being measured in astronomical units (A.U.). Let us first consider the shape of the orbit, then its size.

Looking at the present time, the zero at the center of the plots, one can see that the current eccentricity ($e = 0.01673$) is rather lower than the maximum value attained (0.05788), resulting in the perihelion distance being near its highest value and the aphelion distance near its lowest, meaning that the variations in the flux of sunlight currently received by Earth during an orbit are rather smaller than those which are possible over long timescales. That is, going either backward or forward in time by, say, a hundred thousand years, it is to be expected that more extreme seasonal variations would be in force: just now the contrast in global solar influx between perihelion and aphelion is about 7 percent, whereas the deviation between them is 26 percent when the eccentricity reaches its maximum. Obviously this would have substantial climatic effects.

Now the orbit size. This is *not* simply the semi-major axis (as defined in appendix A). What is shown here is the excess of the semi-major axis over 1 A.U., multiplied by a million. The range of values for the semi-major axis is actually between 1.0000009970 and the sky? The time for Earth to come back to perihelion? The time between the equinoxes, or the solstices?

Here, though, we can leave such details aside. We are interested in a cruder definition of the year, because we are looking for a more substantial variation than the distinctions of a fraction of a day which result from the above alternative definitions. We are wondering whether at some stage the solar year, the time to complete one orbit howsoever defined, might have lasted for only 350 days, say, or perhaps as long as 380 days.

Because another time unit (which we know is varying) is stated there, we need to say what we mean by a *day*. For the purposes of this section, where I am discussing whether the size of the terrestrial orbit about the Sun has changed, by *day* I imply a length of time equal to 86,400 atomic seconds. The question is whether, using an imaginary clock consistently registering seconds and days at that rate, Earth took an absolute length of time equivalent to about 365.25 days to orbit the Sun a million or a billion years ago.

Modern computing power is such that this problem can now be investigated. Astronomers can run programs which follow the paths of all the planets in the solar system, taking into account their mutual gravitational tugs, and see how their orbits change over long periods of time. The results from one such numerical experiment are shown in figure 6, on page 344.

As the extensive caption to that figure states, over the past and future few million years the *shape* of the terrestrial orbit changes somewhat, resulting in more extreme seasonal variations than those experienced in the present epoch. (The changing tilt of our spin axis also has a major influence in this regard.)

The average *size* of the orbit, however, does not change by any significant amount. Cyclic exchanges of gravitational energy between the planets, like two children's swings affecting each other because they both hang from the same tree branch, cause the size of Earth's orbit to change a little, but only by about 1 part in 25 million at most, and within a few thousand years the energy gained is lost again, swapped back and forth between the planets.

The deduction from this is that the absolute length of time which it takes our planet to orbit the Sun is constant, at least on the million-year timescale. On the other hand, one might point out that Earth is about 4.5 billion years

---

1.0000010348 A.U. The oscillations in this parameter are due to resonance effects between the planets, small transfers of energy occurring between them. The long-term average value for the semi-major axis appears to be about 1.000001018 A.U. The essential point is that the size of the terrestrial orbit does not vary by more than about 1 part in 25 million; if you recall that the astronomical unit measures about 150 million kilometers, it follows that this variation is by just 6 kilometers, less than a thousandth of the radius of our planet. The absolute duration of the year, then, measured in atomic seconds, does not alter by any significant amount. Here the intended meaning of the word *year* is the time taken to complete an orbit, and we will not worry about the sophistry which is necessary in appendix B.

(The data plotted here are derived from the calculations of T. R. Quinn, S. Tremaine, and M. Duncan, "A Three Million Year Integration of the Earth's Orbit," *Astronomical Journal* 101 (1991): 2287–2305.)

old, and wonder whether this deduction also applies over such a duration. The answer is that numerical experiments like that producing figure 6 cannot be conducted over periods of billions of years due to various inherent difficulties (for example, the accumulation of round-off errors in the computer codes leads to physical unreality). For this and other reasons it is not clear, in the field of celestial mechanics, that the solar system is stable over timescales of many billions of years. Nevertheless, this is a matter of an inability to demonstrate stability as opposed to any doubt that Earth has followed more or less the same basic orbit for the past few billion years.

All scientific experiments, though, are conducted under circumstances in which the experimenters must make certain assumptions, either explicitly or tacitly. One cannot carry out any experiment in mechanics without assuming that the law of the conservation of momentum is correct (and of course there has never been any demonstration otherwise, so it seems a good assumption). You might ask, then, whether any of the basic assumptions made in such computer simulations of the orbital evolution of the planets might be challenged.

One assumption is that the motions of the planets are affected only by their mutual gravitation, and that of the Sun. In fact, there is a vast complex of comets, asteroids, meteoroids, and dust in space, into which we are continually battering. On a dark night you might see ten shooting stars (meteors) per hour. Does not our accumulation of these things slow us down, like running into the wind? The answer is no, not significantly. The mass of Earth is so large that being struck even by a ten-mile asteroid, like that believed to have finally led to the extinction of the dinosaurs 65 million years ago, slows the planet down by about the same proportion as a speeding automobile has its motion curtailed by hitting a gnat.

It is the mass of the Sun which provides the major factor controlling the movement of Earth. What if the mass of the Sun changes? In computer programs like that mentioned above it is effectively presumed that the Sun has a constant mass. Let us consider whether this is valid.

At Earth's distance from the Sun, the solar flux is about 1.3 kilowatts per square meter. Knowing how far it is away from us, from that you could calculate the total power output of the Sun. Quoted in the first chapter was Einstein's famous $E = mc^2$, and now you can use it. The solar power output is the amount of energy E per second. Using the speed of light ($c = 300$ million meters per second), a few taps on your calculator will render the rate at which the Sun must be losing mass m through nuclear reactions in order to generate that power. The answer is 4 million tons per second.

That is, the solar mass is decreasing at a rate of 4 million tons per second through this avenue, and over millions and billions of years it is obvious that the mass lost must be enormous. But that is on the human scale, in which jumbo jets, skyscrapers, and passenger liners are vast constructions. In astrophysics, one must compare quantities with appropriate standards, in the same way that one does not compare a child's first scribbles with the plays of William Shakespeare. The mass of the Sun is about 2×10^{27} tons (two thousand million, million, million, million tons). Even over the age of the solar system it has lost rather less than one part in three thousand of its mass through the $E = mc^2$ mass conversion.

The Sun also loses mass through the solar wind, the stream of subatomic particles continually flowing outward from its surface and occasionally disrupting our radio communications. It has not all been one-way traffic, though, with larger lumps like asteroids and comets falling into the Sun from time to time. The bottom line to all of this is that over timescales of millions to billions of years the mass of the Sun has not changed appreciably, the terrestrial orbit has remained the same size, and therefore the absolute length of the year (as measured using an atomic clock) has not varied.

For the year to be altered substantially at any stage would require some major catastrophe changing Earth's orbit, such as a collision with a vast planetesimal hundreds of kilometers in dimension. It is thought that a Mars-sized lump struck primordial Earth in its early history, over 4 billion years ago, the debris ejected agglomerating to form the Moon, but since about 3.8 billion years ago it seems that we have not suffered any impacts of such extreme magnitude. The year lasting for about 31.557 million atomic seconds is an unchanging parameter, ticking away for millions and billions of years.

Tidal Drag

Tides are raised by the gravitational attraction of the Moon and, to a lesser extent, of the Sun. It should not be thought that only the waters of the oceans are affected in this way. The solid ground itself also swells up and down with the rotation of Earth under the lunar and solar pulls, although of course to a less noticeable extent. The sea tides are apparent because the fluid flow of water allows it to move sideways, eventually coming up against a landmass where the bank-up makes the difference between high and low tides obvious.

It is the Moon which has the greatest effect. Although its mass is much lower than that of the Sun, it is so close to Earth that the major periodicity of the tides (the lunar tidal day of twenty-four hours and fifty minutes) is that which elapses between transits of the Moon through the overhead meridian, rather than the twenty-four-hour solar day. Nevertheless, the Sun does have an effect. When the Moon is on the Earth-Sun line, at inferior or superior conjunction (the former a day or so before new moon, the latter at full moon), the gravitational pulls of the two act in tandem, producing the most extreme tides, termed *spring tides*. When the Moon is near first or last quarter the gravitational attractions act at right angles, resulting in the lesser-amplitude *neap tides*.

Why are there two tides a (lunar) day? The simple explanation is that the ocean swells towards the Moon on the nearside of Earth because that part is closest to our companion and the gravitational pull is strongest there, making the ocean bulge towards the Moon, whereas on the opposite side of the planet the lunar pull is least and so the ocean also bulges outwards there, but *away* from the Moon. The full scientific answer is somewhat more complicated, but that summary will do for our present purposes.

Our interest here is in the effect of the tides upon the length of the day. As the swell of high tide sloshes around the globe, meeting landmasses every dozen hours and a bit, its progress is impeded by that land and so a frictional effect is applied. Like a spinning ball which you touch with a fingertip, the rotation rate of the mass must slow down. The total frictional force may be huge when it is all added up, but again we need to scale that against the appropriate yardstick, in this case the gigantic angular momentum of the planet. Thus the slowing of the spin rate amounts to just a couple of milliseconds being added to the length of a day each century.

People tend to be familiar with the idea of *linear* momentum, but not so much with *angular* momentum. Linear momentum is why you need good brakes to stop your car. Similarly, an astronaut in space will hurt his or her thumb just as much as down on Earth if he or she inadvertently hits it with a hammer, despite the hammer having no weight in space; its linear momentum and *inertia* are the same in both locations.

For rotational motion, similar considerations apply. Any object may have a *moment of inertia* defined for revolution about some axis, and when it is spinning its angular momentum is the product of that moment of inertia and its angular speed. Tip a bicycle upside down so that you can spin the front wheel by hand. The moment of inertia of that wheel is a measure of its disinclination to start spinning, despite the well-greased axle. Once it *is* spinning, to stop it with your hand (or the brake) you have to take away its angular momentum.

The friction applied by the tides, like the bicycle brake, reduces the angular momentum of Earth. Angular momentum, though, like linear momentum, is a conserved quantity which can be neither created nor destroyed. In the case of the bicycle wheel, when you stop its spinning the angular momentum it possessed is transferred through one route or another to Earth, resulting in the spin rate of the planet changing. Strange to think, but true. When you get into your car and drive west, you speed up Earth's spin and decrease the length of the day. If you drive east, you slow it down. The quantities involved are too small to measure directly, but they can be calculated.

Getting back to the tidal effects, the drag these apply to the entire planet slows down its rate of rotation and thus decreases its angular momentum, but I have pointed out that angular momentum must be conserved. What is the solution to this quandary?

The Lengthening Month

The answer is that the total angular momentum of the Earth-Moon system is what is conserved. Considering this system in isolation (that is, ignoring the huge angular momentum of Earth due to its orbit about the Sun, which we have already effectively shown to be invariant), what happens is that tidal drag causes the spin angular momentum of Earth to be reduced, and to compensate, the orbital angular momentum of the Moon must increase by the same amount.

That angular momentum depends upon the the moment of inertia of the lunar orbit and its angular speed. The moment of inertia of an object has a form Kmr^2 where K is a constant depending upon the shape and various other considerations. The important things to note are the other terms. The mass is m, and since the lunar mass does not change, the distance r from Earth must increase. Due to orbital dynamics this increase in r slows down the angular speed of the Moon a little, but overall the effect is that the orbital angular momentum of our satellite is enhanced as it moves slightly further away from Earth.

The outcome of the slowing of the terrestrial spin rate under tidal drag, causing the day to lengthen and necessitating leap seconds, is therefore the gradual recession of the Moon, resulting in the duration of the month increasing. The increase above the present synodic period of near 29.530589 days is small, only about one twenty-fifth of a second between 1900 and 2100 according to theoretical projections, so that from the perspective of calendar design it does not need to be worried about. On the other hand, our slowing spin rate makes a leap second about every 18 months necessary. To look at it another way, the number of days in a month is actually reducing, because the day length is increasing faster than the month is getting longer, in terms of absolute units of time like atomic seconds.

I wrote that this is based on theoretical projections. One might wonder whether they bear any relation to physical reality. How could the predictions be checked?

One way is through *lunar laser ranging*. The Apollo astronauts left corner-cube retroreflectors on the lunar surface, and physicists at various observatories have used powerful laser systems to direct pulses of light at those mirrors, very precisely measuring the length of time until a tiny fraction of the huge number of transmitted photons makes it back to the Earth-based detector. The round-trip takes about 2.6 seconds, but can be measured to the thirteenth decimal place. This has allowed the rate at which the Moon is receding to be measured with utmost precision: about an inch and a half, or 3.7 centimeters, per year. The month is indeed getting longer.

The Timing of Ancient Eclipses

Such accurate measurements as those mentioned above have obviously only been possible in very recent times. How can we be sure that the understanding obtained is applicable over longer timescales, say the past several millennia over which civilizations have been constructing calendars? The answer comes from the timing of ancient eclipses, most especially total solar eclipses.

Let us refer back to figure 2 in chapter 9, showing the path of the eclipse of A.D. 664 across northern England. That was computed by F. Richard Stephenson of the University of Durham on the basis of many ancient records, allowing the deceleration of Earth's spin to be analyzed.

I mentioned above that the rate of day lengthening is by about two milliseconds per century, but note that this is a *deceleration* (a negative acceleration), not a simple *speed*. There are *two* dependencies upon the time: it is given *per day per century*. Actually, the correct current figure, which is variable, may be closer to 1.7 milliseconds per day per century, and the value is critical. This would mean that one century ago the day was 1.7 milliseconds shorter; two centuries ago the day was 3.4 milliseconds shorter; three centuries ago the day was 5.1 milliseconds shorter; and ten centuries ago the day was 17 milliseconds shorter. Stepping back those ten centuries the size of Earth's orbit was the same so that the absolute length of the year was the same as now, but Earth spun on its axis during a year by an amount equivalent to about 6.2 seconds (365.25 × 17 milliseconds) more than it does nowadays. In terms of an angle in the spin phase, that is about a fortieth of a degree.

Going all the way back to A.D. 664, one has to sum up the contributions for each of the intervening years. The accumulated time slippage is about five thousand seconds, which corresponds to a spin phase of around 20 degrees. This means that if Earth had spun consistently at its present rate, the eclipse path of totality would have been 20 degrees away in longitude, out over the Atlantic, and the eclipse would not have been visible from northern England.

Other eclipse paths may occur with all manner of orientation on Earth's surface, but generally they are only about a hundred miles wide. This allows them to be used as rather precise gauges of Earth's spin phase from the time well before accurate clocks were available. If we know of an eclipse having been observed as being total, say, in Athens in a certain year, but only being partial perhaps in both Thessalonica and Crete, then with modern computations of the orbits of the Moon and the Sun we can work out precisely where the path must have passed and therefore determine the spin phase for that year.

In this way the history of Earth's rotation over the past twenty-seven hundred years has been rather accurately reconstructed, with various applications in geophysical, astronomical, and historical studies. It is found that the long-term tidal deceleration of the terrestrial spin is broadly in accord with the values measured over the past decade or so using sophisticated modern methods, but that it is certainly variable.

There are several reasons for the variability. One is climatic. The example usually given for the conservation of angular momentum is the ice skater spinning on a spot; by spreading her arms she slows her spin rate, while by dropping them beside her body she spins faster. Similarly, the seasonal swelling and shrinkage of the atmosphere affects the rate at which the world turns. There are other consistent trends. One is that during the last Ice Age, which terminated about ten thousand years ago, the overburden of ice compacted the rock and soil layers below. Since that ice melted and ran off into the oceans, the previously compressed landmass has gradually been rebounding. Like the ice ballerina spreading her arms, our planet is slowing its pirouette, part of a cosmic dance lasting for millennia.

The Frequency of Leap Seconds

You may have thought my accentuation of the fact that the day-lengthening rate is a deceleration rather than a simple speed to be tedious, but there is method to my madness. We saw above that it has a practical significance in interpreting past eclipse records. The distinction is also important in modern timekeeping affairs.

Let me ask you this question. If the tidal drag remains the same, so that the day-lengthening rate does not alter, does this mean that leap seconds will continue to be inserted at the same rate? I would anticipate that most readers will answer *yes*. But the answer is *no*. Let us see why.

The atomic second is based upon the length of the astronomical day in 1900, but since then the spin rate has decreased slightly, and the day is slightly longer. The rate of slowdown is near 1.7 milliseconds per day per century. It is now about a century since 1900, and the day lasts for about 0.0017 seconds longer than it did back then. How often do leap seconds need to be inserted? The answer is given by the number of days it takes to accumulate a full one second of delay, which is simply the reciprocal of 0.0017; thus 588 days. That is a bit less than twenty months, and because leap seconds are inserted only at the end of June or December (making eighteen-month gaps with occasional twenty-four-month waits appropriate), and also because the slowdown of our spin is erratic, one can see why leap seconds occur with the approximate frequency which they do.

But now step forward a century. The year is A.D. 2100, and the deceleration has continued to be close to 1.7 milliseconds per day per century. Now, though, two centuries have elapsed since 1900, so that the day lasts for 0.0034 seconds longer than it did in that year which is used as the benchmark for atomic seconds. The reciprocal of 0.0034 is 294, and so leap seconds are required at an average rate in excess of one every ten months. And by A.D. 2400 a leap second would be needed every four months, on average, perhaps one leap second inserted after June 30 every year and two after December 31.

On these grounds one would imagine that somewhere along the line a redefinition of the atomic second will be made, so as to bring it back in line with the time taken for Earth to spin on its axis in some future year to be adopted as the new benchmark. But that has knock-on consequences. The fundamental definition of length, the meter, is given in terms of how far light travels in a certain length of time, so if you redefine the second you will be redefining the meter as well. Radio frequencies are given in hertz, or cycles per second, so that if you change the second you change the frequencies, too. Modern technology brings great advantages, but it brings great complications with it.

The Year and the Month over Geological Time

The matters discussed above pertain to the observations and measurements made in the present epoch using high-technology equipment and analyses of ancient eclipse records in a highly sophisticated manner. Do the results obtained pertain to longer time bases than just the past few thousand years? What about the eons before humans strode the earth? How many days and months would the dinosaurs have experienced in a year?

Imagine that one takes the currently measured recession rate for the Moon and extrapolates backward in time. First, back just a few thousand years: the synodic month gets shorter by about a second or two, not enough to be directly noticeable in ancient observations, and the Moon would have been closer by about a hundred meters, again an insignificant amount compared to its actual distance, which is about 4 million times greater.

But that last simple sum should pique the interest. Multiply thousands by millions and you get billions, and the solar system is only a handful of billions of years old. Reversing the arrow of time would lead to an expectation that the

Moon was actually in contact with Earth about 10 billion years back (that is, orbiting our planet but only just above the surface or atmosphere).

That is clearly nonsense in that Earth and the Moon have only existed for somewhat less than half that time, but it does throw up an interesting question. In reality one cannot just divide the lunar distance by the recession rate to get an idea of its age: the analysis is necessarily much more complicated because the recession rate varies nonlinearly with the rate of slowdown in the terrestrial spin.

When the sums are done correctly, the answer turns up that *if* the tidal drag and concomitant rate of day lengthening had stayed in accord with a proper extrapolation from the present back through the previous eons, *then* the distance to the Moon would have been zero about 2 billion years ago.

Which leaves us with another puzzle. In fact, we have multifarious reasons for being pretty well sure that the Moon has been a separate body in orbit about Earth for at least the past 4 billion years. It follows that the rate of day lengthening and hence the recession rate happens to be rather higher now than in the distant past.

Other geological investigations tell us why. For instance, the present distribution of continental landmasses leads to considerable tidal friction, higher than one might expect for randomly oriented landmasses. Just look at the Americas: they form an impenetrable barrier to the westward-moving tidal swell from the Arctic to latitude fifty-five degrees south, with only the Drake Passage providing a natural gap between Cape Horn and the Antarctic Peninsula. Hundreds of millions of years ago the primordial supercontinents of Laurasia and Gondwanaland were much more compact, allowing the tides to sweep around the globe largely unimpeded, such that the tidal drag was less than at present, and so the rate of change of day length was smaller.

How could we possibly know how long the day lasted, say, a hundred million years ago? The answer comes from fossils. Ancient corals and mollusks, especially marine bivalves, display growth rings which may be interpreted as being evidence of diurnal, monthly, and annual variations in respiration. For example, if a particular set of fossil shells dated to a limestone stratum 200 million years old—from early in the dinosaur era—show a broad variation in ring widths over a cycle which contains 380 distinct rings, then this may be interpreted as implying that there were 380 days in the year at that time. (Let me say it again: the *year* here is of uniform absolute duration; what is being discussed is the number of times that Earth rotates on its axis during that period.) Of course the matter is more complicated than that—for example, a coastal bivalve may have growth rings following the lunar tidal day rather than the solar day—but the idea is simple enough.

The idea may be simple, but it does not follow that the practice is straightforward. In figure 7, on page 354, various evaluations of the number of days in the year made by paleontologists using this basic technique are shown, dating back to about 440 million years ago. Such biological (as opposed to geological) evaluations of the day length can go back only to about 570 million years, because that is the time of the Cambrian Explosion, in which multicellular life suddenly appeared and proliferated. An example is the jellyfish-like fossils first identified in the Ediacara Formation of South Australia. Sedimentary rocks from earlier in terrestrial history bear traces only of monocellular life, the blue-green algae producing stromatolites.

The problem with these paleontological evaluations can be discerned from figure 7. The results published by various researchers seem to follow the trend expected if the tidal drag has remained sensibly constant throughout the past 500 million years (for example, with about 410 days per year at 400 million years ago), whereas that cannot have been the case due to the problem of the Moon colliding with Earth in such a backward-extrapolated model. This seems to be a case of wishful thinking on the part of the investigators: they knew about how many growth rings to expect, and that was how many they counted, but their expectation was based upon a false premise.

The discrepancy is emphasized when the results from modern (past million year) fossil corals and mollusks are considered. The results from those records indicate counts of between 353 and 366 growth rings per year, showing that such counts tend to miss some rings, which tendency would place the more ancient data even higher in terms of days per year. If you based your model of lunar recession on data adjusted in this way to account for "missing" rings, you might find the Moon at zero distance less than a billion years ago!

Turning to geological data, in particular *rhythmites* or layerings found in strata which represent tidal deposition of mineral sediments back even further in time than the Cambrian Explosion, the evidence is to the contrary of the paleontological results, but in accord with the expected day length of, say, 600 million years ago based upon various considerations such as the continental distribution at that time (and given that we know that the Moon has been a separate body for more than 4 billion years). The datum shown in figure 7 results from Precambrian rocks in South Australia, dated to 620 million years ago.

The rhythmites in these rocks have been analyzed by Adelaide geologist George Williams, who finds them to portray a period when there were 400 solar days in the year (plus or minus 7, the uncertainty resulting from the experimental method), 35 more than at present. Similarly, the synodic month lasted for 30.5 ± 0.5 solar days, one more than at present in terms of numbers of days, but actually a lesser amount of absolute time because the solar day was about 10 percent shorter (only 21.9 hours long). Permutating these values, in each year there were 13.1 ± 0.1 synodic months, compared to the present 12.37. The Moon would have been about 3.5 percent closer to Earth at that time than it is at present.

Recently other researchers have similarly determined the numbers of days in the year using even-older rocks from Utah and Western Australia. About 900 million years ago, it seems that there were about 420 days (each lasting 20.9 hours) and 13.5 synodic months in the year, and 2.5 billion years ago there were 466 days (of 18.9 hours each) and 14.5 months per year.

How many days in a dinosaur year? The answer is about 370 to 375, the value reducing substantially between when the first of these giant reptiles evolved about 215 million years ago and when the last of them disappeared 65 million years in the past.

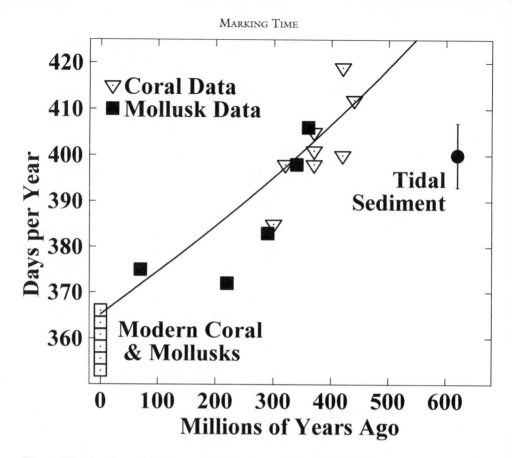

Fig. 7. The number of days in a year has varied substantially over geological time due to tidal friction. We saw in figure 6 that in essence Earth's orbit is of constant size, so that the year has not altered in duration by any significant amount: if there had been atomic clocks available 600 million years ago, they would have indicated a year very near 365.25 × 86,400 atomic seconds long.

Back then, however, Earth was spinning on its axis rather faster, there being about four hundred days (each lasting less than twenty-two hours) in such a year. Shown in this diagram are various evaluations of the number of days in the year derived by counting growth lines in fossilized coral and mollusks (mostly marine bivalves). The scatter in the points results at least in part from experimental uncertainties; for example, modern marine invertebrates have produced the range of values at lower left, with between 353 and 366 growth lines in a year being counted, whereas we know the number of days in a year now. (The above data were taken from Kurt Lambeck, *The Earth's Variable Rotation*, Cambridge University Press, 1980.)

The solid line indicates the number of days in a year which would have occurred if the currently observed trend of slowdown in Earth's spin rate had acted consistently over the past 550 million years, the solar tides being constant but the greater lunar tides when the Moon was nearer leading to the slight curvature of that line.

In reality, erratic variations in the tidal drag would be anticipated, and it is thought that the mean tidal deceleration in the past was somewhat lower than the trend indicated by the line, in accord with the apparent number of days per year in the Precambrian era (the datum at far right, derived by George Williams from tidal deposits of mineral sediments in South

The Aliquot Calendar

Why have I bothered spelling all of this out? In this book we have rather exhaustively traveled through time, through the attempts of different men and women to produce a calendar which follows the celestial cycles and yet is logical and designed for the convenience of human society and industry. Throughout we have seen that religious questions have been of import, often leading to the imposition of calendrical rules which are less than optimal from the perspective of convenience and astronomical veracity.

But our quest for simplicity and logic in the calendar has also been held back by something quite separate from religion. We have been stymied by the fact that the solar day and the lunar month are not an aliquot part of the year. That is, there is not an exact integer number of days in the year defined by Earth's orbit, nor lunations in that year.

My discussion above, though, shows that it has not always been this way. About 1.5 billion years ago there were precisely fourteen lunar months in a year, each lasting for thirty-one solar days, but there was no one around to notice the fact and construct a calendar based on it.

It would be nice to think that in about 200 million years' time there will still be humans living on Earth, and they will be able to devise an aliquot calendar based on a coincidence of precisely twelve lunations each of twenty-nine days in the year, a 348-day year with every nychthemeron (complete cycle of sunlight and nighttime) lasting for 25.2 of our hours. But the reality would not be that way, because the natural trend leads to the twelve lunations per year mark being reached when there is a non-integer number of days in the month, and vice versa.

On the other hand, if humanity does survive for long enough, then planetary engineering is a possibility: perhaps they will artificially increase the tidal drag on Earth to slow it more rapidly, while throwing some major chunks of the Moon into the Sun so as to lighten the former and thus make it recede faster than would otherwise be the case (because a lessened lunar mass would require that it move by a greater distance in order to take up the angular momentum loss of Earth). Then the aliquot calendar would be a possibility, although it seems like a high price to pay.

By that time, of course, humankind may be inhabiting new planets, spread

---

Australia; prior to the Cambrian Explosion about 570 million years ago, the absence of multicellular life means that there are no animal fossils which can be used in such investigations).

It has been suggested that the broad agreement between the fossil data and the backward extrapolation of the current deceleration trend may have been produced by wishful thinking on the part of the experimenters. First, the deceleration many millions of years ago is thought to have been substantially less than the modern value because the distribution of the continental landmasses would have produced less oceanic friction; and second, if that trend is continued for 2 billion years, the calculations put the Moon's orbit on the surface of Earth that length of time ago, whereas we believe with some confidence that it has been an independent body for over 4 billion years.

through the Galaxy, and all with their own days and years; and months, too, if they are orbited by moons. Look at our first stepping-stone in this regard, the red planet Mars. Mars has two natural satellites, the small, rocky lumps we call Phobos and Deimos. The former is so close to the planet that it takes less time to orbit than Mars takes to spin on its axis. There would be two "months," then, for people living on Mars, one of which lasts for less time than the martian day. On that basis, just imagine what a martian calendar would look like. But that's another story.

Chapter 23

❧

SHOULD 2100 BE A DOUBLE LEAP YEAR?

E lsewhere I have discussed the parameters which produce our seasonal climatic variations. Although the changing distance between Earth and the Sun has some effect, it is actually the obliquity of the ecliptic (the tilt of Earth's spin axis) which is the major cause of the seasons.

The basis of the Gregorian calendar reform was a desire to keep the vernal equinox on more or less the same date. To do so, the mean year length produced by a calendar must be as close as possible to the *vernal equinox year* rather than, as is assumed by many writers, the *tropical year*. If one wanted instead to keep the seasons in step, which might be interpreted as meaning the maintenance of all four of the cardinal positions (the vernal and autumnal equinoxes, and the summer and winter solstices) on as close as possible to the same dates, then one *might* use the tropical year, although this is a hopeless task because the lengths of the seasons so defined all fluctuate by several days over timescales of millennia. All this is detailed in appendix B.

The difference between the vernal equinox year and the tropical year is therefore critical in calendrical considerations, and so I have criticized Leroy Doggett for writing, in his chapter on calendars in the *Explanatory Supplement to the Astronomical Almanac* (1992), that "the *tropical year* is defined as the mean interval between vernal equinoxes. . ." although I could have chosen many other books and treatises where the same form of misstatement has been made. The difference is vital.

If you step down to your local library and pull out a copy of the *Explanatory Supplement,* you will find that Doggett continued by saying "it corresponds to the cycle of the seasons." If you thought that I was unfair to Doggett in singling him out, especially given that he is now deceased and unable to defend himself, you will be doubly aggrieved to learn that I am now going to criticize the final part of his sentence as well. You would have good reason in that I have never read any rational, scientific book in which doubt is cast on the use of the tropical year as being the "seasonal year," and indeed those terms are often used interchangeably. The book you are reading is an example: it has been implicitly assumed throughout (until now) that the seasonal year is the tropical year, and vice versa.

I have used Doggett's statement as a rhetorical device, to get your attention:

it is a clear statement, in an official joint publication of the governments of the United States and the United Kingdom, of a fundamentally important "fact" which we do *not* know, without any chance of error, to be the truth. It has been assumed, not proven.

At the back of the *Explanatory Supplement* is a list of many fundamental quantities which have been subject to either precision measurement or definition in a multitude of sophisticated experiments. For example, we know that the mass of the proton and the mass of the neutron are very similar, but distinctly different. Determinations of those quantities have been made, checked, and verified by disparate researchers in many laboratories in different countries. One may then ask whether the same level of rigor has been applied to the seasonal year, and the answer is *no*. At the time that Doggett wrote his chapter, the seasonal year (the recurrence period of the seasons as delineated by a climatic cycle on Earth, as opposed to an assumption of identity with an astronomical period like the tropical year) had not been determined.

Determining the Seasonal Year

How *could* one determine the seasonal year? One way would be to keep records of precise temperatures at one or more locations on a daily basis, and when some years' data had been accumulated, look for the cyclicity.

How many years? Well, let us imagine that we want to show that the tropical year rather than the anomalistic year is the appropriate yardstick to use for the seasonal year. They are both related to the seasons, and we have assumed heretofore that it is the former which provides the cyclicity because it clearly has the dominant influence (the seasons in the Northern and Southern Hemispheres are six months out of phase, and so on). To the fourth decimal place the tropical year is 365.2422 days long, while the anomalistic year is 365.2596 days in duration. Because their lengths vary over century timescales, there is no point in worrying about more decimal places. The difference between those values is 0.0174 days, the reciprocal of which is about 57 or 58. This implies that *if* it were possible to distinguish shifts of only one day in the phase of the climatic cycle, then temperature records stretching over sixty years or so might be adequate.

In reality, any scientist who collects observational data can tell you that there is always noise which blurs the signal which is the quest of the experiment, and in this particular situation everyone will recognize this to be the case: the temperature around any particular date is subject to substantial fluctuations from day to day and year to year. As the saying goes, climate is what you expect and weather is what you get. This means that a longer time base would be required in order to extract the signal from the noise, perhaps a few centuries being sufficient. Also helpful would be the availability of data from multiple separated sites where independent scientists have diligently recorded temperatures for many decades, passing on the baton to succeeding generations.

And that's just what we have in hand: for example, with the invention of the thermometer various English gentlemen started recording daily temperatures from around the middle of the seventeenth century as a fashionable thing to do, a series of observations which have been continued since then through to the present in one form or another. All that is needed is some skilled and imaginative physicist to exploit those data and decipher the underlying cycle, the seasonal year.

Analyzing Past Temperature Records

David J. Thomson (of AT&T Bell Labs at Murray Hill, New Jersey) realized this and set about putting the temperature records from various locations into a suitable format for analysis. The longest-term series of data is a set of weekly average temperatures from central England, as mentioned above. This data set stretches back to 1659, but there are problems in any analysis for cyclicity: being a weekly average (one value for every seven-day block), the reform of the calendar in England in 1752 meant that there was a phase jump of four (that is, eleven minus seven) days in the mean dates for records prior to that year. On the other hand, this discontinuity provides a useful test for the cyclicity analysis.

This sequence stretching back over more than three centuries provided Thomson with his longest time base, but he sought out temperature records from many other locations: three relatively nearby sites in continental Europe (Paris, Basel, and Geneva) from 1760 onward, others elsewhere from soon thereafter (like Edinburgh, Stockholm, and Milan), many in North America starting in 1880–1900 (like Banff, Calgary, and Edmonton in western Canada, and sites in Montana and North Dakota), from San Juan in Puerto Rico, Apia in Samoa, Brisbane and elsewhere in Australia, across the Indian Ocean in Madagascar, and earlier records from New Haven, Connecticut, Toronto, San Francisco, and many other large cities. In all Thomson had temperature sequences from 222 different stations, a veritable cornucopia of data waiting for its secrets to be teased out.

Thomson's results were published in the journal *Science* on April 7, 1995, in a paper entitled "The Seasons, Global Temperature, and Precession." I have some qualms about certain aspects of his analysis, but overall his conclusions seem clear. For most of the temperature records the cyclic period is equal to the anomalistic year and *not* the tropical year. As can be gauged from my preceding comments about the seasonal cycle having been *assumed* by all and sundry to be equal to the tropical year, this was not an expected result!

Various other phenomena were identified in Thomson's detailed analysis, such as the apparent fact that the increasing atmospheric carbon dioxide level due to human activity has caused the temperature cycles over the past sixty years to deviate in their relationship with the cycle of insolation from the behavior witnessed in the preceding centuries. In many ways this is stronger evidence for the existence of an anthropomorphic greenhouse effect than simple measurements of a small upward change in mean temperature over some decades, because of the influence of noise (natural fluctuations). The amplitude and phase of a cyclic phenomenon are much more sensitive discriminants, being less subject to noise.

How Could the Climate Possibly Vary with That Cycle?

It is the cyclicity period found by Thomson, however, which is of most interest to us here. One could ask how the cycle in the temperatures could possibly be governed by the time between perihelion passages (the anomalistic year). Certainly the solar energy striking the entire planet follows a cycle with that period, but how could that determine the temperature cycle at some locality when we know that the intensity of the sunlight at that latitude varies on a cycle equal in length to the tropical year?

Earlier in this book we have used theoretical models which are not entirely realistic in order to move toward an understanding, so let's use another one here. Imagine that the atmosphere of Earth were such that it could transport heat instantaneously, and with perfect efficiency. If that were the case, then the latitude of one's home would not affect the climate you experience. Say you were living at latitude forty degrees north, like Benjamin Franklin in Philadelphia, and it was the summer solstice. On that day you would receive the maximum solar radiation in all the year, but in our current theoretical model the energy absorbed would be instantaneously shared with all other parts of the globe. This would mean that at forty degrees *south*, even though the hours of daylight would be shorter and the sunlight less intense at that time of year, the temperature would be identical to that at its complementary northern latitude.

Under this scenario the temperature would be the same no matter where you were located: equator, North Pole, South Pole, or anywhere in between, daytime or nighttime. That would be the case on any particular day, but as the year progressed there *would* be a change in temperature because the total solar energy received by Earth is highest at perihelion and lowest at aphelion. Thus there *would* be seasons: the whole globe would have summer straddling perihelion in early January, and winter straddling aphelion in early July. This would be the situation, then, *if* the terrestrial atmosphere were perfectly efficient at instantaneously dispersing heat around the whole globe, so that the temperature equilibrates.

In reality, of course, no such thing happens; nevertheless, because the atmosphere does transfer *some* heat between hemispheres/locations, one should expect that the cyclicity of the seasons should contain two contributing frequencies, one being due to the tropical year and the other due to the anomalistic year. Since these are very close to one another it seems plausible that the climatic cycle at various locations (not all Thomson's two-hundred-plus stations show the same cyclicity) could latch onto the anomalistic year, in the same way as a piano with its middle A key mistakenly tuned to 221 hertz can cause a tuning fork for that note, fixed dead on 220 hertz as it should be, to start to resonate.

Will Our Seasonal Cycle Change?

Thomson's paper will not be the last word on this intriguing subject, and there is much still to be understood. In thinking about his result, I came up with an hypothesis which I think is worth considering, and which might be tested using some of the information which has been outlined elsewhere in this book.

My hypothesis is that the temperature cycle for most stations has latched onto the anomalistic year in part because in the present epoch Earth passes perihelion close to the time of the winter solstice; in fact, about twelve days later. The two separate components of the variation in solar energy (changing solar distance and tilt of the spin axis) therefore are not much different in phase at present. Actually, they are close to being at anti-phase for the Northern Hemisphere and in-phase for the Southern Hemisphere. This is only a temporary situation due to precession. Because the date of perihelion shifts by one day in fifty-seven years, about seven hundred years ago the phase relationship would have been perfect (perihelion passage and winter solstice on the same day). My hypothesis is that, all other things being equal, as perihelion slips away from the solstice the climatic cycle (the seasonal year) will latch instead onto the tropical year.

That makes my hypothesis testable, as all good scientific hypotheses should be, except that (1) it may take many centuries until the slippage is sufficient for the decoupling from the anomalistic year to happen; and (2) in the meantime human activities may cause a disruption of the train of events which would otherwise occur (it was noted above that Thomson suggested that carbon dioxide buildup has already resulted in a dislocation in recent decades of the trends portrayed in the temperature records from the centuries preceding). My alternative, of course, is to appeal to earlier times.

The Clue of the Nile Floods

Obviously there are no direct temperature records from before the deployment of the thermometer in the seventeenth century, but the possibility exists that there might be proxies which can indicate to us the length of the seasonal cycle in ancient times. How ancient? I am interested in times when perihelion was more than twelve days *prior* to the winter solstice, and that means earlier than fifteen hundred years ago, and preferably over three thousand years ago.

We have seen that the Egyptians had two techniques for measuring the year at that time and earlier. One was based on the heliacal risings of Sirius (the Sothic year), which occurred 365.2507 days apart (less than the sidereal year because Sirius is far from the ecliptic). The other was based on the annual rising of the Nile in flood, which depends upon the rainfall in Ethiopia. Clearly this is a measure of the seasonal or climatic year, and perhaps a suitable proxy.

But how long was that year? It has often been noted that these two phenomena (the heliacal rising of Sirius and the flooding of the Nile) occurred at about the same time around 3000 B.C., in mid-July. If the flooding followed the tropical year (365.2424 days is about right for the era in question, because it has gradually reduced since that time to the present 365.2422), then because this is 0.0083 days less than the Sothic year the date of the flood would move away from the date of the rising of Sirius by 1 day every 120 years, gradually getting *earlier*. Given the erratic variations in the precise date when the flooding starts, it might take several centuries before the deviation between the starts of the two years (the Sothic year and the flood year) would be noticeable. On the other hand, if the flooding actually followed the anomalistic year (365.2596 days), then the flooding would get *later*, by 1 day every 112 years.

So my hypothesis is testable: did the ancient Egyptians notice the average date of the start of the Nile floods gradually shifting relative to the heliacal rising of Sirius, and if so in which direction (earlier or later)? Did the effects of the tropical and anomalistic years average out for an extended period so as to maintain the inundation at close to the date when Sirius reappeared, allowing the reinforcement of belief in Sothis as a life-giving god? When did the coincidence between the rising of the river and that of the star cease, and when did the flood deviate from the summer solstice and move later in the year, rejoining the heliacal rising of Sirius in late July? I will leave those questions as an exercise to an interested Egyptologist.

The Need for Another Leap-Year Day

So far in this chapter I have skirted around the obvious calendrical question, leaving it as the final substantive matter to be considered in this book. If (1) the seasonal year may actually be identified with the anomalistic year, and not the tropical year as most calendar designers have assumed, and (2) we abandon the Easter question as the major criterion for determining the design of the calendar, instead just insisting that a revised calendar should have an average year length which reflects the climatic cycle, then the average calendar year for which to aim is as close as possible to the anomalistic year, 365.259635 days in the current epoch and not changing by much.

The obvious point to note is that this is more than 365.25 days, so instead of *losing* some leap years, as did the Gregorian reform of the Julian calendar, in fact we need to *add* some. The length of the anomalistic year is only about 1 part in 2,750 different from 365.26 days, which in turn is precisely 0.01 days longer than 365.25, and rather conveniently suggests that we need one extra day every century.

The Millennium/Y2K Bug (in computer date reckoning) and various other considerations have shown us that if there is ever to be another reform of the calendar, it must be planned with a substantial lead time so as to ensure that current software is not just amended, but actually superseded and replaced before the change in dates comes about. If it is a once-in-a-century event, it has to be in the end-century year, and so I suggest that we declare every year ending with two zeroes a *double leap year* containing 367 days. The Gregorian calendar cycles in 400 years, returning at last to the same day of the week. This new "double leap year every century" rule would lead to a full cycle in 100 years, since a century would contain 36,526 days, that being divisible by seven so as to render 5,218 weeks.

Rather than being a Gregorian common year with only 365 days, the year 2100 should contain not only a February 29 but also, to allow the celebrators time to get over their excesses, a January 0. Of course, I will not be around to see the dawn of that day. But I plan to request that, when I die, I am buried with my tongue firmly embedded in my cheek.

EPILOGUE

As one of the great scholars of ancient chronology, E. J. Bickerman, has written, "A calendar is a tool which cannot be justified by either logic or astronomy." The way our calendar has evolved has been almost a random walk, with twists and turns unpredictable. It is a human construct only loosely based upon considerations of the physical heavens, with beliefs about the divine heavens having had a larger part to play in defining the path which has brought us to where we are now, calendrically speaking. As to the future route humankind will take, I very much doubt that questions of science and technology will have a large part to play.

Our various calendars are part of the music and art of human life, not the stark science. Calendar reformers tend to be an indomitable bunch, their philosophy being encapsulated by that which William Shakespeare wrote in *Hamlet:*

> The time is out of joint; O cursed spite,
> That ever I was born to set it right!

The reformer refuses to be dismayed by official reluctance to permit the arguments for altering the calendar to be heard; turning again to the bard, and this time *Richard II:*

> Thoughts tending to ambition,
> They do plot unlikely wonders.

And so it is, I am afraid, for all earnest calendar reformers who argue on the grounds of common sense, efficiency, and logic. As Shakespeare continued a few lines later:

> How sour sweet music is
> When time is broke and no proportion kept.
> So is it in the music of men's lives.
> And here have I the daintiness of ear
> To check time broke on a discordant string;
> But for the concord of my state and time
> Had not an ear to hear my true time broke.
> I wasted time, and now doth time waste me,
> For now hath time made me his numbering clock.

Appendix A

⌘

HOW LONG IS A DAY?

The day is *not* the time it takes Earth to spin on its axis, as most people think. In actuality Earth's orbit around the Sun is not circular, and that affects the length of the day at different points in the year. Also, Earth's spin axis is tilted to its orbital plane, and that in turn affects the day length. These two factors result in the duration of a day changing in a cyclic fashion as the year passes. Only four days each year are really twenty-four hours long (one in mid-April, one in mid-June, one in early September, and the last in late December). Some are longer and some are shorter, but for convenience we define them all to be of the same duration, that duration being twenty-four hours on some form of clock which we know to be accurate. The matter of clocks is addressed in appendix C.

The Solar and Sidereal Days

For now, imagine that Earth's orbit is a circle, and that it spins on an axis perpendicular to that orbit. Further, we will assume that the year (the time taken to return to the same point on that circle) is 365 days long. For our first simple step in defining the day, then, the situation is as in figure A1, on page 366. Note that under these circumstances there would be no seasons, the hours of daylight would always be the same, and if you stood on the equator the Sun would consistently pass overhead at midday, while if you stood at the North or South Pole the Sun would always be on the horizon.

Imagine yourself on this idealized Earth. Each time Earth spins on its axis the stars do a complete loop around the sky. In essence these form a stationary backdrop against which one may measure the spin of Earth.

It takes Earth 365 days to lap the Sun, and there are 360 degrees in a circle. That's not a coincidence: the 360 degrees in a circle derives from the early Sumerian/Babylonian calendar, which had just 360 days, each of them divided into 360 parts, and our degree symbol (as in 90°) is just their representation of the Sun. Earth moves about 1 degree in a day, then.

In figure A1 the positions of Earth separated by ten days are shown. The length of the day is defined by the time between meridian passages of the Sun. In ten days Earth must spin through ten complete rotations (or 3,600°) plus almost another 10°, making 3,610° in all. In just *one* day it spins through an angle of about 361°: at 360° the stars return to their previous positions, but it is not until 361° that the Sun comes back to the meridian because Earth has moved along in its orbit. The length of the day is therefore *not* the time it takes Earth to spin on its axis; it is actually a little bit longer.

The time it takes the Sun to return to the meridian is called the *solar day*. For our simplified description, the solar day would be constant throughout the year. In reality it varies,

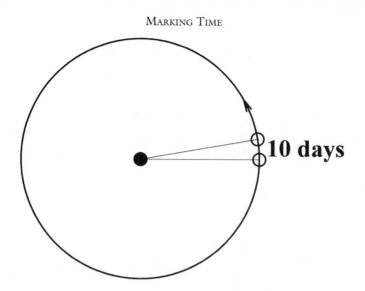

Fig. A1. As an initial simple step, Earth's orbit about the Sun is assumed to be circular, the year 365 days long, and the terrestrial spin axis perpendicular to the plane of the orbit. In 10 days Earth moves through the distance shown. During these 10 days Earth spins ten times *plus* an angle of about 10°, so as to bring the Sun back to the meridian. The length of the day is therefore slightly more than a single rotation of Earth, and in 365 days our planet spins on its axis 366 times.

but we are not worrying about that yet. When I wrote that we would assume here that there were just 365 days in a year, what I was implying was 365 *solar* days. That is, the Sun would cross the meridian 365 times. But how many times would the *stars* cross the meridian during the same period? The answer is 366. That is, Earth spins on its axis 366 times during 365 days; you've only seen the Sun cross the meridian 365 times because the extra terrestrial rotation is taken up by the orbital revolution of our planet around the Sun. The time between the stars returning to their positions—say, the time between successive meridian transits of a particular well-known bright star—is called a *sidereal day*, and it is 365 divided by 366 = 0.99727 times as long as a solar day; in fact, a little less than four minutes short of twenty-four hours.

The time according to the Sun (solar time) and the time according to the stars (sidereal time) therefore proceed at different rates. Sidereal days end progressively four minutes earlier throughout the year, being in agreement only at the vernal equinox (which we will need to define later).

The Effect of Our Noncircular Orbit

We are now ready for our next step up in complexity. For the sake of round numbers, we will maintain the pretense that there are exactly 365 solar days in a year, and we will persist in considering the equator to be parallel to Earth's orbit. Now we will consider the effects of the noncircularity of the terrestrial orbit.

Astronomers have a measure of the deviation from circularity of an orbit, called the *eccentricity*. This word should not be thought of in the same way as it is usually applied to human nature, implying "weird" or "strange." A circular orbit has an eccentricity $e = 0$, while a highly elongated parabolic orbit (like that of a new comet coming into the inner solar system for the only visit when we'll have a chance to see it) has $e = 1$. The planets all have low-eccentricity elliptical orbits. If one counts tiny Pluto as being a major planet, then it has the largest eccentricity, $e = 0.2524$.

Leaving Pluto aside, Mercury has the most eccentric orbit, $e = 0.2056$, followed by Mars with $e = 0.0933$. Venus, however, has an almost circular orbit, with $e = 0.0068$. Thankfully—I write *thankfully* because otherwise our planet would have even greater seasonal variations—Earth also has a low eccentricity, $e = 0.0167$. That's not much of a deviation from a circle, but it is significant in many respects.

Figure A2, on page 368, shows—somewhat exaggerated for clarity—the elliptical orbit of Earth about the Sun. With the circular orbit of figure A1 the speed of the planet would be constant, at 29.8 kilometers per second (km/sec), and that is indeed close to our average speed about the Sun. In our true elliptical orbit, though, our speed actually changes in the same way as a child on a swing is moving fastest when closest to the ground, and slowest when at his or her extreme points. The closest approach to the Sun (which is termed *perihelion*) results in a speed near 30.3 km/sec, while at the furthest point (*aphelion*) Earth is moving at only 29.3 km/sec.

Because orbits are generally not circular, one cannot think of them as having a radius. Rather, the size of an orbit is stipulated by the semi-major axis (denoted a), which is half the distance between the perihelion point q and the aphelion point Q in figure A2. Earth has $a = 1$ astronomical unit (A.U.) by definition, a distance equivalent to a little less than 150 million kilometers (93 million miles).

For Earth's true elliptical orbit there are only two times at which we are moving at right angles to the solar direction: at perihelion, and at aphelion. Between aphelion and perihelion the distance between Earth and the Sun is continually (but not constantly) decreasing. This means that the angle which Earth must spin through to bring the Sun back to the meridian is slightly less than 361 degrees and the days are shorter. Conversely, as the Earth-Sun distance increases while the planet is moving from perihelion to aphelion, the angle which needs to be turned through is greater than 361 degrees, and the day length is increased above twenty-four hours. This is one of the two effects which cause the length of the day to vary during the year.

Figure A3, on page 369, shows the effect of the eccentricity of the orbit upon the length of the day. The deviation is zero when we are at perihelion, which is around January 2 or 3 in the present epoch (twentieth/twenty-first century), and also when we are at aphelion, around July 4 or 5.

The Influence of Our Tilted Spin Axis

Now on to the second effect which varies the day length during the year: the tilt of Earth's spin axis. Up until now we have been assuming that our planet rotates about an axis perpendicular to its orbital plane (the ecliptic). Of course, in reality it doesn't. There is an angle of 23.4 degrees between the spin axis and the perpendicular to the ecliptic. In astronomical terms this is called the *obliquity of the ecliptic* (or just the *obliquity*) and it is usually denoted by the Greek letter ϵ (epsilon). It varies between limits of 22.1 and 24.5 degrees on a cycle lasting forty-one thousand years, so that studies of ancient calendars or megaliths (and the climate) require knowledge of the obliquity in the epoch in question.

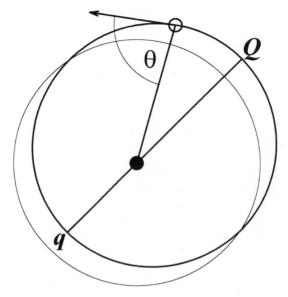

Fig. A2. Earth's orbit is not precisely circular, but actually an ellipse. The circle (lighter line) is shown here so as to indicate the deviation between Earth's true elliptical orbit and a circular path (as used in figure A1). For clarity, the eccentricity of the orbit is exaggerated here (Earth actually has $e = 0.0167$, whereas the orbit shown here has $e = 0.2$). The perihelion point (closest approach to the Sun) is labeled q and the aphelion point (furthest distance from the Sun) is Q. The arrow shows the direction of Earth's motion at a particular instant. Moving from Q to q the solar distance is reducing and angle θ is slightly less than a right angle, while from q to Q Earth is receding from the Sun and θ is larger than a right angle. These small variations affect the angle through which Earth must rotate to bring the Sun back to the meridian, and therefore change the length of the day.

Previously we were considering the effect of the finite eccentricity of Earth's orbit and neglecting the tilt of the spin axis. Now we'll do things the other way around, and resume with the orbit taken to be circular. Figure A4, on page 370, shows the orientation of the spin axis at four points on this (assumed circular) orbit, now viewed at a slant angle rather than in plan view as was the case in figures A1 and A2. These four are labeled *VE, SS, AE,* and *WS* (for vernal equinox, summer solstice, autumnal equinox, and winter solstice, respectively). Despite the fact that I live in the Southern Hemisphere, so that my seasons are six months out of phase with those in the north and the warmer weather occurs around the time defined as the winter solstice, I will maintain that terminology. Indeed astronomical parlance requires this; no matter where you live the terminology is based on

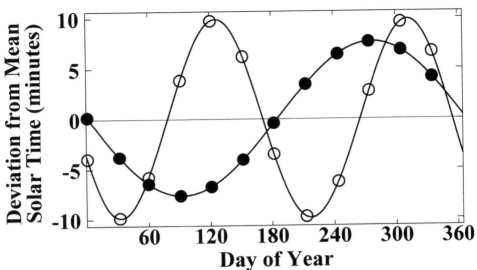

Fig. A3. Deviations from mean solar time (or clock time) of the instant of solar transit through the meridian due to (1) the noncircularity of the terrestrial orbit (solid symbols, located for the first day of each month), and (2) the tilt of Earth's spin axis (open circles). For the former, as our planet recedes after perihelion passage in early January the Sun is delayed in reaching the meridian until Earth passes aphelion in early July, thereafter being earlier in meridian transit than would otherwise be the case. For the latter, the north-south motion of the Sun affects its east-west velocity and therefore the time at which it gets to the meridian.

the seasons in the Northern Hemisphere, and parochial petulance on my part would be pointless.

In figure A4 it can be seen that the spin axis remains pointed in more or less the same direction in the sky throughout the one year represented, that point actually being close to Polaris, the north polestar. The direction of the spin axis moves slowly due to various perturbations. First, it gradually swivels around so as to describe a circle on the sky whose angular diameter is 2ϵ. This is the loop at top in figure A5, on page 372. Looking down from the north the direction of traversal of this circle is clockwise, the opposite direction to the orbital motion of Earth. This gradual movement is called the *precession of the equinoxes*. It takes about 25,800 years to complete, meaning that the north celestial pole was *not* near the star we call Polaris when, say, the pyramids were built in Egypt almost 5,000 years ago. The spin axis never points *directly* toward Polaris, reaching its minimum separation in about the year A.D. 2100. The precession of the equinoxes may be thought of as being similar to the familiar precession of a toy gyroscope, and is due to the tugs and torques imposed on Earth by the gravitational attraction of the Moon, the Sun, and the planets.

The second effect upon the spin axis has already been mentioned: over long periods of time there is a gradual, systematic change in ϵ, the obliquity of the ecliptic. The gravitational tugs of the other massive bodies in the solar system not only cause the orientation of that angle ϵ to swivel around, but they also cause its value to decrease (in the present epoch) by about one-eightieth of a degree per century. This may seem like a small amount, but it can have practical consequences. The tropic of Cancer is at the latitude equal to ϵ. Imagine that you are on a vacation in, say, Mexico, or Taiwan, or at the Aswan Dam in southern Egypt, and you have your photograph proudly taken next to a sign saying "You are standing on the

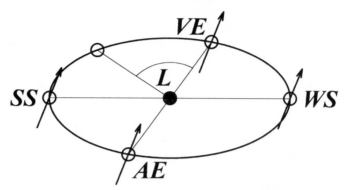

Fig. A4. A slant-angle view of the orientation of Earth's spin axis at different points in its orbit. To a reasonable approximation the spin axis maintains its pointing direction throughout a single year, although over much longer periods of time this moves. The winter solstice (*WS*) is when the spin axis is tipped as far away from the solar direction as possible, and the Sun then attains its maximum southerly declination. At the vernal equinox (*VE*) the spin axis is in the plane perpendicular to the solar direction and the Sun crosses the equator (declination zero). The summer solstice (*SS*) and the autumnal equinox (*AE*) are similarly determined. The position of Earth in its orbit is defined through the solar longitude (*L*), which is measured from the direction of the vernal equinox.

Tropic of Cancer." Even if the entrepreneurs responsible for that sign got its position correct when it was erected, within a year the tropic will have moved fourteen meters south due to the change in the value of ϵ. The Arctic Circle is at a latitude $90° - \epsilon$.

Third, there is a repetitive wobble in the value of ϵ, which is termed *nutation,* caused largely by variations in the torque imposed on Earth by the Moon due to (1) the fact that our natural satellite has a geocentric orbit which is not circular (that is, the lunar orbit about Earth has a finite eccentricity—$e = 0.0549$—in the same way as Earth's orbit about the Sun has $e = 0.0167$); and (2) the nonspherical shape of our planet (it has a pronounced equatorial bulge, and is slightly pear-shaped). The nutational shifts in the locations of the tropics are up to about 280 meters, with a mixture of different cyclic periods of up to 18.6 years.

Finally, there are various small, slightly erratic changes in the direction of the spin axis which are called *polar motions*. The changes in ϵ discussed above alter the borders of the tropics and the Arctic/Antarctic regions, but not the locations of the poles or the equator. The polar motion does move these, the pole spiraling around with an amplitude of a few meters or so; this is caused mainly by redistributions of the mass of Earth. For example, summer in the Northern Hemisphere leads to a heating and therefore slight swelling of the atmosphere, which one would expect to slightly slow down the spin rate of the planet due to the conservation of angular momentum (as when an ice skater stretches out his or her arms to slow his or her spinning on a spot, and brings them in to speed up again). However, this heating and swelling of the air occurs in particular over land, so that the effect is asymmetric and the spin axis is shifted slightly. There are many similar effects at different times of year, such as the filling of lakes, the buildup of ice and snow in winter, and the rebound of the soil as the snow melts in spring. Internal mass redistributions of Earth are also significant, over long timescales: the convective rise and flow of rock in the mantle, for example. On the timescale of a year the erratic wandering of the pole is, as mentioned above, by an amount of order a few meters. This will also contribute such small changes to the true locations of the tropics and the polar circles.

Considering the above influences, one can see that it is not possible to define easily the location of the tropic of Cancer, say, with a precision of better than a few hundred meters at any time, unless one wants to invest lots of time and money in continually calculating and observing how it is moving, with reference to the distant stars and galaxies. Indeed, this is basically how the polar motion mentioned above is determined: the expected terrestrial spin axis orientation is calculated based upon known systematic perturbations, and precise measurements of its actual orientation are made using radioastronomical observations of extremely distant quasars. The difference between the calculated and the observed spin axes give the actual polar wander, which is some meters on the ground and variable in direction and amplitude. Remember that if you ever get to the North or South Pole. Similarly, if someone wants to take your photo straddling the equator with one foot in each hemisphere, you would need long legs to be sure of doing so.

The Motion of the Sun

All of the above effects we can lay aside for the present, since we are concerned only with the apparent motion of the Sun as seen from Earth during one year. From our terrestrial perspective, the Sun seems to move north and south during the year, and from the east to the west during the day. The Sun reaches its furthest northerly point at the summer solstice, which is around June 21. For a Northern Hemisphere observer at a temperate latitude (in Europe or the United States, say) this means that it comes closest to the zenith (the point directly overhead) as it transits the meridian on that date; for an observer on the tropic of Cancer the Sun passes directly overhead at the summer solstice. The word *solstice* is derived from the Latin *sol stetit,* meaning "the Sun stood still," which is what happens: it stops its northerly march and begins to move south again.

Referring to figure A4, the summer solstice is when the Northern Hemisphere is tipped as far over toward the Sun as it can be, and the winter solstice is when the Northern Hemisphere is tipped as far away from the Sun as it can be. Looking now at figure A5, points *A, B, C,* and *D* are on the terrestrial equator, and so are on a plane which we call the *equatorial plane.* This plane may be extended out into the sky (the celestial sphere), and often astronomers just call it the *equator.* This is another place where the unwary may get confused by terminology: usually by *equator* one understands just that circular line around Earth's midriff, as shown in figure A5, whereas in astronomical parlance the term refers to the entire plane on which that circle lies, and which may be extended out into space to an arbitrary distance. It cuts the sky in half, north from south. Perpendicular to the equator (to the equatorial plane) is the plane containing *A, B,* and *P.* The solstices occur when *that* plane also contains the Earth-Sun line. That fact can be used to define the *instant* of a solstice.

What Is the Equinox?

The equinoxes can be defined in a similar way to the solstices (as instants in time), and as we will see later by far the most important juncture to determine is that of the vernal equinox. The discussion both above and below is aimed at making it clear how that equinox is defined. First, though, a slight detour regarding the meaning of the term *equinox* in the present context.

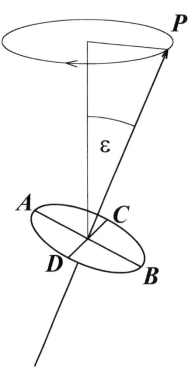

Fig. A5. The tilt of the terrestrial spin axis is called the obliquity of the ecliptic (ε). Gravitational perturbations cause its direction (defined by direction *P*) to precess in the clockwise direction around the circle shown at top, a full loop taking about 25,800 years. Perpendicular to direction *P* is the terrestrial equator, on which the points *ABCD* lie. At the instant of the vernal equinox line *CD* points directly toward the Sun, which is then crossing the equator moving north.

Contrary to popular belief, and contrary to the Latin derivation of the word (meaning "equal night"), the equinoxes are *not* the days on which the lengths of daylight and nighttime are equal. The hours of daylight may be taken to begin when the uppermost point of the Sun rises above the eastern horizon, and end when the last visible part of the solar disk dips below the western horizon. There is also the complication of twilight, which has various definitions: *civil* twilight begins and ends when the center of the Sun is six degrees below the horizon, *nautical* twilight at twelve degrees below, and *astronomical twilight* at eighteen degrees below, and the durations of each will vary during the year because the angle at which the Sun approaches the horizon varies seasonally.

Leaving twilight aside, we proceed with the definition that it is daylight if some part of the Sun may be seen above the horizon. First, the Sun has an apparent diameter of about half a degree. The uppermost point is therefore a quarter degree above the center of the disk, upon which one bases calculations of its position, and that point will rise about a minute before the center does (Earth's rotation rate is about one degree in four minutes).

Second, at the horizon one is looking through a long path length of atmosphere, which has a number of effects. Sunlight at the blue end of the spectrum is preferentially scattered out of the beam (which is why the sky is blue), leaving a reddened sun. The effect of concern here, however, is *refraction,* or the bending of the light path. This is similar to the phenomenon seen if you stick a spoon at an angle into a glass of water: the shaft of the spoon seems to be bent at the water surface, because light passing from air to water and vice versa is refracted. (Refraction is a wonderful thing—without it spectacles wouldn't work!) Returning to the refraction of sunlight in the atmosphere, the effect is that the long path through the air acts like a huge lens, and one can see the Sun rise while in reality it is still below the horizon. That is, if there were no atmosphere the Sun would still be hidden, but with things as they are we can see the Sun a little early. For typical atmospheric conditions this refraction effect moves the image of the Sun by a little over half a degree, and the overall effect (refraction plus the uppermost point emerging) is that at the equinox sunrise is several minutes earlier than would be the case for a point light source (a distant star) observed from an airless planet. And of course the same thing happens at sunset.

Overall, on the day of the equinox the daylight time (defined from the first to last sighting of part of the Sun) is typically six to ten minutes longer than the nighttime. At latitudes near forty degrees this means that the dates when the daytime and the nighttime are both twelve hours are about four days away from the equinoxes; for example, about March 17 in the Northern Hemisphere and March 25 in the Southern Hemisphere for the vernal equinox, and similarly splayed around the equinox in September. On the other hand, the winter solstice *does* have the shortest amount of daylight time, and the summer solstice the most.

Having brought up the question of latitude, and whether Northern or Southern Hemisphere, let me say that henceforward I will generally be making comments which are applicable to the situation at a latitude of forty degrees north. That seems to be an equable value, given that most of my readers will be in North America or Europe. While I might live "Down Under" now, I did live for several years in Boulder, Colorado, which is dead on the fortieth parallel. If you go there and stand on Baseline Road, you will be more or less right on that line. I am not so familiar with the geography of Philadelphia, or Beijing, so I cannot say whether the fortieth parallel is similarly marked in those cities. Two questions for the reader, though: how does the fortieth parallel move with (1) variations in the obliquity of the ecliptic, and (2) polar motion? (Answers: [1] not at all; [2] by amounts up to about the width of the road.)

When Are the Equinoxes?

Now that we understand that the equinoxes are not the times when the days and nights are equal lengths, we can consider how they are actually defined. In fact, like the solstices, they are *instants* of time. Depending on your longitude and the leap-year cycle, the vernal equinox occurs at some instant on March 19–21.

Looking at figure A4, the equinoxes occur when Earth is side-on to the Sun. More technically, in figure A5 we see that this is when the plane containing A, B, and P is *perpendicular* to the Earth-Sun direction. Since the line CD is orthogonal to that plane, it means that CD points directly toward the Sun at the equinox. Astronomers use that fact to define the *longitude* of the Sun, denoted as L in figure A4. (Do not confuse this with the *geographical longitude* used to stipulate locations on Earth.) The solar longitude increases from $L = 0°$ at the vernal equinox, becoming 360° and hence 0° again when a full revolution around the Sun has been completed. The longitude is just the angle between the direction of the Sun and the direction of the vernal equinox, the latter direction often being called the *first point of Aries* by astronomers (although precession has moved it from Aries to the extreme of Pisces, from where it will soon move into Aquarius, hence the "Dawning of the Age of Aquarius" in the musical *Hair* and other astrological babble). Because, as we have seen, the speed of Earth in its orbit varies during the year, the rate of change of longitude L is not constant.

After the autumnal equinox and through the winter solstice to the vernal equinox the Sun has always been south of the equatorial plane, the plane containing $ABCD$ in figure A5 and extrapolated out into the sky. Ever since the winter solstice the Sun has been moving north, and the vernal equinox is defined as the instant at which it crosses the equatorial plane. There are two planes intersecting each other at that angle ϵ, the ecliptic (to which the Sun is constrained) and the equator, and the line of their intersection is CD. The equinox is when this line of intersection is passed. Since the equinoxes and solstices are merely times, one can equivalently represent them as points on the orbit, as is done in figure A4. Because they are constrained to the plane of Earth's orbit, the ecliptic, they all have latitudes (ecliptic latitudes) of zero. Their longitudes are spaced by 90° steps.

Back to the Varying Day Length

We can now resume our discussion of how the length of the day varies during the year. The Sun rises in the east and sets in the west every day, giving it a predominant east-west motion. But between the summer and the winter solstices it is gradually moving south, and between the winter and the summer solstices it gradually shifts north; this causes an effect upon the east-west motion which we use to measure the time of day. Even if Earth had a circular orbit the Sun would still only be on the meridian at noon on four days: those of the solstices and the equinoxes. Between a solstice and an equinox the Sun would seem slow relative to a standard clock (not reaching the meridian until after noon), while in the period between an equinox and a solstice the Sun would be fast relative to the clock (already having passed the meridian by noon). This variation is shown in figure A3. It is clear that the curve for the effect of this tilt of Earth's spin axis undergoes two cycles in each year, whereas the curve for the effect of the shape of Earth's orbit passes through only one cycle in that period.

Apart from the fact that one of these effects varies twice as fast as the other, there is another important point which is obvious from figure A3: they do not have the same start point. In scientific terminology we might say that they do not start *in phase*. For convenience here we may note that the winter solstice is on or about December 22, whereas the date of perihelion is around January 3, so that there is a gap of about a dozen days between them. The Gregorian calendar was defined so as to keep the vernal equinox on or near the same date, and similarly the winter solstice remains on the same date, or very close to it. On the other hand, Earth's orbital orientation precesses such that over many decades the date of perihelion shifts later in the year, a matter we deal with in appendix B: figure A3 is correct now, but if an historian pulls this book from a dusty shelf in several centuries' time, one of the curves will need to be shifted to fit the perihelion date in that epoch.

The combined effect of these two influences upon the position of the Sun is called the *equation of time*. That's a confusing name because it is not usually presented as an equation, as such, but rather as a plot like figure A6, on page 375. In essence, the equation of time is the difference between the time shown by a clock which runs at a constant rate with each day of the year being twenty-four hours long, and the time as indicated by the actual position of the Sun. For any particular place the latter is called the *local apparent (solar) time* (LAT), while the former is the *local mean time* (LMT). These are related, then, as LAT = LMT + equation of time.

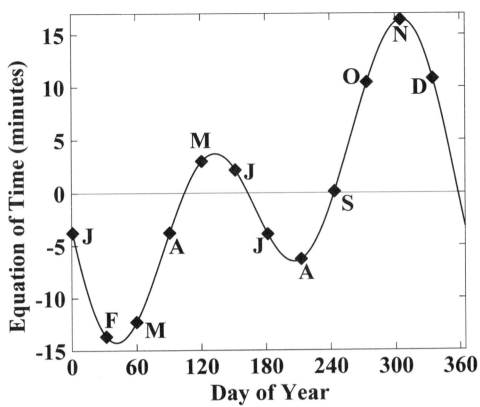

Fig. A6. The two different effects shown in figure A3 combine to produce the equation of time, whose values for different dates throughout the year are shown here (the diamond-shaped symbols represent those on the first day of each month). The equation of time is the difference between the time shown by our clocks, based on the fictitious mean sun, and the time according to the real sun. It is seen that at certain times of year the Sun is up to sixteen minutes away from the position which our clocks presume.

Recapitulation

It is worthwhile to slow down a little here, and recognize exactly what is going on. Although we imagine that our days proceed in step with the movement of the Sun across the sky—as they did many centuries ago before the invention of clocks—in fact we have effectively abandoned the Sun for timekeeping purposes. Our clocks proceed at a rate which is based upon an entirely fictitious entity, which is called the *mean sun*. This is the *average* position of the Sun in the sky, the place it would be if Earth's orbit were circular and the obliquity (tilt of the spin axis) were zero, so that the Sun would proceed at a constant rate along the equator.

The mean sun is only in the same place as the actual sun four times during the year. These are on the dates in figure A6 where the equation of time is zero, around April 16, June 15, September 1, and December 25 (but that is *not* the reason for Christmas being on the last of those days). On the other hand, the equation of time has values reaching four-teen minutes on about February 12 and sixteen minutes on November 4.

One can now see the origin and meaning of the well-known phrase *Greenwich mean time* (GMT). This is the mean time (the time from the mean sun) at Greenwich. The idea of a standard reference time for the world, as GMT served for many years, has advanced and be-come more sophisticated, and so to avoid any ambiguity the term *coordinated Universal time* (abbreviated as UTC) is now used, with a definition similar to that of GMT but based upon atomic clocks rather than the Sun.

The Time Strictly by the Sun

While the equation of time indicates that the difference between LAT and LMT may be as much as sixteen minutes, this does not mean that the length of the day varies by that much from one day to the next. Referring to the plot of the equation of time shown in figure A6, the sixteen minutes is the total amount accumulated over many days. The actual variation in day length is rather less.

We take twenty-four hours of mean solar time as the length of the day—called the *mean solar day*—by which we set our clocks, whereas an *apparent solar day* at any time of year will be defined by the time between successive solar transits, or sunrises, or sunsets, depending upon the marker point we choose to employ. The maximum deviation between the length of an apparent solar day and twenty-four hours of clock time is given by the biggest step be-tween two successive dates in figure A6, and that is about fifty-one seconds.

Setting up a standard timekeeping system and calendar is clearly not so simple as it might at first appear. We have explained what defines a *day,* and learned a little about Earth's orbit around the Sun, but not yet defined what is meant by a *year.* That's our task in appendix B.

Appendix B

∽

HOW LONG IS A YEAR?

There are several different definitions for the year used by scientists, depending upon the particular phenomena of interest, and yet other definitions employed by other professions: accountants have the financial year, while the term *two-year-old* for a thoroughbred horse has a specific meaning in the racing world, because all racehorses in an annual cohort are taken to have the same birthday. So, we must be specific.

In this book we are interested in the evolution of the calendar. If you were to ask some random intelligent person why our calendar is set up the way it is, with the rules for leap years maintaining a certain average length for the annual round, they would likely reply with some phrase like "To keep the seasons in step"—for example, to keep July 4 in the (northern) summer, and Christmas in the winter. As a matter of fact, the definition of the *Gregorian year* (the mean length of the year in the Gregorian calendar) was specifically based upon regularizing the occurrence of Easter, but because the aim was to stop Easter slipping through the seasons, that amounts to close to—but not quite—the same thing. We might therefore state that the period we call a "year" is the period of recurrence of the seasons. But what is that? To answer that question we must begin by considering the causes of seasonal climatic variations.

The Source of the Seasons

We have seen in appendix A that the finite eccentricity of Earth's orbit leads to its distance from the Sun varying, being a minimum in early January and a maximum in early July. Many people seem to believe that this is the prime cause of the seasons, but one can immediately ascertain that such an idea is nonsensical. First of all, the heating would be more intense in January when closest to the Sun, so that summer would be expected around that time. I am sure that Northern Hemisphere readers can vouch that this is not the case. Second, if the seasons were predominantly due to our varying distance from the Sun, they would be in step between the hemispheres, and I can assure you that when it is summer in North America and Europe it is winter in Australia and Argentina.

Actually, the major parameter controlling the seasons is the tilt of our spin axis, the *obliquity* which we have already met. If you imagine an area of a square meter at the top of the atmosphere, oriented so as to be perpendicular to the direction of the Sun, the flux of sunlight hitting it is equivalent to a power of 1.37 kilowatts (kw). That quantity is termed the *solar constant* by astronomers. Not all of that energy reaches the ground, due to absorption and scattering in the atmosphere, but the majority does so. What happens to it then depends upon the nature of the surface material it meets, but averaged over the globe Earth absorbs 63 percent of the incoming light flux and reflects straight back into space about 37 percent.

The absorbed fraction heats our planet, and is eventually emitted back into space as infrared radiation.

The flux is 1.37 kw for an area of a square meter arranged at right angles to the Sun's rays, but if that area were tipped over at an angle of forty-five degrees, then the power available for absorption and heating would be reduced to about 70 percent of the above, or just under 1 kw, and if the area were parallel to the Sun's rays, then clearly no sunlight at all strikes it and the absorption must be zero.

Looking back at figure A4 we can see how this applies to different parts of Earth. At the summer solstice the Northern Hemisphere is tipped toward the solar direction and at any instant it is receiving much more sunlight than the south. At that solstice a square meter on the tropic of Cancer at noon is indeed perpendicular to the Sun's rays. This corresponds to maximum heating for the Northern Hemisphere. At the winter solstice the situation is reversed, and it is the Southern Hemisphere which receives more sunlight. At the equinoxes the incoming sunlight is equably split between north and south.

The actual seasonal changes which occur at any position will depend upon a wide variety of factors, because it is not only the solar radiative heating which controls the weather. If it were so, then meteorologists would have an easy life. To state the obvious, it's colder up a mountain than down in the valley below, even though the sunlight may be more intense above the clouds. Heat is also transported by convection and conduction, through phase changes (snow acts like negative energy), and is stored in large bodies of water so that maritime climates tend to have less contrast between summer and winter. Prevailing winds and currents have major influences.

When Do the Seasons Start?

What do we mean by *the seasons—summer, winter,* and so on? Most of my readers will live at temperate latitudes and so be habituated to four distinguishable seasons, but that is not true for all of the world's population. For most of the people in India and other near-equatorial countries there are only two seasons: the *wet* and the *dry.*

Even if many places do not have four distinct seasons, it is still useful to split the year into four segments one way or another. As a child in England I always thought of the seasons beginning with the first day of a month (March 1, June 1, September 1, and December 1), and I know that people in many other countries do likewise. Astronomically, though—and we've seen that astronomical parameters control the seasons, in broad stroke—it is the solstices and equinoxes which mark the seasons, and so we *could* choose one of these as the start point for our measurement of the year. The solstices and equinoxes are traditionally the *middles* of the seasons, so that midsummer—June 24—was originally set by the date of the summer solstice, the later Gregorian calendar reform divorcing the solstice from that date. Herein we will assume the seasons to start and end with the solstices and equinoxes.

When Does the Year Start?

The marker point for determining the start and end of the day may be chosen arbitrarily—sunrise, sunset, meridian transit, or the construct used most often in civil life (mean midnight)—just so long as we are consistent. Is the same true for defining the year? To show that this is *not* true, let us pick another point in the orbit—perihelion passage—and show how it is *not* a useful juncture to use in stipulating the year length from the perspective of the cycle of the seasons.

Under gravitational perturbations due to the other planets the longitude of the perihelion point is precessing, causing it to advance (move anticlockwise in figure A2, in the same direction as Earth's orbital motion), completing a revolution in about 110,000 years. This is totally different from the precession of the equinoxes, which is moving the equinox backward (clockwise) on the ecliptic to complete a full loop in 25,800 years. It is important to be clear here that there are two quite different motions: one results in the terrestrial orbit swiveling around (line qQ in figure A2 rotates anticlockwise over 110,000 years) while the other results in the orientation of Earth's spin axis twisting around as does a spinning top (the direction of P in figure A5 completes a full turn clockwise around its prescribed path in 25,800 years). Right now we are concerned with the former.

Figure B1, below, shows an exaggerated (for clarity) representation of the precession of perihelion. In that diagram I have shown the longitude of perihelion moving from one passage to the next through an angle of about 5 degrees, whereas in reality it moves through only 360°/110,000, or about 1/300th of a degree. This means that to complete an orbit from perihelion to perihelion Earth must traverse a longitude of about 360.0033 degrees. In the current epoch this takes 365.259635 days. That is a "year," but not the year we need to define a calendar, since it is not perihelion which defines the seasonal change. (But note Chapter 23.)

Astronomers call this length of time the *anomalistic year.* Here "anomalistic" does not mean that it is *wrong* in any way. An *anomaly* in astronomical parlance is an angle which describes the position of an object in its orbit. There are various different ways of measuring

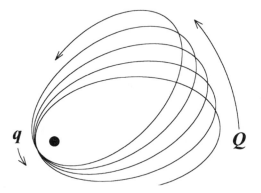

Fig. B1. The precession of an orbit under gravitational tugs from the planets. The terrestrial orbit is actually of very low eccentricity (it is close to being circular), and this, combined with its distance from the other planets and *their* near-circular orbits, results in its rate of precession being very slow, a full rotation taking 110,000 years. To illustrate the concept in this diagram, an orbit with a substantial eccentricity is shown, and the amount of precession from one perihelion passage (transit through point *q,* closest to the Sun) to the next is greatly exaggerated.

those angles, such as the *mean anomaly,* the *true anomaly,* and the *eccentric anomaly.* At perihelion all the anomalies are zero, thus the term *anomalistic year.*

There is another thing we need to say about perihelion. The precession periods given above are relative to the fixed stars. If one instead asked how long it takes between one epoch in which the vernal equinox is at the same longitude as the perihelion point and the next epoch when that is true, the answer is 21,000 years: that is just the reciprocal of ([1/25,800] + [1/110,000]). This means that on our calendar perihelion moves by one day every 57 years or so, obviously affecting the climate. The present proximity of the two is thought to be moderating winters in the Northern Hemisphere in that we are relatively close to the Sun at the winter solstice. On the other hand, one would anticipate that eleven millennia ago winters in the Northern Hemisphere would have been more severe, perihelion occurring during the northern summer (making those hotter than at present).

The Year According to the Stars

During an anomalistic year Earth traverses a longitude of 360.0033 degrees. Let us instead stipulate that the planet moves through *precisely* 360 degrees. The effect of this would be that we have returned to an orientation such that the fixed stars have come back to their apparent positions of a year before. Measuring lengths of time relative to the stars leads to attaching the adjective *sidereal* to a noun, and so this period of time is called the *sidereal year.* It currently has a length of 365.256363 days, but as for all these definitions, (1) erratic perturbations of the terrestrial orbit and spin axis may lead to the actual value at any time being slightly different, this being a *mean* value; and (2) systematic perturbations will make this mean value shift over time spans of centuries and millennia, so that the value is appropriate only for the current epoch.

What is the numerical relationship between anomalistic and sidereal years? The former is one part in 110,000 larger than the latter, of course, because we measure the rate of precession of the perihelion point relative to the fixed stars.

Is the sidereal year the year to use in defining a calendar? Despite the absurd claims of astrologers, the stars do not have a direct influence upon terrestrial affairs, at least insofar as the seasons are concerned. This is *not* the year we are after, then. We need to get back to the equinoxes and the solstices, which are the marker points for the seasons, and see how *they* may be used to define a calendrical year.

The Year According to the Equinoxes and the Solstices

The year length for our calendrical purposes needs to reflect the cyclic period of the seasons. The seasons begin and end with the equinoxes and the solstices. The actual human activities of a year, at least in terms of agriculture and other matters of import to ancient peoples, began in the spring after the winter hiatus. Thus the spring, or vernal, equinox is the appropriate marker point to use in defining the length of the year (even if we choose to celebrate New Year at some other time).

The year counted from the vernal equinox needs a special name, something definitive to make clear what we mean. Here I will call it the *VE year,* and it is not invariant: right now its value is 365.242374 days, whereas at the time of Christ, the beginning of our era, it was 365.242137 days. Note that the sixth decimal place corresponds to almost a tenth of a second, and the difference between those two values is about twenty seconds, which over two millennia would accumulate into a difference of almost half a day, so that if one is to lay out a successful calendar to last for millennia it is necessary to know the length of the year with precision of that order (about ten seconds).

Why is the VE year less than both sidereal and anomalistic years? The answer lies in the precession of the equinoxes. Refer back to figure A4. The sidereal year is the time taken by Earth to move through all longitudes from $L = 0$ degrees to 360 degrees, but the vernal equinox is reached before completing that circuit. In figure A5 the direction of Earth's spin axis denoted by P is swiveling due to the precession of the equinoxes, taking about 25,800 years to complete a clockwise loop. In a single year the position of the vernal equinox regresses by about one-seventieth of a degree in longitude, and thus Earth reaches the next equinox about twenty minutes earlier than it would if there were no equinoctial precession. Thus the VE year is one part in 25,800 *less* than the sidereal year (i.e., [365.256363 − 365.242374]/365.25 is about 1/25,800), whereas the anomalistic year is one part in 110,000 years *longer*, because the precessional motions are occurring in different senses.

What if I had chosen to use instead the autumnal equinox, or one of the solstices, as the marker point to define the beginning and end of the year? Would I have arrived at the same result? The answer is *no*. The lengths of the seasons are *not* equal: for example, the northern spring (the length of time between the vernal equinox and the summer solstice) is not of the same duration as the northern summer (the length of time between the summer solstice and the autumnal equinox), because the speed of Earth in its orbit is not constant, being elevated when it is near perihelion and lowest when near aphelion.

But the location of the perihelion point is changing. In the current epoch, to the nearest couple of hours, in the Northern Hemisphere spring is 92.8 days long, summer 93.7 days, autumn/fall 89.9 days, and winter only 88.8 days long, and those durations are slowly changing as perihelion precesses.

From that it is obvious that the times between successive equinoxes of the same type, or successive solstices of the same type, are not all equal, such that different year lengths based on the autumnal equinox, and the winter and summer solstices, might be defined. As a matter of fact, all of these will fluctuate in length, because the nutation and other wobbles of the spin axis mentioned in appendix A will alter the *instants* at which each of these cardinal positions is reached, from orbit to orbit. Precision calculations show that they fluctuate erratically (but predictably) by up to five minutes or so. Nevertheless, it is possible to calculate mean (averaged over many orbits) values for these different "years." One can quote values to six decimal places, although in reality only at best the fifth is meaningful. In the present epoch (close to the year 2000) these values are:

Vernal equinox year:	365.242374 days
Summer solstice year:	365.241626 days
Autumnal equinox year:	365.242018 days
Winter solstice year:	365.242740 days
Average:	365.242190 days

That average is equivalent to what is called the *tropical year*.

The Tropical Year Incorrectly Defined and Deployed

A surprising thing is that almost all astronomy textbooks, including the official publications of the governments of the United States (via the U.S. Naval Observatory) and the United

Kingdom (via the Royal Greenwich Observatory), have the tropical year incorrectly defined, even though it is the one parameter from which stems the design of the calendar which we all employ!

Take, for example, the *Explanatory Supplement to the Astronomical Almanac,* a huge publication by the two governments above, which is revised every couple of decades. Let me be clear, I love that volume—if I were to be marooned on a desert island with only a dozen books, it would certainly be one of those I would choose—but it is not infallible. In his chapter on calendars the late Leroy Doggett stated that "the *tropical year* is defined as the mean interval between vernal equinoxes," and that is *wrong.* That interval is the VE year, and the error is crucial in many respects.

In fact the tropical year is correctly defined elsewhere in the same volume as being "the period of one complete revolution of the mean longitude of the Sun with respect to the dynamical equinox." This is true because the use of the "mean longitude" implies an averaging over the whole orbit, in the same way as we met the use of the "mean sun" in discussing the length of the day. The tropical year derived from any one orbit (say, that in 1999–2000) will have a value which may differ by some minutes, as the result of influences which we discussed earlier, from a figuratively averaged value which we would call the *mean tropical year;* the "mean" is usually dropped, but understood. This averaging produces the same result as that we saw above when we averaged the four year lengths measured on the basis of the cardinal positions. That is, these are averages with the short-term fluctuations of the order of a few minutes having been smoothed over. Two sets of averaging have been applied to arrive at the result, corresponding to the two uses of "mean" in another description of the tropical year elsewhere in the *Explanatory Supplement:* "The tropical year was defined [in 1900] as the interval during which the Sun's mean longitude, referred to the mean equinox of date, increased by 360°."

Many books—and even articles in quite reputable journals such as *Scientific American*—glibly state that the tropical year is the parameter which is of importance in defining a calendar, and then go on to make absurd calculations about how long the author reckons that it will take the calendar, given a certain leap-year rule producing an average year length, to deviate by one day from the actual (tropical) year length. Take the Gregorian calendar, for example. The mean Gregorian year has a length of 365 days plus 97/400ths (because there are 97 leap days inserted into each four centuries), or 365.242500 days. The difference between this and the current value of the tropical year given above is 0.000310 days, and on those grounds various authors claim that a "correction" of the Gregorian calendar is required about every 3,200 to 3,300 years. There are two serious errors made in such reasoning, however. One is that the tropical year is not constant over millennial timescales; for example, 2,000 years ago its value was 365.242310 days, and if *that* value were used, then an adjustment only every 5,300 years would be indicated. This would again be an erroneous result, due to the second breakdown in the reasoning, which is as follows.

The intention of the calendar reform by Pope Gregory XIII was *not* to make a year which kept all the seasons in step, but actually to keep the vernal equinox on much the same date. In the spirit of the Gregorian calendar, the aim should be to keep the vernal equinox steady, and thus the *year* to use is not the tropical year, but specifically the VE year. If we do so using the above value of 365.242374 days (which would be repeating the first error of reasoning complained about above, because this duration changes over millennia), then the indicated period until an adjustment might be needed is about eight thousand years.

This reasoning, though, is spurious from another aspect: the length of the day is changing such that over eight millennia the accumulated deviation will be lessened by a large fraction of a day.

Of one thing you can be sure: anyone trying to design calendars for use beyond the next millennium has no understanding of the cycles (and variability) of the heavens. And the same applies to anyone making comments about the accuracy (or otherwise) of the Gregorian or any other calendar, unless they acknowledge *precisely* which year is appropriate for that calendar, and do not confuse the tropical year with the VE year. The difference may be small, but it is critical.

Yet More Possible Year Lengths

Astronomers also use other year definitions. First, stellar and galactic distances are often quoted in units of *light-years*. This is the distance which light travels in a year: but which year? The answer is the *Julian year*. The Julian year is the average length of the year in the Julian calendar (leap day *every* four years), which is precisely 365.25 days long. In many astronomical calculations this duration is used for simplicity.

Second, consider your likely response, if asked before you read this appendix, to the question "What defines the year?" You might have answered along the lines of "One orbit of Earth around the Sun." The size of the terrestrial orbit (actually half the distance between q and Q in figure A2) is known as the *astronomical unit* (A.U.). The orbital periods for all planets, asteroids, and comets vary with their orbit sizes, regardless of the shapes of the paths they take (that is, the period does not depend upon the eccentricity). If one asks for the period of an orbit of exactly 1 A.U., one gets what is called the *Gaussian year*, which is 365.2569 days long.

Which Year Is the Calendar Year?

Which of the "years" discussed above would be the best to use in defining a calendar? This question boils down to one of: what is it we really want? The answer to that is simple: we want *stability*. I have noted that all these years can vary erratically by several minutes, but those erratic changes average out to leave only a long-term trend in each. Those trends are such that over the past several thousand years the years based on the summer solstice, the autumnal equinox, and the winter solstice have all decreased, while the VE year has stayed very nearly constant; it follows that the average of the four (the tropical year) has also decreased. Looking into the future, the VE year is again the best behaved, altering by very little over the next few millennia. In terms of stability of year length, it is therefore the VE year which is the preferable duration on which to base a calendar. It also happens to be the basis of the Gregorian calendar.

But enough about years. We have expressed year lengths as numbers of days, having previously discussed day lengths without defining the absolute units in which they may be measured. Our next step is to define the second.

Appendix C

❧

We will define precisely what we mean by a *second* in this appendix. Let us take the *hour* as given, its origin to be considered later. The Babylonians were keen on matters sexagesimal, perhaps because 60 is divisible by 2, 3, 4, 5, and 6, and we get our minutes and thus seconds from them. The units have been passed down to us through Latin, which, I was once authoritatively informed, contributes 64 percent of the vocabulary of English.

In Latin, one-sixtieth of an hour is called *pars minuta* ("small part"), from which we get our noun *minute* (and similarly, the Latin is the origin of the adjective *minute* which is spelled identically but pronounced differently from the noun). If a *pars minuta* is further divided into sixty parts, then a *pars secunda* ("second part") results, hence the origin of our fundamental unit of time, the second. I write *fundamental* because, as we will see, it is the one unit of time which is defined on the basis of physics, giving us not only all other time divisions but also all units of length via the speed of light.

The Ancient Division of the Year

Although the Babylonians used major divisions which we might think of as being equivalent to hours, in fact it was the Egyptians who invented the twenty-four-hour day. This was made up of two blocks of twelve, one being the daytime, the other the nighttime. It was a matter of gradual uptake of this system for time measurement, but basically the twelve seems to have originated in about 2100 B.C. from having ten hours with an extra hour tagged on at each end for twilight; twelve being a submultiple of the Babylonians' base-60 system eventually led to favor. By 1300 B.C. twelve daylight hours and twelve nighttime hours were being used.

During the daytime sundials and shadow clocks (there is a difference: you cannot use a sundial in the tropics because the Sun can pass overhead) were used to split up the time, whereas at night the rising of specific stars or groups of stars on the eastern horizon counted off the hours. The Egyptians divided the zodiac into thirty-six *decans*, or ten-degree zones, each of which was named, although now we know only of those associated with Sirius and Orion.

To us it seems obvious that the length of an hour does not change with the time of day or year (although wet winter afternoon hours may *seem* longer than the hours spent on the beach in the summer), but this was not the case when the hour was invented. The daytime was split into twelve "hours" of equal length, as was the nighttime, but these would be different from each other at any time of year except near the equinoxes, and would alter with the seasons.

Let's use an example to illustrate this, using as an invariant unit of time the minute as measured with a modern clock. Imagine that at a particular latitude—Alexandria, say, the seat of Egyptian learning—sunrise to sunset lasts for 840 minutes (14 of our equal hours) at the time of the summer solstice, so that the nighttime is 600 minutes long, whereas at the winter solstice these are reversed: daytime is 600 minutes, nighttime is 840 minutes. To the Egyptians, at the summer solstice their daytime "hours" would be 70 minutes long, and the nighttime hours each last for 50 minutes, with the reverse being true at the winter solstice. The use of such unequal hours may seem crazy to us, but without an independent clock, how could you tell?

This opens the question of when the first clocks appeared. The wit might note that there seems to be a lot of sand in Egypt, so why didn't they make sand clocks like the egg timers many of us still use for convenience? The answer is that sand is abrasive, and gradually wears down the narrow funnel through which it must flow, altering the flow rate. Egg timers work well so long as you only need to use them once a day.

For continual use a better method is needed; thus the first clocks were water-flow devices called *clepsydras*, employed in Egypt and Babylon from about 1600 B.C. They were also used in India. By arranging for a constant head (uniform water pressure), it is fairly straightforward to design a system from which equal lengths of time could be indicated by equal volumes of water collected, but the problem was that everyone else in the community was using *unequal* hours. This meant that a clepsydra had to be equipped with some device to either vary the flow rate during the year (and allow an interpretation of the different hour lengths for day and night) or have a set of receptacles graduated for different times of year (and day and night). Complicated, is it not?

No matter. The point is that we get our twenty-four hours in a day from the Egyptians, the word *hour* being derived from the Latin and Greek word *hora*, meaning "time" or "season."

The Arrival of Mechanical Clocks

This situation whereby days and nights were split into unequal hours, time being told basically by the Sun, continued for millennia. For an agrarian society, what else is needed? The first mechanical clocks were not constructed in Europe until the end of the thirteenth century, although the Chinese had developed them at least some six centuries earlier, but then lost the art temporarily, the Orient being a region of constant foment.

The essential concept making mechanical clocks a possibility in Europe was the *escapement*, the need for which had been described by a man known only as "Robert the Englishman" in 1271, although the person who succeeded in making the first workable device a few decades later is not remembered. Since it was he who set us on the slippery slope to "Time equals money," it is a shame that we do not know his name. We should commemorate him, perhaps on April Fools' Day.

The oldest surviving mechanical clock, dating from 1386, is that at Salisbury cathedral, in Wiltshire, England. Many of the earliest clocks showed not just the time, but also various astronomical phenomena, incorporating perpetual calendars to indicate church feasts. Giovanni Dondi made the first of these, laboring sixteen years before he completed it in 1364. Although it has not survived, there is a reconstruction of it in the Smithsonian Institution in Washington, D.C. Many other European cathedrals still boast quite remarkable clocks, displaying astronomical phenomena other than simply the time of day, my favorite being that at Lund in Sweden. Giovanni Dondi's father, Jacopo, of Choggia in Italy, is credited with inventing the familiar clock dial in 1344.

The Arrival of Equal Hours and Accurate Clocks

The idea of equal hours was not adopted in Europe until mechanical clocks began to spread in the fourteenth century, and this is the origin of our term *o'clock:* that is, the time according to the newfangled mechanical devices rather than by the Sun, which people had hitherto used as their timekeeper. These early clocks were all driven by hanging weights, and used verge escapements. An improvement, the cross-beat escapement, was introduced in Switzerland by Joost Bürgi in about 1600, starting the tradition of Swiss excellence in clockmaking, but within sixty years the great Dutch scientist Christiaan Huygens invented the pendulum clock.

Huygens was also involved in the introduction of the spring-driven clock. Although this idea was first mooted in the fifteenth century, it was not until Huygens invented the balance spring in 1675 that it became a reality.

At the time there was great interest in the possibility of constructing a clock which would work on a sea voyage, the ship's tossing and swaying obviously making absolutely useless any clock employing a pendulum for ticking the seconds, or a hanging weight for its driving force. A clock which would function at sea would make it possible to determine one's longitude, solving the great problem of navigation, and it is no accident that a small, portable timepiece and a period spent maintaining a lookout on a ship are both called a *watch.* There was considerable pecuniary interest in the development of the first successful marine chronometer. It is in this context that the experiments of Huygens should be viewed, and those of his English adversary, Robert Hooke. There was money to be made.

Huygens's clock was capable of keeping time to an accuracy of about ten seconds per day, many times better than all previous devices. Improving escapements led to even higher precision, as did compensation for temperature-dependent alterations in pendulum length. Other advances obviated the swinging pendulum, and by the mid–eighteenth century John Harrison was taking steps to reduce friction and achieving accuracies better than a second per day, paving the way for safe navigation by ships carrying chronometers based upon his innovations.

For land-based clocks the pendulum still ruled. The next major steps forward were not until the nineteenth century, when variations in atmospheric pressure and thus the air drag were allowed for or obviated by using a sealed chamber. This permitted an accuracy of about one-hundredth of a second per day, even that being exceeded in the 1920s by W. H. Shortt, a railway engineer who built a clock in which interference with the free swing of the pendulum was minimized. His clocks, operated at the Royal Greenwich Observatory until 1942, provided Britain and thence much of the world with its time standard. But changing technology meant the abandonment of the pendulum in favor of electronic systems.

The Electronic Revolution

Quartz crystal clocks were developed from 1928. In view of that it seems surprising that quartz watches were still expensive in the 1970s, but the Swiss watchmaking industry provided much inertia. Even the first quartz systems allowed an accuracy of about a millisecond per day, and thus provided a watershed in timekeeping: for the first time humans had available a tool for measuring time more precise than the spin of our planet. Timekeeping slipped from the hands of astronomers to physicists. The length of the day is actually increasing by about two milliseconds per century due to the fact that the terrestrial spin rate is slowing as a consequence mainly of the drag force applied by the rise and fall of the tides under the gravitational attractions of the Moon and the Sun. Quartz clocks allowed that deceleration to be measured directly.

Until this juncture, any unit of time based upon Earth's rotation could be thought of as being variable in length. If the second were defined as being one part in 86,400 of a day, but the day was the mean solar day which alters as Earth's spin rate slows, then the second must be getting longer. Not by much, but advancing technology made the definition of an invariant second not only possible, but necessary. In 1956 a unit known as the *ephemeris second* was defined, as one part in 31,556,925.9747 of the mean tropical year. But because the tropical year is varying in length, the year employed was that at the start of 1900.

That is fine as a definition on paper, but how could it be put into actual use? The answer is that in the 1950s the first cesium clocks were being brought into service in laboratories in Britain and the United States, and these are accurate to about one-millionth of a second over a year (a fractional uncertainty of around one part in 10^{14}), and led to their direct use in providing a reference measurement for the second: "The second is the duration of 9,192,631,770 periods of the radiation corresponding to the transition between the two hyperfine levels of the ground state of the cesium-133 atom in zero magnetic field." That is what a second is; it is totally independent of the spin and the orbit of Earth.

This definition, then, is divorced from any astronomical considerations, but it is *based* on the length of the tropical year back in 1900. Paradoxically, this second, which is called the *SI second* (SI stands for *Système International d'Unités*, or International System of Units)— or perhaps the *atomic second* is more intelligible—is founded upon a reference length of time in an epoch before the atomic age even began!

We have decided to state that a *day* is 86,400 of these atomic seconds long, but in reality a mean solar day, at the end of the twentieth century, is about two milliseconds longer than that and gradually, but erratically, deviating further from that count. The *day* which was the unit for the lengths of all the *years* we discussed in appendix B, but which we could not define there, is comprised of 86,400 atomic seconds.

Ultraprecise Time Determinations

Timekeeping worldwide is now based on the numerous cesium clocks maintained in France, Germany, the United States, the United Kingdom, Japan, Canada, Korea, Russia, Italy, Australia, and Switzerland, comparisons showing a relative wander of one microsecond or less per year (one part in 10^{14}, corresponding to one second in three million years). Even this precision is now being superseded, with hydrogen maser clocks having an accuracy one hundred times better (one part in 10^{16}), although only sustainable over periods of about a week.

For example, the U.S. Naval Observatory, which provides the time standards for the United States, employs sixty cesium clocks and ten hydrogen maser clocks spread between twenty controlled vaults, and these are all intercompared every hundred seconds so as to maintain a stable and reliable time standard good to a billionth of a second per day. The timescale from such clocks is called *international atomic time* (abbreviated TAI, not IAT, because the abbreviation is from the French, the ultimate responsibility for international time standards having moved from London to Paris some decades ago).

Recapitulating Universal Time

The time frame which, by international agreement, is used for civil timekeeping around the world is coordinated Universal time (UTC). It is based on TAI. Different time zones maintain round offsets from UTC. For example, time zones in North America are integral numbers of hours behind UTC.

The terms *Universal time* (U.T.) and *Greenwich mean time* (GMT) are ambiguous, being used sometimes to refer to U.T.1 (a scientific timeframe which varies with Earth's irregular and slowing rotation) or some other ill-defined scale, and sometimes in connection with UTC, so that it is best to avoid the use of U.T./GMT in modern-day affairs. UTC is the time which is given by broadcast time signals.

The day, the year, and the second have all now been defined. The fundamental division of time we have yet to meet is the month.

Appendix D

❦

I am no poet, as the following lines demonstrate:

> With cutlery silver
> Sliced citrus bright orange
> And grapes lustrous purple
> They feasted mid-month.

One might think that, even if it scans, my quatrain has no rhyme or reason, but that would only be half correct. It has a reason, and that is to impress upon you the fact that the word *month*, like *silver, purple,* and *orange,* has no rhyme in the English language. Or at least none I've ever been able to find. Perhaps that is why we speak of purple *prose* rather than *poetry.*

Another way to say that *month* has no rhyme would be to say that it is discordant, and in this appendix we will see that in designs for our calendar the discordant month has certainly disrupted matters. The only feasible way to make things fit has been to change the lengths and forms of the months according to the Moon.

The problem with the Moon, from the calendrical perspective, is that it does not have a period which is an aliquot part of the year. That's a technical term, which just means that the month as imposed by the Moon does not divide evenly into the year as defined by Earth's motion about the Sun.

In this discussion we will just think in terms of *solar years* and not worry about their precise length. More important now is to think about the length of the *month.* Here we are not concerned with the *calendrical months,* which are arbitrary definitions according to human whim. Nor are we worried about usage of *month* to mean a period of precisely 30 days, or perhaps four weeks, such as when a rental agreement stipulates payment "monthly on each fourth Monday"; that is not of interest for present purposes. Our quest here is not for some month based upon human tinkering, but for the natural cycle of our celestial neighbor.

The Astronomical Months

It will not be a surprise to hear that astronomers have a number of ways of defining the *month,* with an appropriate adjective appearing in the name of each. Four of these months have periods of 27 days plus a fraction:

Tropical month (equinox to equinox):   27.32158 days

Sidereal month (fixed star to fixed star):   27.32166 days

Anomalistic month (perigee to perigee):   27.55455 days

Draconic month (node to node):   27.21222 days

The *tropical month* here is the time between alignments with the direction of the vernal equinox (the first point of Aries). In a fit of masochism one might split that tropical month up into four unequal months based on the times of passing the lunar equinoxes and the solstices: these would each be varying quite quickly because, as we see below, the lunar perigee is precessing rapidly, although their average would remain almost constant.

The *sidereal month* is quite straightforward, being the time taken for the Moon to complete an orbit compared to the celestial sphere, but we are not concerned with the stars.

The *anomalistic month* is more interesting. The size of the lunar orbit around our planet (more strictly, about the barycenter, the center of mass of the pair) is 384,400 kilometers. The geocentric distance of the Moon fluctuates about this value because its orbit is noncircular, the closest approach to Earth being termed *perigee* and the furthest recession *apogee*. The eccentricity is about 0.0549 (although under perturbations due to the Sun it varies between 0.044 and 0.067), so that perigee is at a distance of about 363,300 and apogee at 405,500 kilometers. During this month the lunar distance varies by plus and minus 21,100 kilometers, but note that for a particular observer the *daily* variation in distance is several thousand kilometers (up to the terrestrial diameter of 12,750 kilometers), being lowest when the Moon is overhead and highest twelve hours later. Note that although the difference between the anomalistic and sidereal *years* was only 0.00327 days, producing a 110,000-year precession period, the difference between the anomalistic and sidereal *months* is 0.23289 days, resulting in a precession period of only $27.32166 \times 27.55455 / 0.23289$ = 3,232 days, or 8.85 years. The advance of the lunar perigee is rapid due to (1) its relatively high-eccentricity orbit; and (2) the major perturbations produced by the attraction of the Sun.

The *draconic* (or *nodical*) *month* is the time between passages of the Moon through its node on the ecliptic (transits through Earth's orbital plane). The lunar orbit is inclined at 5.15 degrees to that plane on average (the inclination varies between 4.96 and 5.32 degrees), and as a result it can appear up to that angle away from the path of the Sun across the sky. The draconic month is of most interest for predictions of eclipses (that's where the word *draconic* stems from: the ancients thought that a dragon ate up the Sun during an eclipse).

To get an eclipse, Earth, the Moon, and the Sun must be very close to collinear, and thus the Moon must be at a node. Above, it was noted that the lunar perigee advances (moves anticlockwise in the plane of the ecliptic), and that is reflected by the anomalistic month being longer than the sidereal month. To the contrary, the draconic month is shorter than the sidereal month by 0.10944 days and the node of the lunar orbit moves clockwise or *regresses*, taking $27.32166 \times 27.21222 / 0.10944$ days, or close to 18.6 years, to complete a rotation. This is the origin of the 18.6-year nutational cycle mentioned in appendix A.

Which of these months is of use to us in delineating a calendar? The answer is *none of them*. For the month in a calendar we are interested in the brightness cycle of the Moon, and that actually has a periodicity of 29.53059 days. This is called the *synodic month*; it is the mean time between two full moons. It may also be called a *lunation* or simply a *lunar month*. The reason that the synodic month is longer than the period it takes the Moon to orbit Earth is similar to the reason why the day is longer than the time it takes Earth to rotate on its axis: as the Moon orbits us, our planet is moving along its own orbit around the Sun, and in order for the three to be aligned again the Moon must complete more than one circum-terrestrial loop.

This is illustrated in figure D1, below. At positions A and B the Moon is opposite the Sun in the sky, and at those times Earth and the Moon have the same solar longitude. Such a condition is called *opposition* or *superior conjunction;* it is the time of full moon. If the Moon were half that loop away, between Earth and the Sun, the condition is called *inferior conjunction* (and sometimes just *conjunction,* with *opposition* being the norm above), and it is soon the time of new moon. I wrote "soon" there because many texts confuse conjunction or dark of moon with the new moon: new moon is when the slender lunar crescent is sighted just after sunset, having been missing for about three days straddling conjunction.

The Bright of Moon

Because we are discussing the dependence of the brightness of the Moon upon its position relative to the Sun, it is appropriate to mention here a matter which many people, especially those who have never camped out far from city lights, appear not to appreciate. At first and

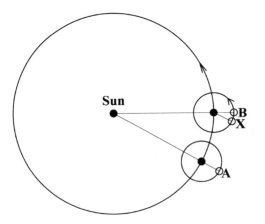

Fig. D1. The orbit of the Moon about Earth is shown in greatly exaggerated fashion in this diagram, and also as a circle for simplicity. At both A and B the Moon is full, and the time between these (which averages to about 29.53 days) is called the *synodic month.* At position X the Moon had completed one sidereal month since A, but it still had some distance to go before reaching full moon, which occurs when Earth and the Moon are aligned with the sun in terms of longitude (if all three were also in the same plane—if the Moon were at a node—then a lunar eclipse would occur). This means that the synodic month, the cycle of lunar brightness which is of interest for human affairs, is the longest of the lunar months.

last quarter (when the Moon is at right angles to the solar direction, which is termed *quadrature*), half of the lunar disk visible from Earth is illuminated by the Sun, the rest being in shadow. At full moon essentially all the disk is illuminated.

If I were to ask you now what the ratio is between the moonlight coming to Earth at full moon and first/last quarter, you would likely answer *two* based upon there being twice as much of the lunar disk reflecting sunlight in our direction. In fact, the answer is *nine:* the intensity of moonlight at full moon is nine times as high as at the quarters.

The reason is that the lunar surface is rough, not smooth and shiny. Imagine that 10 percent of all photons of light are absorbed each time they strike a lunar surface segment, which is rough and granular. At the quarters, light must be scattered through 90 degrees so as to go off in the direction of Earth. Some photons will require only one reflection to do this, but most will bounce around between the dust grains comprising the lunar soil, perhaps 10 percent being absorbed on each bounce. If twenty bounces are required on average to direct a photon toward Earth, then only $(1 - 0.1)^{20}$, or about 12 percent, would survive.

At full moon, though, a large proportion of photons will be directly backscattered toward Earth. Even if photons average two bounces, 81 percent would still survive. The situation is more complicated than this simple explanation, but this is broadly the scenario. The significant brightening at full moon is called the *opposition effect*, and it is also displayed by other objects such as asteroids. This tells us that the Moon and asteroids have rough, dusty surfaces, due to the churning effects of meteoroid impacts: they are not like shiny ball bearings or the mirror balls seen in discotheques.

This might seem just an interesting aside, and you might be wondering what this has to do with the calendar. Here is the answer. The single factor which has had the dominant effect in defining our calendar has been the desire of the Christian churches to regularize the date of Easter. The extensive ecclesiastical rules are detailed in chapter 13, but as a simple statement *Easter Day is the first Sunday after the first full moon after the vernal equinox.* This puts Easter close to a full moon, but following it. One of the original motivations for that— many centuries ago—was to provide sufficient light to allow early Christians making an Easter pilgrimage to Jerusalem to travel at night: not only is the full moon nine times brighter than the quarter, but also it is available all night. Try camping out at both full moon and new, and you'll notice the difference as you try to find your way to the nearest tree to micturate (relieve your bladder) in the middle of the night.

The Calendrical Influence of the Moon

In discussing the various types of month above, I gave values for their lengths with some accuracy, but in every case these were *average* durations, although I did not specifically say so at the time.

The averaging must be done, figuratively speaking, over hundreds of months—in fact, over some *decades*—for the values to have any real meaning, because there are many factors which lead to short-term alterations in the precise periods from one orbit to another. The theory of the motion of the Moon is one of the most complicated problems in celestial mechanics, and it has occupied the careers of many great scientists from Newton's day to this. If the Hawking rule cited at the start of this book were true—that every equation halves the potential sales of a book—then no book would ever have been published on the lunar theory; actually, many have, but they are not best-sellers. The equation one needs to employ if one wants to *calculate* (as opposed to *observe*) the precise position of the Moon at any instant has over fifteen hundred terms.

The determination of the mean synodic period does not actually result from timing the Moon for thousands of years, but from precise, shorter-term observations which allow us to make the required calculations. The ancients derived the lengths of the synodic month and the solar year using repeating intervals between eclipses over centuries.

For calendrical purposes we are only really interested in the synodic month, the time between similar lunar phases (between full moons, say). In his book *Questioning the Millennium*, Stephen Jay Gould wrote that "the Moon takes 29 and a half days to circle the earth (29.53059 days, to be more precise) . . ." In this and in several other points he is incorrect. One could forgive the use of "circle" even though the lunar orbit is not circular, but there are other errors in that statement. As we have seen in figure D1, the synodic period is *more* than the time taken to circle/orbit Earth, and the 29.53059 days is *not* precise: it's an average over many months, so that full moons/new moons do *not* occur as regularly as clockwork. That mean value is given to five decimal places, whereas the time span of a lunation can vary in the first decimal place, ranging between 29.2 and 29.8 days, up to an eight-hour deviation in either direction from the average.

Why Does the Synodic Month Vary?

The reason for the variation in the length of the synodic month may be understood from figure D1, with a little imagination.

In that diagram the lunar orbit around Earth is shown as being circular, whereas in fact it is elliptical with an eccentricity of near 0.055. If that orbit were circular, *and* Earth's heliocentric orbit were circular, then the synodic month might be of constant length. The synodic month, we saw, is longer than the sidereal month by an average of $29.53 - 27.32$ days, or about 2 days and 5 hours.

This is the mean interval it takes the Moon to move between positions X and B in figure D1. But if perigee lies between X and B, then the actual time taken will be shorter than that average, while if apogee is in that gap, then the time will be longer, because the Moon moves slowest at apogee.

We have already seen that the perigee/apogee of the lunar orbit takes 8.85 years to cycle around all positions, so that they are not fixed in orientation, but the revolution of Earth around the Sun in a year results in both perigee and apogee passing through that arc XB on that shorter timescale. Apart from that, Earth's varying speed in its orbit also affects the time taken for the full moon to come around again. The details are not important here: the point is that the lunation is quite irregular in length, varying by over twelve hours during a typical solar year, providing other difficulties for calendar makers beyond the fact that the synodic month does not fit neatly into the year.

On top of all of this, the month is getting longer. The Moon is gradually receding from Earth, extending the former's orbital period. Around the year 1900, the mean synodic month lasted for 29.5305886 days, but by 2100 it will have lengthened to 29.5305891 days. Not much—about one twenty-fifth of a second—but it is a significant difference if one is considering the precise long-term motion of the Moon.

Fitting the Month to the Year

Having come a full circle from our starting point in this appendix, we can now consider the relationship between the lunar month and the year. For the sake of this argument, we may take the year to be 365.242 days long, and the month 29.5306, deliberately employing six-figure accuracy in each case. Twelve times the latter figure renders 354.367 days (but remember that in any particular year twelve lunations may last for a time substantially different from this calculated mean value). This is 10.875 days short of the solar year. For simplicity, let's take that to be 11 days. How can that be incorporated into a calendar?

One way is to base your calendar solely on the lunar phases. This is called a *lunar calendar*. An example is the Islamic calendar, which is defined to be twelve (observed) lunations long, so that it is (1) variable in length, but usually 354 or 355 days in duration, and (2) shorter than a solar year by about 11 days, so that the seasons arrive progressively earlier in this lunar year, proceeding through a complete cycle in about 33.6 solar years.

A second way is to maintain twelve months based on the phases of the Moon, each lasting either 29 or 30 days (so as to average near 29.53), and every few years insert an extra (thirteenth) month in a year. This is called a lunisolar calendar. Many contemporary societies use such systems, examples being the Hebrew and Chinese calendars. The general principle of adding a month (or some other specified period) is called an intercalation. If this were done every third year with a 30-day month being added, then the three years would still be short of about three days according to the Moon. This could be accommodated by having more months of 30 days rather than 29, but then the calendar would get out of step with the Moon.

An alternative is to use a longer intercalary cycle. The number of synodic months in a year is about $365.242/29.5306 = 12.3683$. Having one intercalary month every three years is equivalent to making the decimal part there equal to 0.3333. One wants to get closer to 0.3683. A better fit is obtained using $3/8 = 0.3750$, implying three intercalary months in an eight-year cycle. Convergence onto 0.3683 can be achieved by using longer cycles—for example, $7/19 = 0.3684$—and even that can be improved upon. In general, short repeat times are employed, and the $7/19$ cycle has been favored by many cultures: this is the Metonic cycle, as discussed in chapter 5.

The third basic form of calendar abandons the phases of the Moon and defines month lengths which, on average, are longer than the mean synodic month. The calendar then follows the Sun only, and so is called a solar calendar. The Gregorian/Western calendar which we all use is the prime example, stemming from a desertion of the lunar month by the Romans prior to about 400 B.C. (see chapter 6).

The Movement of the Moon

A final point I would mention regarding the Moon is that, given that it affects us in so many ways, it has always been a surprise to me that so few people seem aware of how it moves across the sky. I do not mean its daily motion, but rather its monthly path.

Thinking back to when we discussed Earth's orbit around the Sun, the apparent effect is that the Sun moves eastward among the constellations by about 1 degree per day so as to complete a 360-degree revolution in a sidereal year. The Moon completes such a revolution in a sidereal month of about 27.3 days, implying that it has to move through 13.2 degrees per day, again in the eastward direction.

That is compared to the stars. Compared to the Sun, the mean motion is the difference between these rates, about 12.2 degrees per day, and because 1 degree is equivalent to about four minutes of time, this means that the Moon rises successively almost fifty minutes later from one day to the next.

At new moon it is glimpsed very soon after sunset and just before it itself dips below the western horizon. The amount of time required after conjunction before the new moon can actually be seen depends upon a number of parameters, including the latitude of the observer, the time of year, and the location of perigee in terms of both celestial latitude and longitude; this is not merely an academic matter because many religious junctures (such as the declaration of a new month in the Islamic calendar) depend upon the sighting of the new moon. The next day it sets almost an hour later, and four or five days after that it is setting near midnight, until eventually it is setting as the Sun rises, and of course rising (as the full moon) as the Sun sets.

The thing which seems to confuse people the most is why the Moon appears to move backward (eastward) across the sky from night to night. Their confusion is based upon the realization that the Moon is orbiting our planet in the same sense as Earth spins, and the same direction as we orbit the Sun: anticlockwise as viewed from the north/above. So shouldn't the Moon move forward/westward across the sky?

My way of explaining things is this. From Cape Canaveral satellites are launched generally toward the east. Those in orbits only a few hundred kilometers up, like the space shuttle, complete an orbit in about ninety minutes at a speed of seven or eight kilometers per second, traveling west to east: they seem to cross the sky in the direction opposite to that in which the Sun and the stars move. The further up you go, the longer the orbital period and the slower the speed, until at an altitude of about thirty-six thousand kilometers the orbital period of twenty-four hours exactly matches the spin rate of Earth, and the satellite remains above the same point on the equator (if it has been placed into a zero-inclination orbit). Such satellites are said to be *geostationary*, and are used for communications and broadcasting.

Further up still, the orbital period must be longer than twenty-four hours, and the satellite would appear to slip back compared to Earth's spin, thus moving from east to west during each day, but not by a full 360 degrees. In a twenty-five-hour orbit a satellite would move through about 345 degrees, and thus from one day to the next would appear to have moved west by 15 degrees. The Moon has a 27.3-day orbit, and from one day to the next it appears to have moved west by almost 347 degrees. And that's equivalent to having moved *east* through 13 degrees.

The Lunar Day

The above has a possible calendrical significance which is easy to miss. When we were considering the *day*, we saw that our unit with that name is based upon the time, averaged over the year, between the Sun's crossings of the meridian. That is the *mean solar day*, or twenty-four hours.

On the other hand, if you are interested in the stars, as are astrophysicists, you would need to know how much time elapses between successive passages of the stars through the meridian; that is the *mean sidereal day*, a little over twenty-three hours and fifty-six minutes.

In the same way, one could define a *mean lunar day*, being dependent upon the time between meridian transits of the Moon. From the earlier discussion one can see that this is about twenty-four hours and fifty minutes in duration.

Who on Earth would use such a day length? The answer is: *any seafaring peoples*. We often say that there are two tides a day, or something similar, but the "day" implicit there is the *lunar day*. If you look up times of high tides in a newspaper or an almanac, you will find that they are separated by about twelve hours and twenty-five minutes. Broadly speaking, the ocean bulges toward the Moon on the point closest to our celestial neighbor, where its gravitational attraction is largest, and also bulges on the opposite side of Earth, where the attraction is lowest. (A proper detailed description of the cause of the tides, including the phase lag behind the lunar direction and the role of the Sun in producing neap and spring tides, is more complicated than that, but this will suffice for present purposes.)

If you depended upon the sea for your livelihood, then it is the tide-raising Moon which is more important to you than the Sun, and so you would use a day which is twenty-four hours and fifty minutes long. The Eskimos of Greenland formerly did this, as did other societies living beside the sea, such as those in Polynesia and Hawaii.

Julius Caesar Rhymes Again

I cannot resist making a final comment about Julius Caesar in connection with my unrhyming litany of silver, orange, purple, and month. In chapter 6 you will see how he dealt with the *month,* giving us the calendar month lengths we use today. As a Roman senator he wore a toga with a wide *purple* stripe; when he was made dictator for life from 44 B.C., he took to wearing an all-purple toga, thus enraging the other senators. He also had his profile stamped on the *silver* coins of the realm.

But *orange?* In culinary terms everyone will have heard of a Caesar Salad, but there is another recipe associated with his name: the delicious blend of orange juice, milk, vanilla essence, and ice (plus a raw egg if you are so inclined) invented at a Los Angeles lunch counter in 1926. Although most familiar now through the chain of diners which takes its name from the drink, the *Orange Julius* ensured that my four discordant words were brought into association at last, two millennia after Julius Caesar's death. "Waiter, an Orange Julius and a Caesar Salad, please!"

SELECTED BIBLIOGRAPHY

Listed below are various scholarly (as opposed to popular-level) texts to which the reader might refer as a start toward delving deeper into any particular aspect of calendar and timekeeping affairs. It is not intended to be a complete reference list. Several other suitable volumes are also mentioned in passing within the main text of this book.

Achilis, E., *Of Time and the Calendar,* Hermitage House, New York, 1955.
The last of Achilis's books about calendar reform and the World Calendar Association.

Aveni, A. F., *Empires of Time: Calendars, Clocks and Cultures,* Basic Books, New York, 1989.
Aveni writes accessibly on many of the human aspects of time.

Bickerman, E. J., *Chronology of the Ancient World,* Thames and Hudson, London, 1968; revised 1980.
Excellent description of how dates in the ancient world are derived; includes the B.C./A.D. dates for many civilizations and king/consul lists.

Colson, F. H., *The Week,* Cambridge University Press, Cambridge, U.K., 1926.
Fascinating but dated scholarly account of how the week came into being, along with related timekeeping matters.

Dershowitz, N., and E. M. Reingold, *Calendrical Calculations,* Cambridge University Press, Cambridge, U.K., 1997.
The authors give algorithms allowing conversions to be made between dates stipulated in most common ancient and modern calendars.

Fraser, J. T., *Time, the Familiar Stranger,* University of Massachusetts Press, Amherst, 1987.
For background information on matters of time, look up this and any of Fraser's several other books in a good library.

Howse, D., *Greenwich Time and the Longitude,* Oxford University Press, Oxford, 1980; revised and published by Philip Wilson, London, 1997.
The best book on how astronomy has contributed to timekeeping, navigation, and the determination of longitude, through to almost the present.

Jones, C. W., *Saints' Lives and Chronicles in Early England,* Cornell University Press, Ithaca, N.Y., 1947.
Discusses many aspects of the period A.D. 600–1000, when our calendar's evolution was influenced by events in Britain.

Macey, S. L. (editor), *Encyclopedia of Time,* Garland Publishing, New York, 1994.
One could spend days picking gems out of this very extensive volume. If you have a specific query about timekeeping and related matters, this is a good place to start.

Metford, J. C. J., *The Christian Year,* Crossroad, New York, 1991.
A nice description of the religious matters which shape the internal structure of the year.

Michels, A. K., *The Calendar of the Roman Republic,* Princeton University Press, Princeton, N.J., 1967.
Scholarly treatment of what we know about the pre-Julian calendar used in Rome, including the many festivals and other significant days.

O'Neil, W. M., *Time and the Calendars,* Sydney University Press, Sydney, 1975.

A succinct but erudite account of calendar systems. Does not fall into the traps that catch out many other authors, but still contains a few errors of fact.

Poole, R., *Time's Alteration: Calendar Reform in Early Modern England,* Taylor and Francis, London, 1998.

An erudite discussion of the debates which led to the eventual change of the calendar in Britain (and the colonies) in the eighteenth century.

Poole, R. L., *Studies in Chronology and History,* Clarendon Press, Oxford, 1934; reprinted 1969.

A fascinating set of scholarly papers concerning calendar development through the Middle Ages.

Salzman, M. R., *On Roman Time,* University of California Press, Berkeley, 1990.

Concerned with one of the few Roman written/illustrated almanacs of which we still have facsimile copies, this one from A.D. 354. If you want to see what a late Roman calendar actually looked like, see this book.

Seidelmann, P. K. (editor), *Explanatory Supplement to the Astronomical Almanac,* University Science Books, Mill Valley, Calif., 1992.

To discover the true astronomical basis of timekeeping, this is the major reference source. There are errors within its covers, however, in particular in the statement of the definition of the tropical year, as I have described in the text.

Stephenson, F. R., *Historical Eclipses and Earth's Rotation,* Cambridge University Press, Cambridge, U.K., 1997.

See this book for examples of how ancient eclipses are computed and used to figure out how the terrestrial spin rate has slowed over past millennia. Stephenson has also written extensively about the application of other astronomical methods in chronology—for example, in using comets to date past events.

Whitrow, G. J., *Time in History,* Oxford University Press, Oxford, 1988.

Another good source of background reading, although fallible.

Zerubavel, E., *The Seven Day Circle,* Free Press, New York, 1985.

Excellent modern account of how the week evolved.

INDEX

Page numbers in *italics* indicate figures or maps